REPRODUCTION

Reproduction

Jack Cohen, DSc, FIBiol

Senior Lecturer in Animal Reproduction,
University of Birmingham, England

BUTTERWORTHS
LONDON - BOSTON
Sydney - Wellington - Durban - Toronto

THE BUTTERWORTH GROUP

ENGLAND
Butterworth & Co (Publishers) Ltd
London: 88 Kingsway, WC2B 6AB

AUSTRALIA
Butterworths Pty Ltd
Sydney: 586 Pacific Highway, Chatswood, NSW 2067
Also at Melbourne, Brisbane, Adelaide and Perth

SOUTH AFRICA
Butterworth & Co (South Africa) (Pty) Ltd
Durban: 152–154 Gale Street

NEW ZEALAND
Butterworths of New Zealand Ltd
Wellington: 26–28 Waring Taylor Street, 1

CANADA
Butterworth & Co (Canada) Ltd
Toronto: 2265 Midland Avenue, Scarborough, Ontario, M1P 4S1

USA
Butterworth (Publishers) Inc
Boston: 19 Cummings Park, Woburn, Mass. 01801

First published, 1977

ISBN 0 408 70798 4

Library of Congress Cataloging in Publication Data

Cohen, Jack.
Reproduction.

Includes bibliographies and index.
1. Reproduction. I. Title.
QP251.062 591.1'6 76-47512
ISBN 0-408-70798-4

Printed in Great Britain by Chapel River Press, Andover, Hants.

PREFACE

This book has its origins in courses on embryology and reproductive biology at the University of Birmingham, UK. There have been changes in emphasis in biology teaching at all levels, away from 'classical' teaching about organisms, anatomy, phylogeny, and natural history and towards the understanding of physiological processes like digestion, osmo-regulation and excretion; evolution and behaviour studies, ecology and biochemistry are replacing comparative morphology and systematics as hallmarks of the 'compleat biologist'. Recent biochemical discoveries have uncovered the biochemical nature of genetic transmission, but not yet of the genetic control of development; it is in this area that biological reductionism, the belief that 'biology is only complicated chemistry' has most adherents, and much positive achievement. Many modern school biology syllabuses emphasise the biochemical component to the exclusion, sometimes, of all else but a cursory treatment of mitosis and meiosis and a mandatory lesson or two on the hygiene and responsibility of human sexuality, dressed up as reproductive biology of the human being.

Emboldened by student interest in my rather different approach to reproductive biology, and desirous of having my arguments, examples and prejudices made more generally available, this book was conceived. Its gestation has been traumatic in many ways, notably for the failure of an original intention to share authorship with Dr T.D. Glover, then of Liverpool University and now Professor of Veterinary Anatomy at the University of Queensland.

A broad approach has been taken, for subjects of great relevance to the reproduction of animals have classically been available only as separate texts. Genetics is usually only marginally related to gametology, and reproductive endocrinology is often tied to behaviour studies or to gametology without assembling both under the reproductive umbrella. Larval forms, senility, offspring care and developmental genetics rarely form parts of a university biology syllabus, but are usually relegated to the less formal parts of biological education or left out altogether. Embryology, too, is rarely seen as a part of reproductive biology; this demonstrates especially how adherence to ancient or modern teaching rituals has split biological relevance into different mental compartments. The assortment of all these reproductive topics into one book, however clumsily, should emphasise that all are part of the reproductive biology of animals.

A guiding principle to both author and publisher has been price, for in these days of escalating costs no purpose is served by a book, however beautifully produced, whose covers remain unworn on the bookshop shelf.

The chapters are of three kinds. Some, for example those on sperms, eggs, fertilisation and developmental biology are close to my professional expertise and I have found difficulty in selecting relevant material. Others, notably the chapters on the arithmetic of reproduction and on sexual congress, give a fairly orthodox view in what I hope is a new way without raising the hackles of too many professional readers. A few chapters, including those on larval forms, life cycles and especially reproduction in mammals are much thinner, with conjecture replacing solid foundation in many places. For all but the last I have to plead the ease of theorising in the absence of information to disprove hypothesis; I am not expert in these areas but feel that they should be represented here, however inadequately.

For the Reproduction in Mammals chapter I have been given freedom by a series of five excellent books, with this title, by Austin and Short. I could not begin to treat this subject fully in so short a space, so have attempted the bare minimum of information; I make no apology for the evolutionary series from viral protein to Christianity with which I finish mammalian peculiarities. Cultural reproduction fitted best as a climax to the mammal chapter.

J.C.

ACKNOWLEDGEMENTS

My gratitude is due, firstly, to all the students at Birmingham University who have received, or suffered, developmental and reproductive biology in my courses, and especially to those who have criticised by other than examination performance. My colleagues Brendan Massey and Brenda Murray have helped work out and give these courses, and have read the manuscript in an early version; many of their comments have resulted in changes for the better. Lawrie Finlayson has also been patient enough to read and annotate most of the manuscript. I find it impossible to ascribe origins of my ideas to individual conversations; the following friends and colleagues have contributed thoughts, examples, ideas: Jack Llewellyn, Barney John, Alex Comfort, Otto Lowenstein, Jim Berreen, Hugh Wallace, John Jinks, Isaac Asimov, James Blish, Christopher Zeeman, Jack Grainger, Owen Harry, Phil Dyer, Dennis Wilkins, David Jones, Anne Maclaren, Shirley Toogood, Russell Coope and Bill Potts.

Thanks are due to Butterworths and, especially to Glen Hughes for her patient midwifery and ante-natal care. My wife drew many of the line drawings, and Chris Harris produced all of the others. Photographs are mine except where indicated; Kevin Tyler, my research student during the writing of this book, not only suffered fairly quietly but also produced many of the photographs and much constructive criticism.

CONTENTS

NOTE TO THE READER

I have assumed that you have some knowledge of basic biology, including the processes of cell division, mitosis and meiosis. This is not meant as an elementary biology text but as a book of examples and ideas in reproductive biology; I have attempted to show which ideas are mine by quoting references wherever manuscript readers have asked for them, for whatever reason. These are given as superscript figures rather than the more usual (Smith, 1962) form, because most readers will not be concerned with most references and numbers are less intrusive. They are organised by chapter to avoid large numbers and to enable me to comment occasionally. I have always given a second reference if the first is not in English, and have avoided the primary reference for its own sake, choosing instead secondary references if they give context, are more readable, or if I have not seen the original. Page numbers are given in the reference list.

There is a *Glossary*, because my use of even familiar words (perhaps especially of familiar words) may require explanation; and there are many technical terms for which definition in the text would annoy those readers who disagree in detail; these too are indexed.

There is a *List of Organisms* which are mentioned as examples, for it is tedious to classify each example in the text, yet the reader should know that, for example, *Perameles* the bandicoot is a mammal rather than an insect or a fungus. Many student essays give classical examples without the writer doing more than transferring a name from the textbook to his paper; this list should at least direct attention to a small part of the animal kingdom. Unfortunately, access to this list can only be made via the index, so two steps are necessary. Please take them as often as you need.

This book is not aimed at a specific readership. I hope that students of biology will find that it draws together many threads they have met in other contexts. Veterinary, nursing and medical students will find more about reproduction here than in some of the specialised or general text books. I had biology teachers in mind when I drew specific examples (for example Siamese fighting fish) and I would like to think that this book makes 'reproduction' a teachable part of the syllabus. I know that my professional colleagues will find some parts of this book to be controversial; but I hope parts will please and excite them to. This is not a treatise, nor a textbook circumscribed by a past syllabus. I believe that most people concerned with reproduction in animals or people will find ideas and examples which illuminate this central area of biology.

If you are a professional into whose field I have trespassed, please treat my suggestions as critically as you would those of your colleagues and let me know my sins, whether of omission or commission. If you have other suggestions, I will be grateful; even if no further editions of this text appear, generations of students at Birmingham may be grateful too.

LIST OF ILLUSTRATIONS

LIST OF TABLES

1

INTRODUCTION

1.1 The Scope of Reproductive Biology

Reproduction is the fundamental process of all living things. The other so-called characteristic properties, irritability, metabolism and so on, may be absent; but if an individual reproduces, it must be alive. Surprisingly, the study of reproduction has not yet become a distinct academic discipline. It forms an uneasy intersection of genetics, embryology and ecology and is the basis of animal and crop production. Endocrinology has recently contributed greatly to the study of mammalian reproduction, sometimes to such an extent that many modern books with 'reproduction' in the title have little more than endocrinology to offer. But by and large general knowledge of reproduction is still typified by anecdotal books like Wendt's *Sex Life of the Animals*[1]; they are often fun to read, but are basically nineteenth-century natural history. It is my aim to bring some facts, and some theories, of reproduction into a form that is accessible to criticism. I concentrate on animal reproduction because my training is in animal biology, but some of what follows applies, of course, to all living things.

Reproduction takes a variety of forms among living things. On this planet, the molecule deoxyribonucleic acid (DNA) is used by nearly all creatures for the transfer of information from one generation to the next, and in recent years elucidation of the biochemistry of genetic mechanisms in bacteria and viruses has given us many tools with which to attack the comparable processes in higher organisms.

It has slowly become clear, however, that models built on these prokaryotes (organisms without a nucleus, like bacteria) are not entirely applicable to the eukaryote (nucleated) cell, and still less to the organism of which the cell is a part. We have had to change our ideas about the phylogenetic relationships of bacteria and higher organisms; it has been discovered that mitochondria and chloroplasts, probably centrioles and possibly the centromeres of chromosomes, have their own DNA rings which are responsible for their replication. This has meant that we can no longer think of the cell as an 'evolved' bacterium. It is of another order of complexity. It probably arose as a community of different prokaryotes, and its properties depend on the different properties of these 'evolutionary symbionts'[2] at least as much as upon their common structures. The chromosomes, in particular, are not comparable to the DNA rings of bacteria but are very complex structures, with several kinds of histones and a spectrum of other proteins as a necessary part of their structure and function.

For these reasons it is all the more unfortunate that a doctrinal philosophy of biology has arisen whose cardinal bases are that 'the cell is the unit of life', derived from studies of the prokaryote *Escherichia coli*, and that 'DNA is God and RNA is his prophet'[3], a concept arising from studies with bacteria, their viruses and biochemistry. As we shall see, these concepts have contributed immensely to our detailed understanding of mechanisms of heredity and development, but this very success of molecular biology has directed attention away from many other important aspects of the biology of reproduction. This attitude has led many to believe that most of the problems of reproduction have been solved or are being resolved by the methods of biochemists; even if all is not quite understood yet, they say, then it *should* and *will* be understood in these 'molecular' terms. I believe this 'reductionist' attitude to be at least misleading, and often mistaken. For example, none of the principles of Mendelian inheritance have been simplified as a result of the elucidation of the biochemistry of the genes which are their basis.

In the following chapters we will consider some of the important processes which contribute to an understanding of the reproductive cycle. But we must not restrict our enquiry only to those subjects in which recent technology has given us new insights. Neither, of course, should we fail to illustrate these areas in as much detail as we feel the new insight warrants.

This book has a further, and perhaps more pretentious aim: it should replace fact by theory. 'Theories destroy facts' said Medawar[4] epigrammatically. We may take a simple but classical example to see what is meant by this. Before Newton, let us imagine, people had to remember many categories of events independently: apples fell down, people fell down, dropped objects fell down. It was harder to walk upstairs than downstairs, water ran downhill, the undisturbed surface of molten lead was flat enough to mould glass panes, the orbits of the planets were elliptical, and flames pointed up; some or all of these required independent memory. But once Newton said 'Every object...............attracts every other object..............' they all fell into one category, filed in the same pigeonhole and individually retrievable from the central theory by the application of a question. The individual facts were rendered trivial by the generalisation, and so were 'destroyed' as individual items to be remembered.

Biology still has few of such unifying principles, and in consequence has many facts to be remembered. But we must not pretend that the problems are not there in areas which are not covered by such unifications as we have. Part of the reproductive field has been generalised by the discoveries of the structure and some of the functions of DNA, and in the same way mammalian reproduction has benefited from the explanatory powers of endocrinology. But we must not assume that DNA-based hypotheses are potent in a broader context; nor can knowledge of the endocrinology of reproductive processes in mammals be a substitute for anatomical, behavioural and ecological theories. The study of mammalian

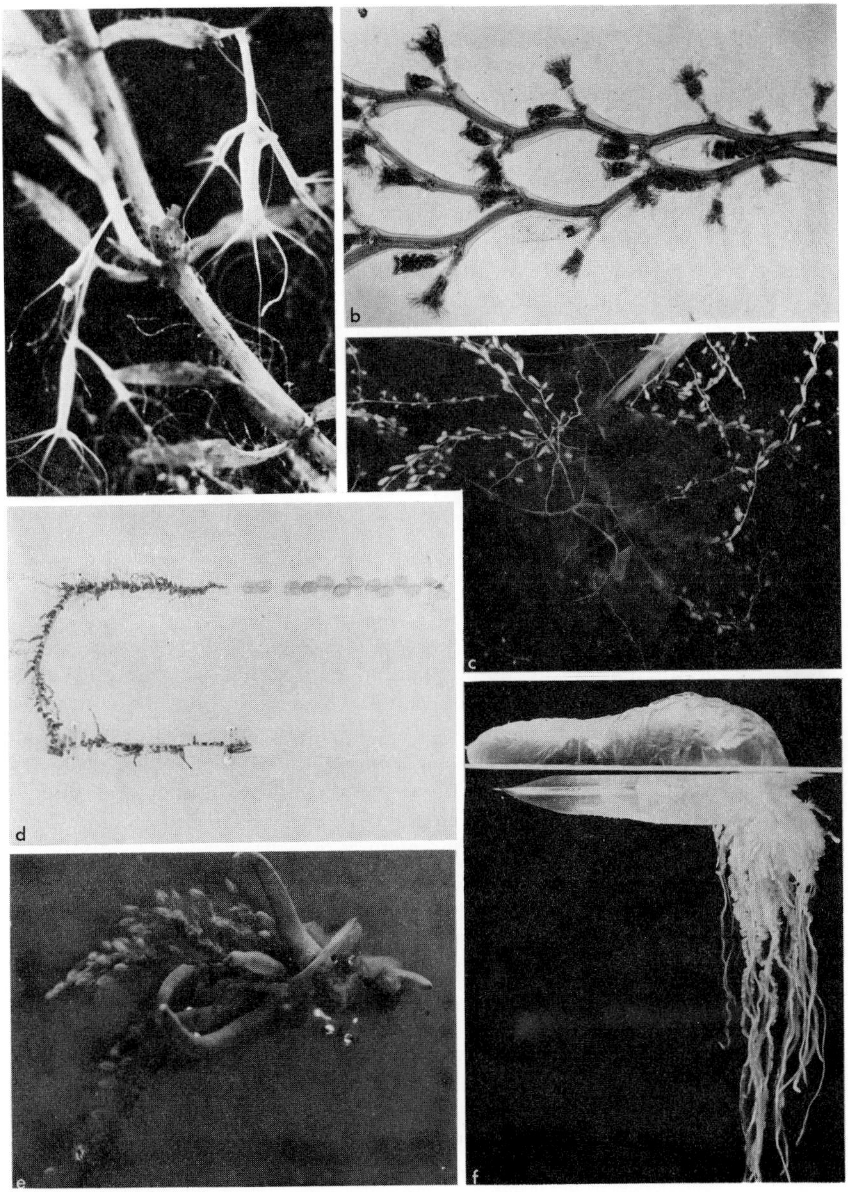

Plate 1 VEGETATIVE MULTIPLICATION AND REPRODUCTION IN THE HYDROZOA (a) Living Hydra, *showing asexual reproduction by budding; the two specimens each have two buds, and are pendant from a waterplant (Photo by C. Royston). (b) Part of an* Obelia *colony (compare* Figure 13.2*). (c) A large* Obelia *colony with dying-back at its centre (compare* Figure 1.1*) (Photo by P. Dearn). (d) A siphonophore (possibly* Halistemma, *museum specimen). Swimming bells (modified medusae) pull the long (20 cm) colony along, trailing the chain of other 'persons', which may break off and lead a free existence after producing medusoid 'persons' for their locomotion (Photo by T. Morris). (e) Another siphonophore (possibly* Anthophysa, *museum specimen). Muscular bracts protect the abbreviated (5 cm) colony, which has 'persons' (polyps) specialised for feeding, defence and reproduction (Photo by T. Morris). (f) The Portuguese man of war* Physalia. *This individual is a complex colony of varied 'persons', including one forming the float (Photo by T. Morris)*

reproduction has, however, been completely taken over by this endocrinological attitude, as any study of the literature will amply justify; more than 70% of research papers in the field are endocrine in bias, which grossly exaggerates the part that endocrinological processes play even in the reproduction of mammals.

Previous attempts at an overall view of reproduction have either dealt phylogenetically with reproductive mechanisms, just as one might list anatomical features without much attempt to relate them to function, or they have been weighted heavily towards the vertebrates and especially the mammals. Marshall's *Physiology of Reproduction*[5], although it suffered somewhat from both these tendencies, nevertheless forms a broad basis for our concept of the field even 50 years later. This small volume cannot hope to be compared with such a great work, and indeed that is not the intention. But it is hoped that it will give modern readers a few genuinely new insights into reproductive processes, and thus serve to stimulate as well as to instruct.

1.2 Definition

Reproduction is the 're-production' of a group of organisms; it is not multiplication nor mere replication. Nor is it breeding. Even when a family has produced children whose developmental programmes have made them resemble the parents, reproduction has not yet occurred. Imagine a 1920s family photograph of a granny, mother, father and three children, succeeded by a 1950s photograph similar in all details but 'stepped down' one generation. This is my picture of reproduction. A population of parents producing another, similar but later, population of parents is reproduction. So is an egg producing another egg. That behaviour in a population which contributes more or less directly to the emergence of a later population, resembling the first, is reproductive behaviour. Some of it, of course, may *reduce* the number of entities involved, but this shouldn't worry us; all sexual reproduction does that, at fertilisation.

Human reproduction gives us some problems, because human sexuality is so potent a part of human relationships and serves both procreation and recreation. Much human breeding seems to be 'unintentional', a by-product of the recreation, so human sexuality in its social and individual contexts must be considered in any general textbook about reproduction. Human reproductive behaviour is modified by Papal pronouncements, public health practice, the cost of 'family planning' advice or materials and in fact by all human activity. Equally, of course, the reproductive behaviour of all other organisms is enmeshed in their vegetative lives, but because we know less of this interaction than in the human case, and because it is probably simpler too, abstraction of their 'truly' reproductive behaviour is easier for us. It may be (p. 66) that much of human cultural behaviour springs from reproductive, or at least sexual, bases. This should not make us leave human sexuality to the 'specialist' or to the 'sexologist'. We should attempt to see it in the context of the reproduction of human organisms and cultures, even though we are conscious that the complexity of its nature forces a superficial view.

1.3 Kinds of Reproduction

The activities of organisms have been divided into **vegetative** and **reproductive** activities, but the distinction is often difficult to draw. Vegetative activities include growth, feeding, locomotion, and in general the physiological activities which characterise the life processes of individual organisms. Reproductive activity includes not only the actual production of eggs, sperms and progeny but also those activities, such as copulation and lactation, related mainly to the start of a new generation.

In some organisms the vegetative process of growth itself results in reproduction. In the branching colonies of *Obelia*, for instance, the oldest part of the hydrorhiza dies away; as each early branching is reached by this decay the colony becomes separated into two (*Figure 1.1* and *Plate 1*, p. 3). Sea anemones, too, commonly lose parts of the basal disc as they creep over irregular rocks. Such 'pedal laceration' often results in a chain of anemones across a rock surface, the tiniest immediately behind the parent and progressively larger specimens further away. The budding of *Hydra* is another common example of vegetative reproduction (*Plate 1*). Because wound healing and regeneration are part of the vegetative activities of so many organisms, the regeneration of torn starfish into several new individuals is also considered to be a vegetative reproduction; no new structures or special organs are required. This is the hallmark of **vegetative reproduction**.

Other organisms have special processes, organs or resistant stages, whose production is associated with special reproductive events which are, however, not sexual. Such **asexual reproduction** is very common in many phyla of the animal kingdom, including the chordates; the urochordate salps and sea squirts have some of the most remarkable methods of asexual production of new individuals (*Figure 4.9*, p. 88). All vegetative reproduction is after all, asexual; we can find examples, starting with the mitotic multiplication of some protozoans, and budding of *Hydra*, which provide us with all grades through the stolon-borne new individuals of salps to the gemmules of sponges. These latter, and such comparable structures as the statoblasts of ectoproct *Bryozoa*, are cell aggregations within a resistant shell formed by the organisms in times of stress, and capable of reorganisation to produce a new individual when circumstances improve. This is usually what is meant by asexual reproduction, but the complex linear budding of polychaete worms and the production of buds on a special stolon are also covered by the term. Examples are shown in *Figures 1.1* and *4.9* (p. 88) and in *Plates 1* and *2* (pp. 3 and 15).

Sexual reproduction is the commonest form of reproduction found among animals. Its major advantage seems to be that one individual may combine genetic innovations (**mutations**) from two ancestors not closely related to each other. Asexual reproduction only allows the production

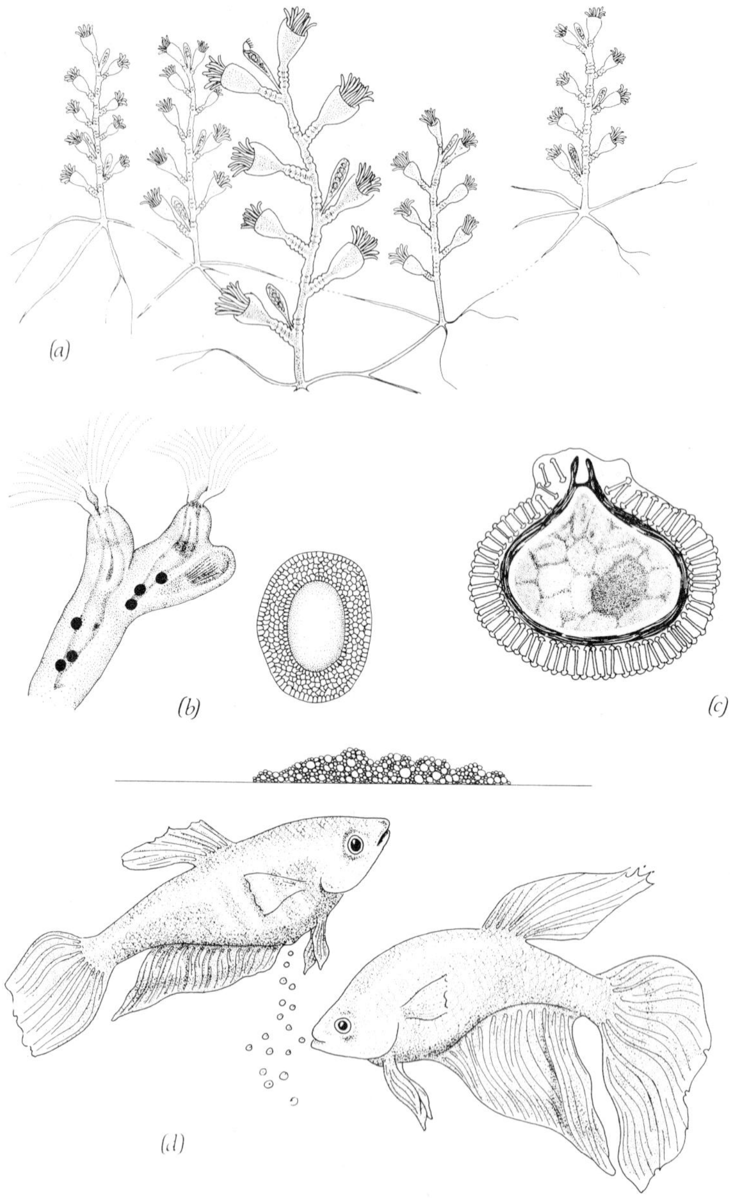

Figure 1.1 VEGETATIVE AND ASEXUAL MULTIPLICATION (a) An Obelia *colony, from a fixed specimen; the decay of the strands of hydrorhiza on the substrate results in separation of the colonies. Compare* Plate 1. *(b) 'Polyps' of a freshwater ectoproct, from life; note the statoblasts inside, released at death. One statoblast is shown enlarged. (c) The gemmule of a freshwater sponge. The wall has special spicules, amphidiscs, and the nutrient-laden cells within hatch from the 'neck' after the gemmule has been freed by the death of its parent sponge. (d) A pair of Siamese fighting fish,* Betta splendens, *at the nest. Compare with* Plate 5

of '**clones**' of individuals, each clone consisting of those individuals descended from one ancestor. If that ancestor has a genetic innovation, then only its descendant clone possesses it; it cannot spread through the population. Therefore combinations of useful characters take much longer. But sexual reproduction has as its central theme **amphimixis**, the actual fusion of living things from different hereditary stocks, resulting in a **zygote** with a new heredity, which may combine innovations from different ancestral lines. Sexual fusion actually temporarily diminishes the number of entities concerned, because two **gametes** become one **zygote**. As each parent usually produces more than one gamete, the parental population can nevertheless be reproduced numerically. In a totally asexual species, only one individual of each previous generation contributed to any one present individual. Sexual organisms, in contrast, have more ancestors in previous generations; so that they have two parents, nearly always four grandparents, generally eight great-grandparents and so on until most of the population is included. The Adam-and-Eve view that sexual ancestry *reduces* is clearly a special case, true only sometimes in the early stages of expansion of a population.

1.4 Varieties of Sexual Reproduction

The characteristic fusing entities, the gametes, are **sperms** and **eggs**, each with a haploid number (n) of hereditary determinants (usually taken only to be chromosomes, but possibly including flagellar apparatuses, chloroplasts, or other organelles). The fusion product (**zygote**) is diploid ($2n$). Meiosis, the kind of cell division which halves the number of chromosomes received by its products, usually precedes fusion, but there are many organisms, like most fungi, where fusion is followed by meiosis. Animal gametes (sperms and eggs) are usually direct meiotic products, but this is not normally the case in plants.

There are many variations on this theme. Protozoa often show meiosis, following the fusion of gametes (*Figure 1.2*), in the production of spores containing the haploid trophozoites (growth form) of the next generation, so the dominant vegetative form is haploid. In some protozoans, like the eugregarines and the adeliid coccidians, trophozoites form an association (**syzygy**) prior to the divisions which form the gametes (*Plate 3*, p. 43). It is sometimes uncertain where meiosis comes in the cycle, and the organisms in syzygy are called **gamonts**. Many protozoans do not have sperms and eggs which are very different, but may produce apparently equivalent gametes (**isogametes**) as in *Chlamydomonas rheinhardi.* The gametes of *Paramecium* and other ciliates do not leave the parent organisms, but travel through the cytoplasm of the conjugant as haploid nuclei, and never form separate 'cells' (*Figure 1.2* and *Plate 3*, p. 43). But these are all exceptional. In nearly all animals larger eggs are fertilised by smaller sperms. Most eggs carry food reserves, and most sperms are motile, usually flagellate with a posterior flagellum (*Figure 1.2*).

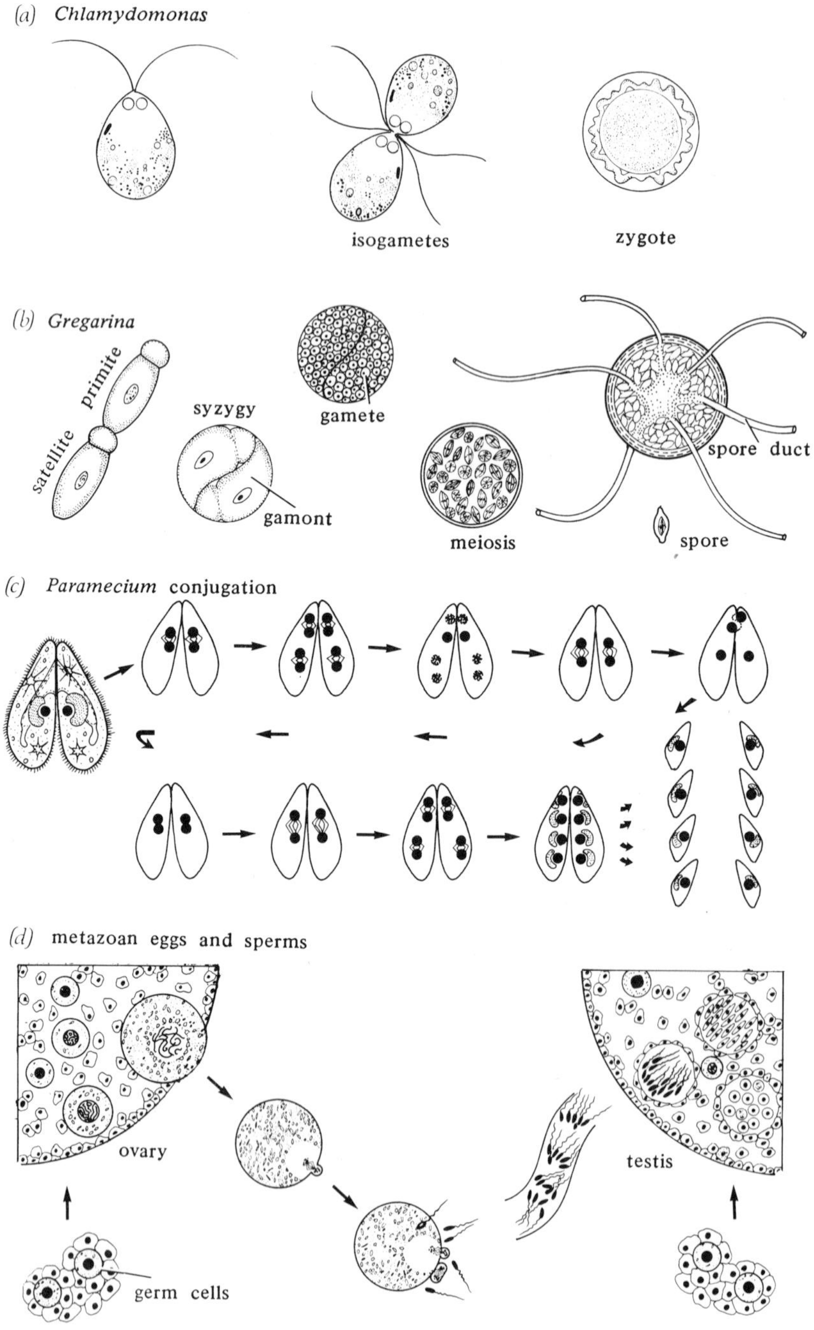
(a) Chlamydomonas
isogametes
zygote
(b) Gregarina
primite
satellite
syzygy
gamete
gamont
meiosis
spore duct
spore
(c) Paramecium conjugation
(d) metazoan eggs and sperms
ovary
testis
germ cells

1.5 Cell Division and Cell Multiplication

The egg is usually regarded as a cell, but this can be misleading and is rarely useful[6]. Most eggs are first recognisable as early **oocytes** as they enter meiosis, and they *are* usually fairly typical cells at this stage. The nucleus usually changes very early into a condition unlike that in any other cell (the 'lampbrush chromosome' state) when all parts of all chromosomes seem to be active. While this is going on the cytoplasm is increasing in volume and becoming spatially organised. At ovulation its organisation is such that the characteristic development of the new organism will start by cleavage of the egg, and the organisation of the early embryo will appear *without* specific instructions from the fertilisation nucleus. The cells of cellular organisms, or the acellular protozoans, *multiply* by mitosis, the progeny being equivalent to each other. But eggs *divide* into parts which differ in important ways, so that nuclei included in them can show *different* parts of the hereditary programme. The organisation of the egg is much simpler than that of even the early embryo, but to the extent that this organisation is a transform of the young embryo, a simplified plan, the egg is less comparable to one cell of the embryo than to the whole embryo. The egg divides into cells which are different; eggs rarely increase their substance during cleavage. So to compare the egg to a cell is rather like comparing a village to a city block; even though the village may become a city the comparison is not valid.

There is a question of phylogeny here, as well as of semantics. If early metazoans were colonial protozoans like *Volvox*[7], then cleavage may be indeed a matter of multiplication (even though it doesn't look like it because eggs do not usually grow during cleavage). But if the early metazoan resulted from specialisation of the parts of the cytoplasm of a protozoan which became multinucleate[8], then cleavage is a true division of the egg and the egg is best regarded, like the protozoans, as non-cellular rather than unicellular.

Figure 1.2 FORMS OF SEXUALITY (a) Chlamydomonas *vegetative form; isogametes fusing; and thick-walled zygote. (b)* Gregarina *trophozoites in tandem (primite and satellite); encysted together, (syzygy); formation of gametes mitotically; meiosis after fusion of those gametes; spores each containing eight haploid sporozoites, within the old cyst wall from which residual cytoplasm has built the everted spore ducts. Compare* Plate 3. *(c)* Paramecium*; conjugation of two compatible mating type trophozoites; micronuclei divide twice, meganuclei disappear; three of each four nuclei degenerate, and the survivor divides; one from each organism (♂ nucleus) passes into the partner's cytoplasm, where it fuses with the remaining (♀) nucleus; each fusion (zygote) nucleus then divides three or four times, and pairs form micro- and meganuclei of four or eight new paramecia. (d) Male and female gonads of a metazoan; germ cells invade the young gonad (ovary or testis) where they divide mitotically many times before their ultimate progeny, oocytes or sperms, are released. The oocyte then completes meiosis, the haploid pronuclei fuse and the zygote again segregates somatic and germ cells*

1.6 The 'Germ Plasm' Idea

Sexual reproduction in the metazoans depends upon a very early difference in the fate of parts of the egg. In nearly all animals which have been investigated, those parts of the egg which will come to form the gametes of the next generation are segregated from the rest as the **'germ cells'** at a very early stage of development. The rest of the organism develops, and these cells then invade the **gonad** and take up residence there. Sometimes, as in frogs, fishes, and some annelids this germ cytoplasm can even be recognised in the egg after only one or two cleavages. Weissmann, working with the nematode *Ascaris* in which all cells but the germ cells show fragmentation of the chromosomes, first suggested that metazoans consisted of a potentially immortal 'germ plasm' which produced, and was housed by, the mortal 'soma' or body in every generation[9]. He worked at the end of the nineteenth century, and his idea caught on, not only with biologists but with other scientists and with newspapers. The 'monkey gland' treatments of Voronoff (injections of animal testes) derived from his theories. Weissmann's ideas lost favour, not only because of the distorted publicity given them, but also because no biologist was able to demonstrate 'segregation of the germ plasm' except in *Ascaris*. Attention had turned to the new 'chromosome' studies, and to the new experimental embryology, and Weissmann's germ plasm theory was coupled with Lamarckian ideas about the inheritance of acquired characters, although it was, of course, exactly opposite (his 'determinant' theory, however, had Lamarckian overtones). Then in about 1925, Bounoure[10] discovered 'germ cells' in the frog, and since then, they have been found in all major phyla and in all five classes of vertebrates. So Weissmann's ideas have again gained general acceptance, and in consideration of metazoan reproduction we must follow

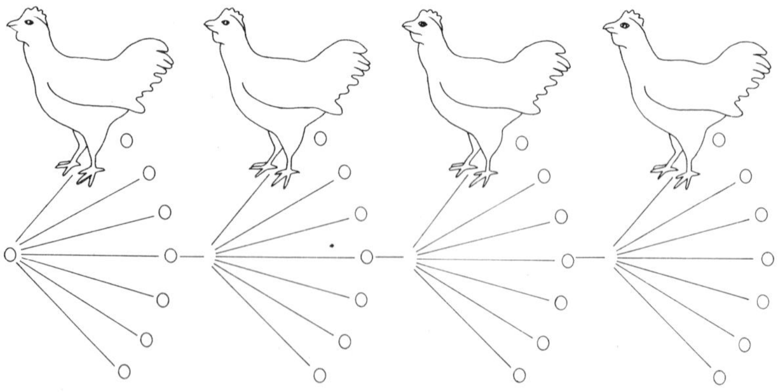

Figure 1 3 GERM LINE AND SOMA For the embryologist there is no problem deciding which came first, the chicken or the egg. The chicken is obviously only the egg's device to make another egg. (S. Butler, 1861). The egg, however, makes both the chicken and the eggs it will lay; the only heredity in this diagram is horizontal, and the chickens' bodies and most of their eggs have no genetic future

both threads: *the germ cells make gametes which fuse making a zygote, and this goes on to make more germ cells which make more gametes; the zygote also makes a soma, or body, which carries the germ cells and whose properties determine whether or not these germ cells will actually have any chance of meeting with others.*

REFERENCES

1. Wendt, 1965
2. Jinks, 1964
3. Asimov, c.1967
4. Medawar, 1963
5. Marshall, 1910, 1922, 1956
6. Cohen, 1971
7. Hyman, 1940
8. Hadzi, 1963
9. Weissmann, 1904
10. Bounoure, 1939

1.7 Questions

1. What is the difference between 'breeding' and 'reproduction'?

2. How can mutations 'spread' in a population of asexually reproducing organisms?

3. What are the effective differences between cell division and cell multiplication, and where are these processes found?

4. In what sense is it useful to consider metazoan parental bodies as out of the direct line of descent?

2

ARITHMETIC OF REPRODUCTION

2.1 Axioms

'Reproduction' is not synonymous with 'breeding'. Breeding is only the production of progeny, reproduction is the replacement of one population by its progeny. A cat has not reproduced when it has had kittens; only when two kittens have replaced the parents as breeding adults is reproduction said to have occurred. Animal populations do not show the 'natural increase' in numbers which we might expect from simple calculations. They do, of course, fluctuate but by and large they remain constant over long periods of time. A few, like man and the starling, are increasing in numbers fairly rapidly. Some, like the housefly and the great whales, are probably on the decline. Yet other species, perhaps most notably the locust and the lemming, have cycles or waves of population explosion. But when the numbers of offspring produced in the species, which tell us the potential for explosive population growth, are compared with the rates of change of population size, the *constancy* of population is impressive. It is clear that there are many controlling factors which maintain a 'balance of nature'. For example prolific herbivores are reduced in number by carnivores, who themselves are reduced in numbers when herbivores become scarcer, e.g. after a year of poor grazing. The fact that animal species have so contrived the scale and timing of this pruning of the population that long-term stability is achieved, is remarkable. The stability itself leads to what might be called 'axioms of reproduction'.

2.1.1 AXIOM 1 – TWO PARENTS PRODUCE TWO PARENTS

This is an alternative formulation of the observation that, except for rare examples and occasions, the numbers of breeding organisms in a species remain approximately constant from year to year. It is, of course, a general statement that relates to mean numbers; the axiom is not intended to distinguish between breeders within the species, in their relative contributions, but only to state the overall case. A particular two parents one year may finally produce more or less than two successors. But if the overall population remains approximately constant, there can be no alternative to this rule. To put it another way, although the variance of offspring number for any population of parents may be high, the mean is about two per parent pair. The numbers of offspring per parent in some cases may have a 'normal' distribution, so that the mean is a useful statistical parameter. But even if the numbers of offspring per parent form, for example, a Poisson series with very different contributions by different parents, the parental number still only replaces itself in most species; so the average, two offspring from two parents, must still occur.

Stability of animal populations through time is usually explained by variable, number-related environmental effects, e.g. predators which find numerous prey, or dependence of a species upon limited food resources. It must be realised, however, that the way of life of the organism itself determines the nature of this stability and its relationship to outside forces. So the reproductive strategy of a species is that structure upon which these forces impinge; the interaction determines, for example, at which stage of the life history most of the reduction to parental numbers occurs.

2.1.2 AXIOM 2 – NEARLY ALL SEXUALLY PRODUCED ORGANISMS FAIL TO BREED

Breeders produce tens, hundreds, thousands or millions of progeny during their lives. Because animal populations remain fairly constant, only the same number as the numbers of parents will breed in their turn. Sixteen eggs laid in her life by a female starling, or 50 million eggs laid by a female cod, on average only result in just over two breeding starlings or two breeding cod. All the other offspring die before they achieve parenthood.

Breeding organisms nearly always produce more than one offspring during their lives; the young are smaller than their parents and must grow before they themselves can reproduce. Because only a tiny minority survive to breed, nearly all deaths are vegetative failures. So most deaths result from failure to feed, from predation, from disease or high parasite load, or from developmental or environmental accident to **juveniles**. Very nearly all failure of zygotes to contribute genes to the future results from this vegetative failure; that is to say, a failure to grow up.

2.1.3 AXIOM 3 – MOST JUVENILE DEATHS ARE DISCRIMINATORY

At first sight it seems that this cannot be so. It seems obvious that there are many juvenile deaths that are dependent totally upon accident and in which, as far as one can determine, the victims are random representatives of their species. But other species may avoid such accident by adoption of a slightly different life style. Three fair examples might be: the destruction of nestling birds in a forest fire (but some birds *do* breed in isolated trees, or out of the dry season); the killing of all fish fry by pollution or drying up of streams or ponds (but many species do have behavioural, physiological or anatomical adaptations for this eventuality); or the accidental engulfing of planktonic larvae by baleen whales (but many such larvae live at depths, or inshore, where such whales are rare). It is probable that *any* juvenile caught in this way would be killed even though, as indicated in our examples above, other species may avoid death from these causes.

A moment's reflection, however, suggests that such *random* cases, in which all organisms die, are exceptional as causes of juvenile death. More usually, morphological, behavioural and physiological differences among progeny result in differential survival in a stressful, and competitive, environment. The classical example, since Darwin[1] read Malthus[2], is selection by food deprivation. Those juveniles better at finding, killing, digesting, storing or utilising food will survive their brethren even if marginally, and will benefit by the subsequent lack of competition; often they will also have benefited by eating their siblings.

2.2 Chance or Selection?

Having listed three axioms of reproduction, let us now consider the selection of breeders from the juvenile population. Starvation may not be the main selective factor for many species; differential tolerance of temperature extremes, oxygen and water lack and parasite load probably largely determine which juveniles survive. In some forms most juveniles do seem to survive and, notably in teleost fishes, birds and mammals, competition for territory or even mates may be the final arbiter (*see* Chapter 13). Classically, random factors in the environment have been 'blamed' for the deaths of these juveniles[3] or even for the later selection of breeders. For example, it is said that 60 million oyster larvae must enter the difficult and dangerous world of the plankton for 200 little oysters to come out, or 22 adders must be born for six to survive their first year, and so on. Nature is often profligate, we say, in gambling such long odds on an outside chance. We see organisms such as mammals and birds, who produce few offspring and look after them better, as betting on the favourite at short odds. But we must remember that other species (betting on 'outsiders' at long odds) pay off just as well; two breeders reproduce two breeders. The 'cost' of 20 million cod eggs or 10 000 yolkier salmon eggs or 20 yolky dogfish eggs, that is the production of eggs by fishes of comparable size, is not very different (*see Table 2.1*).

The immense numbers of offspring produced, especially by marine forms like the cod, plaice or oyster, must be seen in context. We usually interpret 'profligacy of nature' as a kind of exuberant insurance against failure in a hard world, but this interpretation is based on a hidden assumption made explicit in our oyster and adder examples above. This assumption is that variety among the young is small, so that effectively every one of them could come through if 'luck' favoured it. The opposite assumption is that the few survivors are the *only* ones which could have survived even in the best conditions because their siblings all carry genetic defects. Clearly the truth is somewhere between these two extremes: some juveniles are just unlucky, but some are indeed defective.

Each death, however, cannot be scored in this way. Deaths are not totally random with respect to the organisms' properties, *or* perfectly

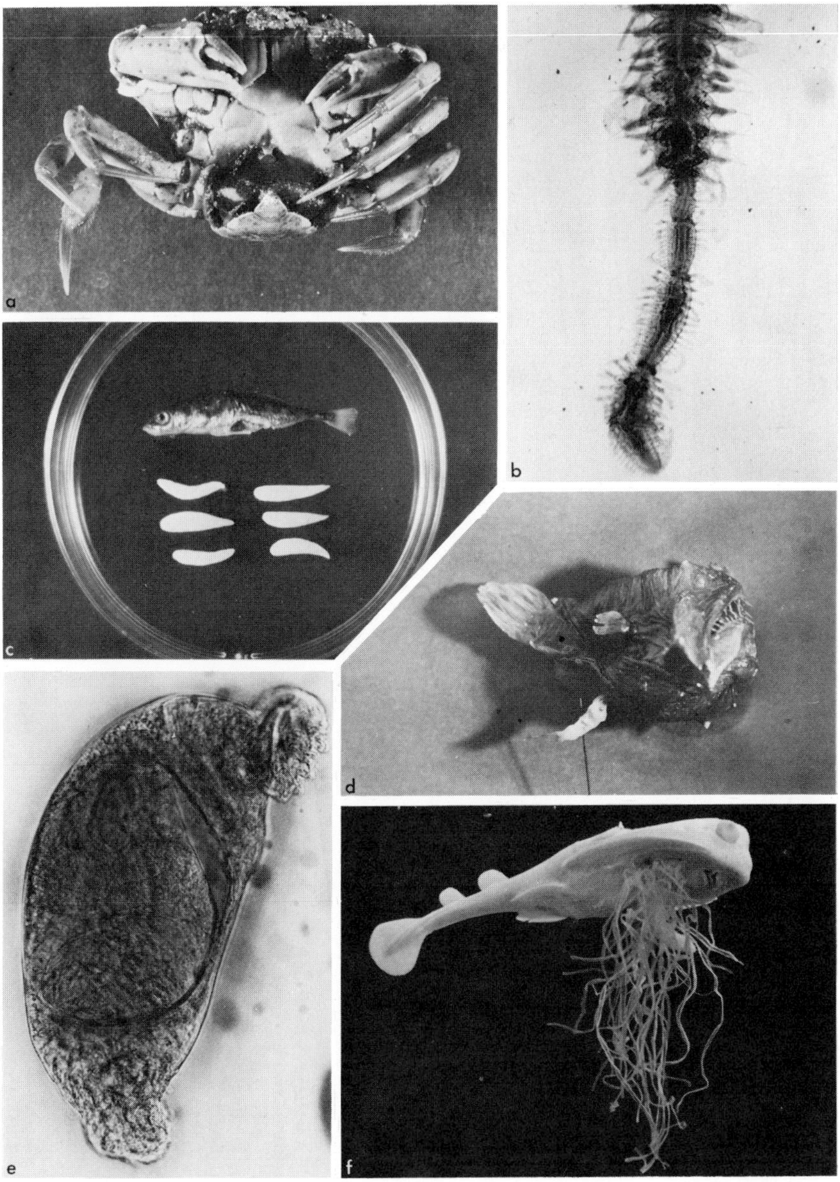

Plate 2 REPRODUCTION AND PARASITISM (a) A shore crab infected with Sacculina *(compare* Figure 10.6*) (Photo by T. Morris). (b) The posterior end of a polychaete worm, with a budding string of juveniles. The most posterior has eyes and is nearly ready for independent life. (c) Plerocercoids of* Schistocephalus *and the stickleback from whose peritoneum they were taken (see* Figure 4.11*); the stickleback seemed healthy (Photo by K. Tyler). (d) A female angler fish (*Borophryne*) with a parasitic male; this is a preserved specimen (courtesy of Dr H. Greenwood) and the male is held up into the focal plane with a pin. (e)* Gyrodactylus *with one young in the uterus and a further juvenile inside that. The haptor of the mother is at the top, of the daughter at the bottom, of the grand-daughter at the top (Photo by M. Shaw). (f) An embryo* Torpedo*, with long gill filaments for oxygen and nutrient uptake from mother's oviduct (Photo by T. Morris)*

Table 2.1 Numbers and care of offspring[4,5,6,7]

Organism	*Common name*	*Egg status (o,ov,v)*	*Yolk provision*	*Parental (or sibling) care and feeding*	*No. of eggs or young per brood or season*	*Total production per breeding female*
Homo sapiens	Woman	v		++++ M	1	13 ?
Capra hircus	Goat	v		+++ M	1–5	20 ?
Oryctolagus cuniculus	Rabbit	v		+++ M	8 (1–13)	25 ?
See Table 16.1 for more mammals						
Anas	Mallard	o	++++	++	9	25 ?
Sturnus vulgaris	Starling	o	++++	++	4.8	16 ?
Turdus migratorius	American robin	o	++++	++	3.4	10 ?
Larus argentatus	Herring gull	o	++++	++	2.5	7 ?
Parus major	Great tit	o	++++	++	10.3	30 ?
Perdix perdix	Partridge	o	++++	+	14	50 ?
Alligator mississipiensis	Mississippi alligator	o	++++	+	29–88	300 ?
Thamnophis s.sirtalis	Garter snake	ov	++++	–	28 (6–51)	100 ?
Anolis carolinensis	American 'chameleon'	o	++++	–	8–10	20 ?
Anguis fragilis	Slow worm	ov	++++	–	7–19	40 ?
Chelone mydas	Green turtle	o	++++	–	700	3000 ?
Chrysemys picta	Painted terrapin	o	++++	–	4, 5–9	40 ?
Ambystoma maculatum	Spotted salamander	o	++	–	100–250	500 ?
Rana pipiens	Leopard frog	o	++	–	3500–6500	12 000 ?
Rana catesbiena	Bullfrog	o	++	–	10 000–25 000	100 000 ?
Hyla regilla	Pacific tree frog	o	++	+	730–1250	3000 ?
Sminthillus limbatus	Cuban painted frog	o	+++	+	1–3	10 ?
Acipenser ruthenus	Sturgeon	o	++	–	182 000–1 million	6 million ?
Anguilla rostrata	American eel	o	++	–	5–20 million	5–20 million
Cyprinus carpio	Carp	o	++	–	0.5–2 million	10 million ?
Gadus morrhua	Cod	o	++	–	2–20 million	50 million ?
Mola mola	Marine sunfish	o	++	–	300 million	1000 million ?
Perca flavescens	Yellow perch	o	++	+ ?	10 000–40 000	100 000 ?
Pleuronectes platessa	Plaice	o	++	–	2–5 million	20 million ?
Scophthalmus maximus	Turbot	o	++	–	9 million	72 million ?
Salmo salar	Salmon	o	+++	– ?	7–10 000	7000 +
Mustelus laevis	Smooth hound	v	+	–	7–15	35 ?
Scyliorhinus canicula	Dogfish	o	++++	–	8–20	40 ?

Organism	Common name	Egg status (o,ov,v)	Yolk provision	Parental (or sibling) care and feeding	No. of eggs or young per brood or season	Total production per breeding female
Echinus esculentus	Sea urchin	o	+	–	1 million	3 million ?
Crassostraea virginiana	Oyster	o	+	–	0.5–1 million +	6 million ?
Mytilus edulis	Mussel	o	+	–	0.5–5 million	4 million ?
Helix pomatia	Roman snail	o	++	– ?	40–200	400 ?
Apis mellifera	Honey bee	o	++	++++ (M)	120 000	0.5 million
Vespula vulgaris	Wasp	o	++	+++ (M)	22 000	50 000 ?
Eciton burchelli	Army ant	o	+	+++ (M)	1 million	6 million ?
Anomma wilverthi	African river ant	o	+	+++ (M)	40 million	240 million ?
Bellicositermes natalensis	Termite	o	+	+++	13 million	100 million ?
Musca domestica	Housefly	o	+++	–	75–100	75–100 ?
Periplaneta americana	Cockroach	o	+++	–	200–1000	200–1000
Macrosiphum pisi	Pea aphid	ov	++	–	50–100	7–20 generations/season
Cyclops viridis	Green cyclops	o	++	+	20–160/2 weeks	2–400 ?
Daphnia longispina	Waterflea	ov	++	+	4–35/4–6 days	2–300 ?
Homarus americanus	Lobster	o	+++	+	8500	20 000 ?

o = oviparous
ov = ovoviviparous
v = viviparous
M = milk

selective, discriminating against defect. Nearly every death must have causal components of both classes, namely random and selective. When most die but some survive, which is what usually happens, we can easily see this as a discrimination between organisms and thus as a selection of a biased population. When all individuals die the situation obviously cannot be part of the normal reproductive strategy; and where none die, even in the long term, it is also irrelevant to the present issue except in special cases, discussed further in Chapters 13 and 14. Despite this argument, that each death has both accidental *and* selective components, it is probably permissible to use percentage fatality as a measure of the 'severity' of selection. (Haldane[8] used

$$\log_e \frac{\text{percentage survival of organisms in one group}}{\text{percentage survival of organisms in another}}$$

as a measure of different 'intensity of selection' in two groups, but here we are concerned with single populations at different stages of the life history.) If the stress is simple, for example if oxygen content of the watery medium falls to the level where some tadpoles die of anoxia, and the tadpole population varies in only one way, say length of gills, then this use of percentage fatality to measure severity of selection is permissible, whether or not competition for oxygen occurs. This is still true if both short *and* long gills are less efficient, or if average gill length is least efficient, or indeed if variety in gill length of one individual is being selected for or against. But as variations in different larval structures, and in more than one environmental parameter, are involved in the successive losses from a population, it is less clear that percentage fatality is an accurate or useful way of measuring selectivity. In our simple example, survivors may have required long gills, but *then* have needed darkest pigmentation *and* flattest shape. Because of gene linkage and the effect of one gene on several characters, the distribution of genes among the individuals may have more effect than the severity of each stress upon the percentage survival. This percentage will not then be a measure of 'selectivity', but of original genetic assortment i.e. of how many individuals happened to combine the necessary properties.

In larger populations of larvae, however, the chances of 'lucky' combinations (as of all other combinations) increase, and so percentage fatality is again more representative of the degree of environmental stress. So for small populations and simple stresses, *or* large populations and complex patterns of mortality, we can probably consider percentage survival as a useful indication of the extent of selection, if a regular small survival rate of a particular stage of the life history is a characteristic of the species. The enormous numbers of hatchlings produced by marine forms like oyster and cod clearly fall into this category. There have been many attempts to culture these eggs, particularly of oysters and flatfishes, and it seems reasonable to suppose that the experimental design precluded environmental hazards as far as possible. Nevertheless, less than 1% of plaice eggs (and far fewer for other flatfish) have achieved a length of 5 cm in culture[9]. Oysters have not been successfully tank-cultured, but many oyster 'farms' have attempted the culture of larvae in enclosed bays with ample food. Nevertheless less than one egg in 10 000 produces a settled oyster[10]. We must therefore assume that even the 'best' environment imposes stringent selection in these cases i.e. that most of the zygotes have built-in defects.

It is very difficult in this connection to define a 'defect'; we must digress and consider Wallace's concept of **'hard'** and **'soft' selection**[11]. Some organisms, for example plaice larvae with blind guts, die whether they are competing with others or not; this is recognised experimentally as **non-density-dependent mortality**. Other organisms, for example blind cave fish or heavy-metal-tolerant plant varieties, do very well in non-competitive situations but regularly lose out to competition by eyed fish[12] or normal wild plants[13] when competing in the normal habitat. The first kind of selection, for *internal* reasons which are lethal in any conceivable environment, is called by Wallace 'hard' selection, and forms the central core of Whyte's[14] arguments that many developmental routes from the egg lead not to adults but to embryonic or juvenile death; these organisms carry the **'lethal' genes** of the geneticist. The other kind of selection, seen only in more stringent circumstances, usually with competition from siblings, is called 'soft' selection, because its extent varies and its pressure may be relaxed. It leads to **density-dependent mortality**.

'Soft selection' is closely related to another problem of selection, which I will illustrate with an imaginary example (*Figure 2.1*). Suppose we have a species of oyster 50% of whose eggs hatch, releasing 20 million veliger larvae into the plankton from each female. We know that in the normal course of events only two of these (on average) would become breeding

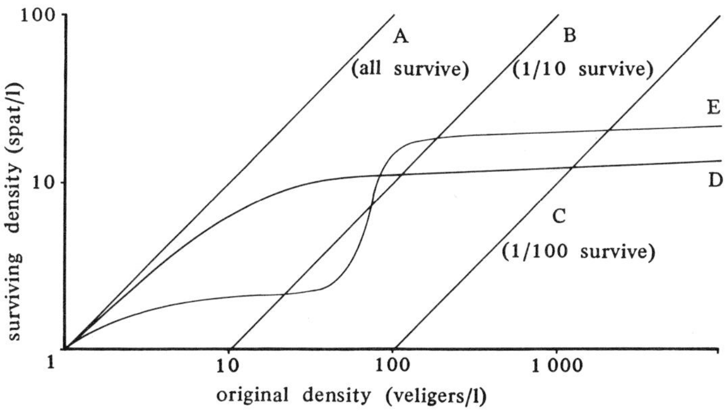

Figure 2.1 DENSITY DEPENDENCE OF SURVIVAL An imaginary experiment with cultured oyster larvae in tanks could result in line A: all larvae survive to become spat ('baby' oysters); or B: 1/10 of any *larvae become spat, whatever their starting density; or C: only 1/100 of larvae survive to spat; or D: only 10 spat per litre can be produced, whatever the original density, because this is the 'carrying capacity' of the system, defining final or 'bottleneck' density. Any real situation, especially in natural conditions, must be more complicated and may even resemble line E if early high density gives protection or better food acquisition*

adults; suppose also that we know that 2000 usually settle as 'spat' (young oysters). We might estimate the contributions of hard and soft selection to this reduction in numbers by culturing the veligers in oxygen-rich sea-water with plenty of food, at various densities (*see Figure 2.1*). Survival of a constant proportion independent of density up to carrying capacity would show that all except this proportion are defective and are eliminated by 'hard' selection (lines B or C). The survival of all larvae, up to a certain density well below the apparent carrying capacity, with progressive density-dependent losses above this, would be strong evidence of 'soft' selection (line D). But another interpretation of these lines is possible, the 'accident' hypothesis. Those who believe that accident is responsible for most larval deaths would have no difficulty in explaining lines B or C, or line D. Constant proportions of survivors, as in lines B or C, could result from any of a great range of accidents to which *all* are very susceptible, but 1/10 or 1/100 escape: from high susceptibility to cosmic rays; from failure to feed because a diatom gets stuck cross-wise; from internal damage from ingested radioactive particles; etc. I have purposely chosen fairly absurd examples, because even these would be very difficult to disprove. Proportional survival of essentially acciden-tal hazards, which have an equal chance of killing each organism, is an attractive alternative hypothesis to hard selection. If hard-pressed its pro-ponents can even postulate developmental accident *inside* the organisms should the survivors be shown to differ morphologically or biochemically from those which succumb.

Line D can also be explained as due to 'accident' in a number of ways: there could be a 'trace element' whose concentration was only sufficient to maintain 10 veligers per litre; waste products secreted by all organisms could only diffuse away quickly enough at this low density (i.e. 'carrying capacity' was over-estimated); disease spreads between the organisms above a certain density; etc. Here I have chosen less absurd-sounding hazards. It is very difficult indeed to distinguish accidental survival of a minority from selection of that minority, and I know of no naturally occurring case in which selection can be shown to operate in every generation to discrim-inate in favour of the élite which survive to breed.

The same problem is seen in discussions of the relative contributions of genetics and environment to, for example, IQ[15], and its roots are to be found not only in the old 'nature versus nurture' arguments but also in a pair of letters to *Nature* about 1930, which I believe biased much later argument in favour of 'accident' and against selection.

The botanist Salisbury[16] wrote in 1928 of some observations he had made of quadrats in a wood. He noted all tree seedlings which sprouted in his area over some years, and observed that nearly all of them pro-duced the pair of cotyledons, but that very few survived to produce four pairs of leaves; if they did survive, he thought their chances of attaining sexual maturity were good. The observation that juvenile mortality, imply-ing selection, had such a major effect on population structure puzzled Salisbury, who as a good neo-Darwinist believed that natural selection acted predominantly by discriminating between parents with different genes, and permitting some to produce more progeny than others. He proposed that a more important and previously unconsidered role of natural selection was to permit survival of only a tiny proportion of young plants, and to discriminate between their **juvenile** abilities: 'The mortality and therefore the operation of natural selection is almost entirely confined to the juvenile stages of development.' This was heresy, and called forth an answer from R.A. Fisher[17] the noted statistical geneticist. Fisher made a very strange statement: 'The selective elimination of certain individuals at stage C is for this reason as effective in modifying the genetic constitution of the species as the selective elimination of 10 times as many individuals at stage B, or of 100 times as many at stage A, or of 1000 seeds in their initial condition.' That is, if 1000 eggs produce 100 larvae of which only 10 survive to become adults, the death of 10 eggs is equivalent in evolutionary terms to the death of one larva, and the death of 10 larvae to that of one adult. He called this idea the 'Principle of Reproductive Equivalence' and seemed unaware that its basic assumption was that there *was* no selection during development, that *any* larva was just as likely to survive. Of course, if there *was* any selection for any larval character, each genetically different larva would have a different chance of survival, and the only 'equivalence' would be between the larva which had the genetics to survive and the adult it became; the failed larvae are not 'equivalent' to *any* adults! Such was Fisher's reputation, however, that the 'Principle of Reproductive Equivalence' is still quoted as *evidence* against juvenile selection[18]!

We have examined the question of selection of larvae as part of the foundation for the introduction of heterosis and genetic load, as these two genetic concepts relate directly to our subject. But the treatment must necessarily be brief, and the reader is referred to works such as Wallace's book on genetic load[11] for more detailed discussion. **Heterosis** is defined as that situation in which heterozygotes for a particular pair of alleles contribute more progeny to future generations than do homozygotes for either allele. The **genetic load** carried by a population is the proportion of unviable individuals *necessarily* produced by this population when it reproduces. Both are explained in the next section.

2.3 'Hybrid Vigour'

It has been commonly found that a heterozygote for two alleles is a 'better survivor' than similar organisms which are homozygous for either. The name given to the phenomenon is heterosis, and a classic example is the sickle-cell gene, a haemoglobin variant in man. The normal allele, which most of us possess, contributes normal haemoglobin to our red blood corpuscles. The mutant allele produces haemoglobin which differs in only one amino acid, but which is less efficient as an oxygen transporter and causes the biconcave disc-shaped cells containing it to collapse into sickle shapes when stressed. The homozygote individual for this allele is weakly, very anaemic and dies in childhood unless given considerable medical support. The heterozygote individual has some corpuscles that possess each kind of haemoglobin, and although it may be that in extreme circumstances he is marginally less athletically able than normal homozygotes, this is entirely speculative at present. The condition was, however, common in malaria-endemic regions of Africa, because the homozygous mutant and the heterozygote are very much less likely to maintain an infection of the malarial parasite, which lives in the red blood cells. In these endemic areas, it can be seen that heterosis applied. Many parents, sometimes up to 60% of them, are heterozygotes for this particular gene, so their progeny are of three kinds; half of these are heterozygous like their parents (Ss) one quarter homozygous for normal haemoglobin (SS) and one quarter homozygous for sickle-cell anaemia (ss). The homozygous sickle-cell types were weakly, and in these primitive conditions never survived to breed; the homozygous normals also frequently failed to mature because they caught malaria. Only the heterozygotes could mature. This is a very dramatic instance, but the phenomenon is very common indeed, and has been clearly demonstrated in insects[19], fishes[20], snails[21], birds[22] and mammals[23]. Sometimes the reverse occurs, where the heterozygote is less likely to survive than *either* homozygote, but this is probably much less common. A frequent explanation is that the heterozygote usually has more flexibility, that is to say, it has more possible answers to environmental challenge. This pertains in the case of sickle cell anaemia[23] where some haemoglobin carries oxygen well, while other haemoglobin protects from malaria. This heterozygote advantage is usually claimed to be the basis of 'hybrid vigour', in spite of its one apparent immense drawback as a reproductive tactic; four offspring must be produced for two parents to result. So for every gene demonstrating heterosis, at least a quarter of the offspring are less likely to survive.

There is now evidence that many enzymes, and other proteins, are made from several sub-units often as dimers, tetramers or other polymers. Where the sub-units are all similar because they are coded from the same DNA sequence, only one kind of protein is assembled. But heterozygotes may produce two kinds of sub-units from the two different alleles, which can be assembled in pairs into three kinds of protein. Indeed, more than two units are commonly required to form each protein molecule; four is a common figure, as in haemoglobin and the enzyme lactate dehydrogenase (LDH), and here the existence of two kinds of sub-unit produces five kinds of assembled tetramer (*Figure 2.2*). This allows great quantitative versatility in enzyme activity for such a heterozygote, whereas the homozygote is limited in this respect. This versatility may be expressed in time or space, and is now considered to be a major part of the explanation of heterosis. Indeed, many organisms have 'built-in' heterosis for many important enzymes and other proteins. They have several 'copies' (duplications) of the gene in each haploid chromosome set, each producing a slightly different sub-unit for the polymeric protein[24]; each of these may also be heterozygous in the diploid condition but, even if not, a variety of proteins of similar function are produced (*see Figure 2.2* for LDH). For haemoglobin, embryo, fetus and adult always have different combinations of sub-units giving different oxygen-carrying properties, i.e. there is temporal versatility; for LDH, different tissues have different proportions, giving a range of properties in the different parts of the organism at the same time.

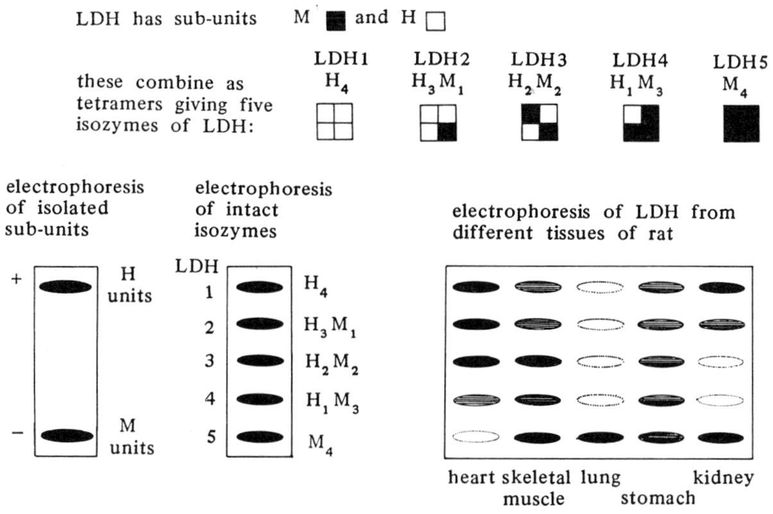

Figure 2.2 THE ISOZYMES OF THE LACTATE DEHYDROGENASE (LDH) ENZYMES OF VERTEBRATES LDH is a tetramer, with two kinds of sub-units, M and H from different structural genes. These associate randomly in the cell, giving five combinations. Skeletal muscle has mostly M, so LDH 5 predominates; heart muscle has mostly H, so LDH 1 predominates. Electrophoresis separates the sub-units, with H moving anodally. Extracts of various tissues behave as shown, indicating the relative activities of the M and H genes, associated with requirements for enzymic activity at different redox potentials. Kidney probably has two populations of cells, at least

If those juveniles which survive to parenthood are heterozygotes for one heterotic gene, only half their progeny are heterozygotes like them, and will survive stringent 'soft' selection. But if two gene loci are heterotic, only a quarter of the progeny are like them, and finally $(½)^n$ is the proportion of surviving progeny for n heterotic loci under 'soft' selection, in a stable population.

This must be qualified, however. Genetic strategies which produce non-viable products are said to generate 'genetic load'; a major contribution to this genetic load is the segregational load exemplified above and described below in more detail. Few heterotic genes show such a dramatic effect as sickle-cell anaemia. Usually, the heterozygote is only marginally more likely to survive than one homozygote, the other homozygote often being rather more handicapped. These three relative handicaps are usually quantified by contribution to the population of breeding animals at equilibrium. If the proportions of SS, Ss and ss (in the sickle-cell example) are x, y and z then the frequencies of S and s gametes which produced them must have been $x + y/2$ and $y/2 + z$. These gametic frequencies are usually represented as p and q, (where $p + q = 1$), and will produce by random mating p^2 of SS, $2pq$ of Ss, and q^2 of ss. This equation, $p^2 + 2pq + q^2$ for the proportions at equilibrium of homozygotes and heterozygotes in a population producing p and q gametes carrying two alleles, is the **Hardy-Weinberg equation**. It tells us that there is no 'tendency' to change proportions of gametes in successive generations for simple mathematical reasons (*see*

Figure 2.3). It also tells us the proportions in which we would expect to find the three classes in a population under constant 'stabilising' selection, provided that the proportion of zygotes remains constant for those characters being assessed until breeding occurs (this is usually true for 'adult' characters but not, of course, for larval ones). The situation is much more complicated, however, in successive generations after a change in selection either because of a new allele or because of a change in the environment, and the reader is referred to a text such as Wallace's[11] if he finds the following arguments inadequate.

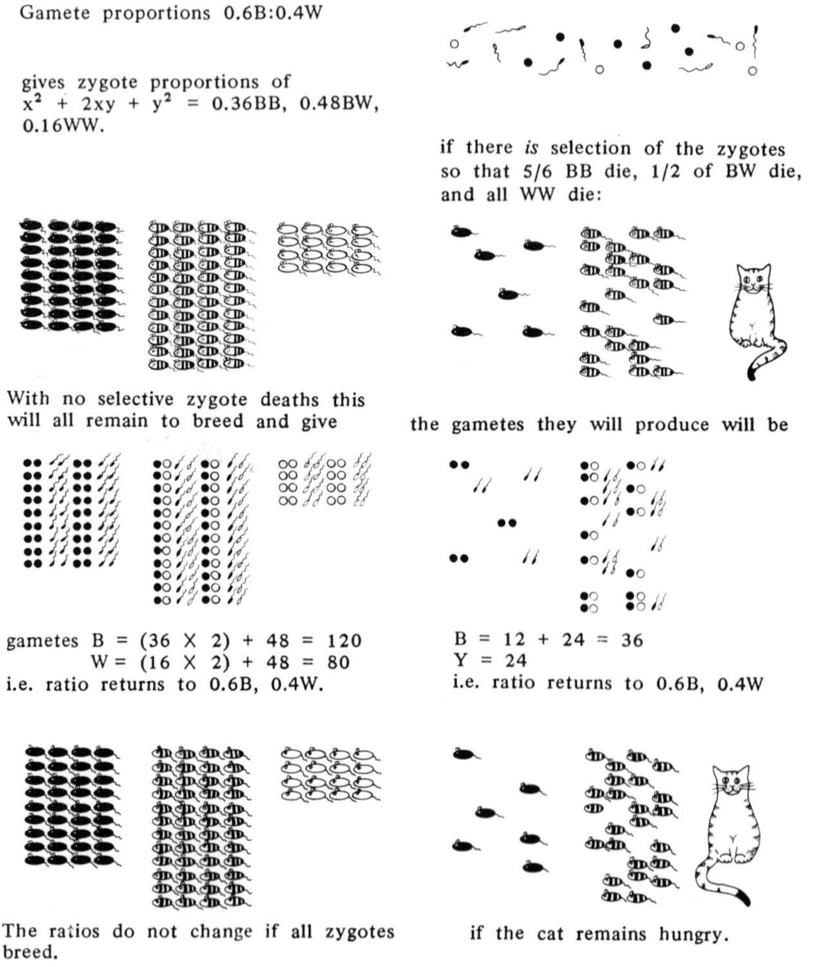

Gamete proportions 0.6B:0.4W
gives zygote proportions of
$x^2 + 2xy + y^2 = 0.36$BB, 0.48BW, 0.16WW.
if there *is* selection of the zygotes so that 5/6 BB die, 1/2 of BW die, and all WW die:
With no selective zygote deaths this will all remain to breed and give
the gametes they will produce will be
gametes B = (36 X 2) + 48 = 120
W = (16 X 2) + 48 = 80
i.e. ratio returns to 0.6B, 0.4W.
B = 12 + 24 = 36
Y = 24
i.e. ratio returns to 0.6B, 0.4W
The ratios do not change if all zygotes breed.
if the cat remains hungry.

Figure 2.3 THE HARDY–WEINBERG RATIOS On the left is seen the theoretical case, with no selection, immigration, etc. The ratios of black, striped and white mice repeat in each generation, and produce repeating ratios of gametes. On the right is shown a more realistic situation, equally repetitive (balanced) but with selection. Note that if selection is removed then the allelic ratios will be maintained, converting the right hand example to the left hand one in this special case but not usually

The rapidly increasing complexity of this kind of analysis, combined with the unlikelihood of selection remaining constant as the phenotypic proportions change (at least partly because each organism forms part of the environment of its peers), renders this kind of exercise impractical even with simple models and computers to play with. There have been many attempts at a theory of gene frequency change under heterotic selection, but it is probably best to use the simpler mathematics of the heterotic contribution to the 'genetic load'.

Because of heterosis, then, populations cannot 'breed true'; in every generation variety must be engendered from breeding heterozygotes at any locus, less well fitted or less adaptable than the parental genotype. This is the major contribution to 'genetic load' carried by that population, and is called the **'segregational load'** because it is produced by the Mendelian segregation of alleles from the heterozygotes. Other contributions to the 'genetic load' are **'mutational load'** and **'incompatibility'** (e.g. rhesus factor). Genetic load is a convenient, but in certain respects misleading, concept which amalgamates all sources of genetic variation, and even if deaths are not considered it can be taken as a measure of the number of fertilisations required to generate the specific genotypes found in a breeding population. It is a concept in direct contradiction to Fisher's Principle of Reproductive Equivalence of course.

We may be guided about which to believe by digression into a consideration of **'relative fitness'**, often wrongly called 'fitness'. A population in which a given generation stage, say the breeders, is constant in number and genetic make-up from generation to generation is said to be in **balanced equilibrium.** The group within such a population with highest gene transmission, which in heterosis would be Ss, is assigned a relative fitness of 1; the other groups, in our case SS and ss, are assigned fitnesses $1 - s$ and $1 - t$. Then the overall balanced equilibrium population which is involved in a repeated formation of p and q gametes, $p^2 + 2pq + q^2$ breeders (*or* zygotes *or* any other stage), is

$$p = \frac{t}{s + t}$$

and

$$q = \frac{s}{s + t}$$

and the 'genetic load' on the population (the loss of individuals required to maintain a balanced heterosis) is

$$\frac{st}{s + t}$$

There are many assumptions in this simplification. The interested reader should refer to Wallace for argument and detailed accounting of many interesting reproductive situations. Or, if 'genetic load' does not appeal as a view of selective deaths, John Maynard Smith's book[25] on evolution gives alternative formulations.

For example, if 20 heterotic loci, all with $s = 0.05$ and $t = 0.2$, exist in a population of fruit flies, then

$$\frac{0.01}{0.25} = 0.04$$

fail to breed and 0.96 survive for *each* locus, and $(0.96)^{20} = 0.44$ is survival for all 20 loci; so the load is just over 2, and 2 fertilisations are required to replace each breeding adult.

Both Fisher[26] and Haldane[27] gave their reasons for believing that populations cannot tolerate many heterotic loci; both considered that the 'wastage' involved would cause too large a depletion of the population. But observation of organisms such as plaice and oysters shows that apparent wastage of the magnitude required does in fact occur in each generation, and that the larval population is large enough to support it even taking into account 'accidental' attrition. For example, even if $s = 0.5$ and $t = 1$ for 20 loci, 1 in 800 larvae survives this segregational attrition, this 'soft' selection. Only 800 fertilisations would be necessary to reproduce each breeding adult. 'Accidental' death could still account for more than 99.9% of deaths of juvenile plaice. $st/s + t$ is a maximum for the load; each death is assumed to lose only one homozygosity from the population. But, of course, each dying individual will usually have more than one homozygosity.

There is now new evidence for the existence of very large segregational genetic loads in wild populations. Recent studies of adults from wild populations of fruit flies[28], eels[29], mice[30] (and men[31]) and many other organisms by a new technique, electrophoretic separation of enzyme variants, has revealed very high and consistent levels of heterozygosity at many loci. This argues both for heterosis, and for high s and t, at these loci. Proteins are extracted from individuals, separated electrophoretically, and then the enzymes concerned are identified by their specific reactions, usually giving colours; or other specific proteins may be detected with specific antibodies. For monomeric proteins homozygotes show one spot, heterozygotes may show two; for polymeric enzymes, e.g. tetramers like LDH, more than the normal five kinds are found (*Figure 2.2*). This method is so sensitive that 30 different enzymes of one fruit fly could be analysed. This method gives only a *minimum* heterozygosity, because it is to be expected that many mutant enzymes are electrophoretically very similar to the normal, and so finally give the same point. Although very many mature animal populations, that is to say populations of breeding individuals, have been analysed in this way, it is unfortunate that very few juvenile populations have been investigated. This is obviously a very useful experimental tool for investigating the genetic structure of these juvenile populations, because 'direct' genetic investigation by breeding is restricted to those organisms which survive to breed, or to those few kinds of organisms in which all juveniles can be brought to sexual maturity and bred from. So we will await genetic analysis of the survival of breeders from juveniles, although the high heterozygosity of wild breeding populations leads us to expect a very heterotic, and therefore selective, survival; such analysis should tell us whether, and to what extent, the dying juveniles represent genetic load as well as accident.

2.4 Gametic Redundancy

There is another example of the 'Profligacy of Nature' in which we have good evidence for a high level of production defect. This concerns the numbers of sperms offered to eggs, the extent of sperm wastage at copulation. This has been expressed as a redundancy,

$$\frac{\text{number of sperms offered}}{\text{number of eggs fertilised}}$$

and varies enormously between species[32]. Fruit flies and some other insects have a sperm redundancy under 100, while many mammals, like man, ejaculate hundreds of millions of sperms for each egg in the female genital tract at the time. The classical explanations of this wastage have involved inefficient utilisation of sperms, the relatively enormous genital tracts of female mammals being supposed to require a high concentration of sperms in the vagina for a few to diffuse to the fertilisation site[33,34]. This is exactly comparable to 'accident' explanations of numbers of cod or oyster eggs, and is as difficult to disprove.

There is, as would be expected, a very good correlation of sperm redundancy with size (weight) of animal, $r = 0.7$ for 32 species[35]. But there is a higher correlation between sperm redundancy and a production event during sperm manufacture, i.e. crossing-over between meiotic chromosomes, $r = 0.81$ for 24 species; there is little correlation between chiasma frequency and weight, 0.21 for 29 species, and all of this correlation could be explained by the other association[35]. Analysis of the sperm redundancy and chiasma frequency figures strongly suggests that about 20% of cross-overs result in an 'unacceptable' sperm, so that for man with about 50 cross-overs in the production of each sperm, only about one sperm in one million would be expected to be free of these errors and acceptable for fertilisation (*Figure 2.4*); only about 400 of these acceptable sperms would be produced in a normal ejaculate. If this is valid, the female tract must select these sperms, all others being rejected and destroyed because of production defect; there is now some evidence that this may be so[36].

How does this relate to the juvenile survival we have been considering? It does so in two ways. Firstly, it shows that reproductive processes, characterised by 'like produces like', need not be free of considerable error, what the communications engineer would call 'noise', provided that this is filtered out within the same reproductive cycle so that it is not additive (*Figures 2.3* and *2.5*).

Secondly, we must recall that sperms are not alone in having cross-overs during their production. So do eggs, and we find that the effective eggs of mammals come from a very large population of oocytes nearly all of which degenerate, leaving only a small number for ovulation[37];

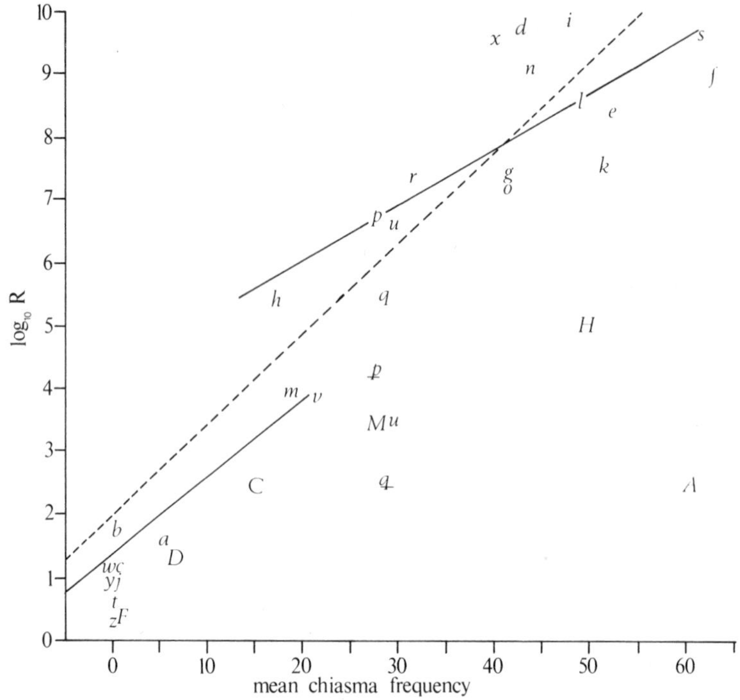

Figure 2.4 GERM CELL WASTAGE AND THE NUMBER OF CHIASMATA DURING EACH MEIOSIS Each point is the figure for one species. Small letters refer to sperms-per-egg and chiasmata during spermatogenesis of, a. Anopheles, *b.* Apis, *c.* Asplanchna, *d.* Bos, *e.* Canis, *f.* Capra, *g.* Cavia, *h.* Cricetulus, *i.* Equus, *j.* Drosophila, *k.* Felis, *l.* Homo, *m.* Locusta, *n.* Macaca mulatta, *o.* Macaca nemestrina, *p.* Mesecricetus, *q.* Mus, *r.* Oryctolagus, *s.* Ovis, *t.* Pseudococcus, *u.* Rattus, *v.* Schistocerca, *w.* Steatococcus, *x.* Sus, *y.* Dahlbominus, *z.* Aculus. *Capital letters refer to female germ cell wastage.* *C.* Carabus, *D.* Drosophila, *F.* Forficula *nurse cells per oocyte; H. woman and M. mouse oocyte atresia. A is the anomalous axolotl,* Ambystoma, *sperms-per-fertilisation.* p, q *and* u *are minimum* in vitro *sperm-per-fertilisation for hamster, mouse and rat*

this redundancy too, is what would be expected from the number of cross-overs if about a fifth of cross-overs 'fail'. Mammals are peculiar, however, in that they finally produce very few offspring, having reduced egg and sperm numbers from the very large numbers of meiotic attempts, presumably by selecting 'acceptable' ones. This would be expected, because in the mammals, like other viviparous forms (or indeed any species which invests considerably in each offspring, like sharks and birds) there is not time or space in the uterus to 'test' all fertilisations for this possibly chiasma-associated error, so it is filtered out before fertilisation. Many other errors, like mutation and non-dysjunction, get through this filter, of course; the filter only rejects one class of 'unacceptability'. 'Chiasma-associated unacceptability' in the gametes of some marine forms, though, need not be filtered out before fertilisation. It can remain in the fertilised eggs, and contribute to the variability of the larvae without affecting the parents' overall breeding efficiency; this is so only if there is sufficient prodigality to include losses from inevitable accident, and from genetic load arising from other causes, as well as this possible chiasma-associated load. Indeed, by serving to introduce an additional (mutational?) source of variety it may increase the chances of *some* offspring surviving to breed.

2.5 Accident and Selection

Let us now summarise the overall sexual reproductive strategy in terms of contributions to the arithmetic from accidental and selective factors. We will express them as percentage survival or as ratios, whilst appreciating that each death has both causes. For simple selections of small populations, and complex selections from large populations, this approximation is probably valid.

First, take a population of individual breeders in a population subject only to stabilising selection. A few germ cells in each of the females will multiply by mitosis to produce e oocytes which enter meiosis, usually early in the life of the female. Sexual congress occurs f times in her life and she produces E eggs on each occasion (*see Tables 2.1* and *2.2* and *Figure 2.5*).

Throughout the adult life of the male, germ cells multiply by mitosis and produce spermatogonia, which in turn produce spermatocytes, which then undergo meiosis. There is minor wastage during this process, but it is probably small (less than 90%) in most species[38]. Of the enormous sperm production, however, only a small proportion may actually be offered at a potentially fertile sexual congress (see example in *Table 2.2*). Let the number of sperms in such an emission = S (*see Table 2.3*).

S sperms are offered at sexual congress to E eggs. Sperms usually have a very high apparent wastage, which we should at this stage not judge 'accidental' but label redundancy,

$$R = \frac{S}{E},$$

the number of sperms offered to each egg. This can be seen to have at least three components (*see Figure 2.4*):

1. The proportion of 'acceptable' sperms, without production error (perhaps 'chiasma-associated error') is possibly small[35]; at least the reciprocal of this proportion must be ejaculated so that one 'acceptable' sperm is available. Let us call this true redundancy the ratio of

   $$\frac{\text{all sperms}}{\text{acceptable sperms}}, r;$$

   so $1/r$ is the proportion of acceptable sperms, for reasons implicit in the process of spermatogenesis, whether 'chiasma-associated' or not.

Table 2.2 Possible arithmetic of .reproduction for a pair of organisms and their progeny. (Compare *Figure 2.5.* and see text)

Stage	*Process*	*Entering*	*Competent to pass process* $1/r$	*Design permits* $1/n$	*Accident permits* $1/m$	*Completing*	R =	*Comment or example*
Spermatocytes		10^9						
	Spermatogenesis 10^9 could give 4×10^9 spermatids		1 ?	1/2 ?	1/2 ?	10^9	4	This is *not* sperm redundancy but spermatogenic redundancy
Sperms								
Oogonia	Oogenesis	10^6						
Ovulations			1/100	1	1	10	100	Atresia, supposedly all 'chiasma-associated' i.e. C $\triangleq$ 20 ($0.8^{20} \triangleq 1/100$)
Sperms (all)		10^9		1/10		10^8 per copulation	10	Say 10 copulations, 9 with infertile females; 10 adequate sperms find each egg, i.e. n = 100; therefore m = 10. But sperm redundancy R (*Figure 2.4*), i.e. S/z for one copulation = $10^8/10^4 = 10^4$
Sperms (ejac)	Fertilisations		1/100	1/10	1/10	10^4		
Eggs		10^4					1	
Zygotes		10^4						
	Development		1/1.25	1	1/5	1600	6.25	
Hatches		1600						
	Larval life		1/4	1	1/4	100	16	
Metamorphosis	Metamorphosis	100	1/2	1	1/2	25	4	Both a stage and a process – difficult and dangerous. If arthropod, this is low for R_{meta}. Juvenile life is a time of competition, i.e. 'soft' selection.
	Juvenile life	25	1	1/25	1/2	10	2.5	
'Territory gained'		10						Competition again (in many organisms by hormone status)
	Adult life	10	1/1.25	1/2	1	4	2.5	
Breeder selection		4	2	1	1	2	2	Competition again (hormones in mammals, plus a variety of behavioural tests in many creatures).
	Breeding	2						
Reproductive status	Ageing	2						Completion of individual life history of one of the few breeders.
Senile		1 ?				0		

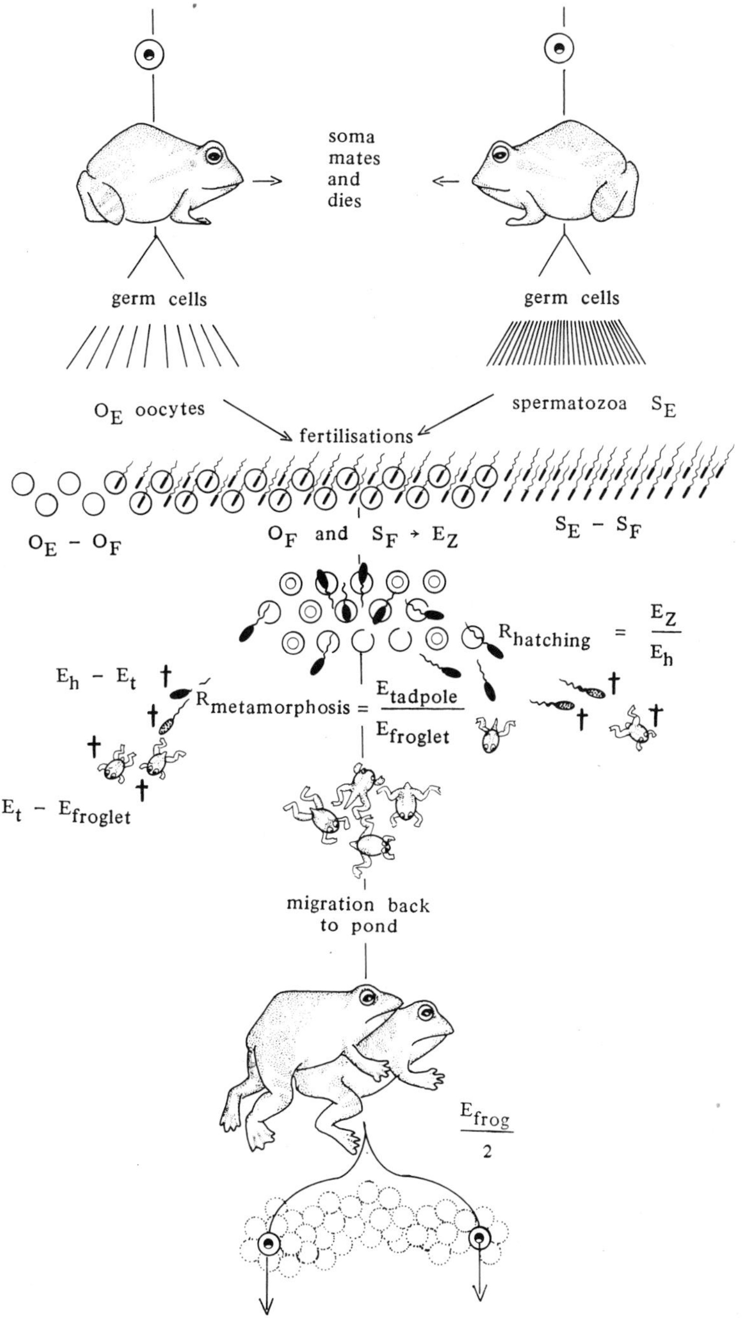

Figure 2.5 REPRODUCTION OF FROGS, TO ILLUSTRATE TABLE 2.2 *There are all kinds of cheats and simplifications in this diagram; please use it only as an aid to understanding the components of* Table 2.2. *† means 'dies'. These individual deaths cannot be assigned to* m, n *or* r, *because these factors apply in a strategic, not an individual, sense; only* R *for each stage is shown. But if the reader wishes to imagine some tadpoles and froglets as carrying a lethal genotype (contributing to* r*) others being desiccated, suffocated, starved or eaten because of the frog's breeding strategy (*n*) while others accidentally die (*m*) this may make* Table 2.2 *more comprehensible. But it is a dangerous simplification, because* m, n *and* r *are abstractions, proportions of a global measurement of attrition and should not be related to individual deaths*

Table 2.3 Sperm redundancy in some animals

Organism	*Common name*	*No. of sperms*	*No. of fertilisations*	*Source*
Hydra	Hydra	75	1?	Zihler, 1972
Limulus	Horseshoe crab	c.60 000 *minimum*	Per fertilisation	Brown and Knouse, 1973
Aedes	Mosquito	c.1500	c.57	Jones, 1968
Apis	Honey bee	6 million	<100 000?	Mackensen and Roberts, 1948
Dahlbominus	Parasitic 'wasp'	300	54	Wilkes, 1965
Drosophila	Fruit fly	c.4000	<700	Peacock and Erickson, 1965
Reticulotermes	Termite	c.9561	<200	Weesner, 1969
Aculus	Peach silver mite	50	23	Oldfield, Newell and Reid, 1972
Loligo	Squid	8 million (X25?)	?	Austin, Mann and Mann, 1964 Drew, 1911
6 Echinoids	Sea urchins	100–1000 *minimum*	Per fertilisation	Branham, 1972
Lytechinus	Sea urchin	60 *minimum?*	Per fertilisation	Timourian, Hubert and Stuart, 1972
Echinus	Sea urchin	1000	Per fertilisation	Tyler, 1962
Coturnia	Quail	20 million	10	Wentworth and Mellor, 1963
Gallus	Domestic fowl	2800 million	10	Nelsen, 1953
Megaliea	Kangaroo	32 million	1	Sadleir, 1965
Bos	Bull	4000 million	1	Spector, 1956
Equus	Horse	8400 million (in 70 ml)	1	Spector, 1956
Homo	Man	350 million	1	Spector, 1956
Mus	Mouse	3 million	6	Snyder, 1966
Myotis	Bat	300 million (in 0.05 ml)	1	Altmann and Dittmer, 1964
Oryctolagus	Rabbit	150 million	8	Mann, 1964
Rangifer	Reindeer	2 400 000	1	Dott, 1971
Sus	Pig	25 000 million (in 250 ml)	9	Spector, 1956

2. Only a *very* well designed species could produce only one acceptable sperm for each egg. Animal sperms do not usually exhibit chemotaxis, so it appears that collision with the egg is a chance affair. The number, n, of acceptable sperms required in any ejaculate to fertilise an egg can be regarded as a measure of the biological hazards of the journey: n must start for one to finish. There are few ejaculates, too, in which all sperms are physiologically effective. Many sperms lack tails or heads, are crooked or otherwise abnormal anatomically; others are simply immotile for a variety of reasons. Let us suppose all these problems result from 'construction faults' during spermatogenesis. Only a proportion of these, therefore, can fertilise and this situation must also be part of the overall redundancy that is programmed into the organism's reproductive tactics. So

$$R \text{ (sperm redundancy)} = mnr \ ,$$

where m is necessary provision for the unexpected, and is found from actual R by dividing by supposed r and supposed n.

The number of fertilised eggs, Z, is usually only slightly smaller than E, and not all these zygotes hatch, or even begin to cleave. We may draw lines anywhere, but embryo → larva → juvenile → adult → breeder all usually represent reductions in the total number from the original egg population. Part of this attrition is doubtless attributable to 'accident' (m) but part is also attributable to internal factors related to homozygosity, mutation or other genetic load factors, including the 'chiasma-associated unacceptability of *Figure 2.4*, whatever this is. The degree of attrition, R, from any one 'stage' X to the next stage Y, for example from fertilisations to hatches (R_H, *see Table 2.2*) or from adults to breeders, has distinct components m, n and r, already introduced in the special case of the sperms offered in an ejaculate:

m – where $1/m$ are physiologically and morphologically free from *accidental* damage and $1 - 1/m$ are *accidentally* (independently of genotype) prevented from completing the stage. Common poisons or oxygen lack are examples of irregularly but commonly occurring adverse conditions that could intervene to prevent $1 - 1/m$ completing the stage. Parasite infection is another possibility of the unpredictable attrition factors subsumed under m.

n – where only $1/n$ of perfect specimens are able to complete the stage, for non-genetic reasons that are associated with the particular reproductive strategy adopted. For example there may not be enough implantation space in mammals or lack of territory for grouse (p. 238) and these may be found quite regularly. n interacts with r as a measure of 'soft' selection: m covers insurance for accident, real 'profligacy'. n will be larger in really competitive situations: sperm numbers are high when several males can inseminate a female.

r – where only $1/r$ are genetically (developmentally) competent to complete the stage. r may be seen as 2 factors, r_x normally from the egg genotype and r_y from the sperm. $1 - 1/r$ will fail to complete the stage even in 'ideal' circumstances throughout all the stages of development, because here the causes of attrition are internal genetic causes ('hard' selection i.e. genetic lethals). r could also be taken to include extrachromosomal factors such as lack of competent mitochondria in protozoans, and inherited cortical damage to eggs, as well as the 'unacceptability' of gametes that might be associated with some cross-overs.

Thus, for example, pathological polyspermy would fall into the category of m or n, depending on whether it is occasional, or regular and always occurring in a specific proportion of cases. m, n, and r have all been described by their reciprocals, as fractions completing the stage, but are better considered as factors of R, that is, as the number of individuals needed to start for one organism to complete the stage; i.e. if $1/n = 1/30$ then 30 must start for one to finish. *Table 2.2* gives all these factors, in sequence, for a breeding pair. Redundancies may also be expressed as m, n, and r for any step of the reproductive situation, e.g. for human pregnancy:

$$R_{\text{preg}} = \frac{\text{fertilisations}}{\text{births}},$$

and is probably[39] between 3 and 12 in different circumstances, say 8; possibly $m = 2$ (malimplantation, disease e.g. tubal infection and various accidents abort half) $n = 2$ ('natural' and induced abortion as part of reproductive strategy, inviable mosaics from non-dysjunction at first cleavage, polyspermies etc. remove half) $r = 2$ (half of the remainder have genetic anomaly, e.g. trisomies or nullosomies from meiosis of egg or sperm, and are resorbed or aborted);

$$2 \times 2 \times 2 = 8 = R_{\text{preg}};$$

one birth results from eight fertilisations. It will be seen that n and m are sometimes not clearly separable, and that r is often difficult to dissociate too.

2.6 Reproductive Arithmetic in Context

Much of this chapter may seem to have overstated the obvious; but this was considered necessary because of many earlier assertions denying the importance, or occasionally indeed the existence, of reproductive losses. Saunders, for example has stated that: 'Almost without fail, each egg produced in the right environment does form a new individual that, in turn, makes sperms or eggs that begin another generation'[40]. Or Sadleir[41]: 'unlike numbers of invertebrate species in which many members of the species are asexual and play no part in reproduction, every individual in all vertebrate species takes part in reproduction at one time or another'! It is clear from the context and the earlier half of the sentence that Sadleir means 'breeding' by his word 'reproduction' – and he is of course wrong. Or, more subtly, Haldane excluded the possibility of loss of about 60% of the populations (as homozygous rejects) even as a tenth of the number of adults spread over many generations[27]; he saw this scale of loss as clearly absurd ***and*** contrary to observation. But this loss is, of course, trivial compared with 2/40 million *normal* survival in each generation for a cod, and not excessive for fruit flies, which may have a survival of only about 1/40. Even wild mice may have only about 1/12 survival after birth[30,42], i.e. 91.5% loss before breeding: 60% as segregational genetic load as part of this 91.5% would not seem excessive even in one generation. Wallace accepted the criticisms, from Kimura and Crow, of his suggestions of eight alleles at each of 5000 heterotic loci (each with marginal heterosis) in *Drosophila*[11]; they pointed out that this would result in only one or two individuals per thousand surviving, and Wallace agreed that this was untenable. We should note, though, that it is probably[25] closer to the real situation than the 1/10 loss supposed by Kimura in another context[43].

Finally, Lewontin and Hubby's discovery[28] of high heterozygosity in wild fruit fly populations dropped a bomb in population genetics whose fire is still burning[44]. This is probably because many geneticists seem reluctant to accept the unavoidable fact that nearly all sexual organisms fail to breed because they die young.

This chapter has taken the opposite viewpoint from these geneticists, and sees Wallace's figures above as not totally unreasonable. 1/1000 surviving is a little extreme for *Drosophila*; females usually lay 150 eggs or so, but a very minor adjustment to his number of alleles or loci would modify his figure by an order of magnitude. Biological opinion on this issue seems polarised. On the one hand embryologists like Saunders, persuaded perhaps by the elegance of developmental processes, relegate most death to accident, and indeed seem not to admit its extent; the same may be said of those geneticists who follow Haldane and Fisher. Wallace and other modern geneticists, notably Lewontin, see much juvenile failure as selective (genetic load) and the surviving breeders as a special small part of the zygote population, whose own genetics must, however, reproduce the diversity of the zygote population in each generation. There is a great difference of philosophy here. The first view sees 'like producing like', well adapted to the environment so that, unless accident befalls a developing organism, it will live to breed. The second view sees 'like' producing mostly 'unlike', the unlike failing to grow up, and a proportion of the like also failing; those which survive to breed are a proportion of the like, carrying a **'balanced genome'** which decrees that the majority of the next generation will be 'unlike' too. *In the first view wild parents differ, and their contribution to the future differs accordingly, the classical neo-Darwinist view. In the 'balanced'* *view all wild parents are very alike, because they have been selected stringently from a diverse set of zygotes; any pair of them must produce this diversity again.*

Perhaps it is our built-in reluctance to consider harm coming to our own juveniles which prevents us from viewing dispassionately the almost total contribution by juveniles to the food web of this planet. It would be a much less interesting ecology if only senile animals provided fuel for other species.

There is such a diversity in reproductive strategy among animals, however, that the two views represent ends of a spectrum. Most mammals and birds, some beetles, and treefrogs take the first approach and produce few but viable offspring. Plaice, cod, oysters and perhaps bullfrogs take the second, more profligate, approach. Most organisms produce some variety, and so lose some offspring.

2.7 Human Reproductive Arithmetic

The grand exception is, of course, the human species. Other exceptional circumstances were excluded in our reproductive axioms (p. 12), where we considered only populations of stable size; the expansion phase of organisms invading new habitats is not covered by the axioms, nor are other rarities and anomalies such as longer-than-annual cycles. In many respects the arithmetic of human reproduction resembles the logarithmic expansion phase of a colonising species (*Figure 2.6*), and two rather complex explanations can interact to account for this logarithmic increase during all of human history.

This successful breeding history spans very different life habits; hunting and food-gathering was succeeded by agriculture, by urbanisation, and so on; indeed these life styles co-existed so that towns were supported by agriculture, raiding barbarians gathered spoils from towns. A complex interactive cultural ecology was formed, whose major property is the continuous invention of new habitats, new niches. This scenario has been historically documented by Darlington[45], who views the mobile expertise-carriers as the medium of cultural diversification. Fremlin[46] also described the history of man as continuous invention, and showed that technology *could* allow the increase in numbers to proceed disastrously, up to the physical limits. Such views explain the directions of human diversification, and increase of numbers, as a constant invention of new habitats to expand into. Fremlin documents these as loss of erstwhile constraints: little predation after the invention of stone tools and fire allows some increase; then winter (or dry season) starvation is alleviated by storage of food in settlements; agriculture follows, allowing greater numbers again; but these may equally be seen as invention of new ways of life – habitat diversification.

The achievement of success in other organisms has usually led to a reduction in breeding rate and to a balanced reproductive strategy. The downhill bicycle race of man's incontinent increase and diversification contrasts with the apparently conservative reproductive strategies of some

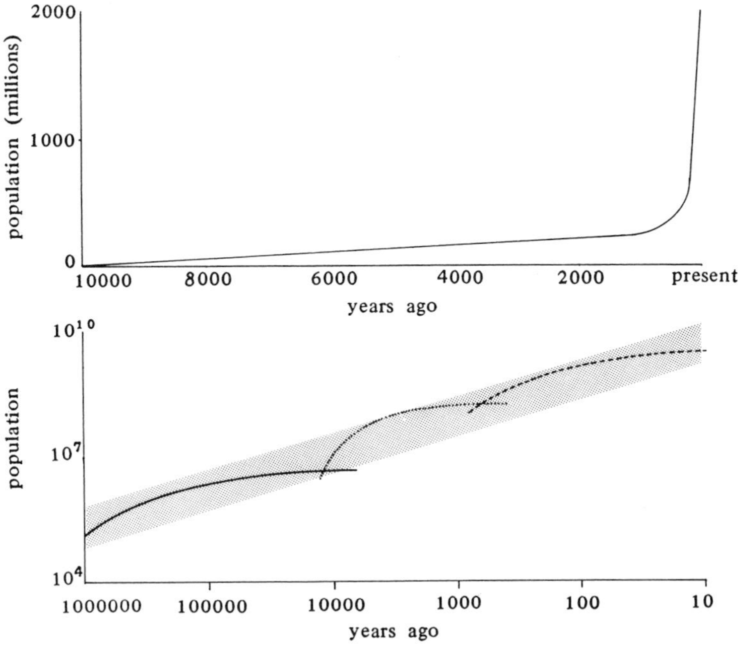

Figure 2.6 HUMAN POPULATION INCREASE (after Deevey, 1960). The upper graph shows an arithmetic plot of human numbers during our history, according to certain assumptions; note that wherever *one stands on such an exponential curve, yesterday looks placid and tomorrow is an emergency. If there is an emergency, it is not because of the shape of the increase but because we now approach a limit. The lower graph, plotted on log/log scales, exemplifies this. Deevey believed that three population surges occurred (toolmaking, agriculture, science/technology) but his figures are so uncertain that one straight line is an alternative interpretation, as shown in stipple. See also* Figure 14.2 *for a micro-view*

other innovators, dolphins, elephants and gorillas. This can be seen as a direct consequence of the involvement of continuous sexuality in his cultural patterns, so that sexuality, and therefore breeding, characterise the dominant individuals in most human cultures. The biblical concept of 'natural increase' may, then, be the natural result of a coupling of our breeding mechanisms with our speciality, which is varied cultural interaction between versatile individuals. Like some tumours, our mode of life and our increase may be coupled, so that balanced reproduction is only possible in the presence of stringent external controls. Technology is committed to the removal of such controls, by public health measures like the eradication of disease vectors, by more efficient food production and by use of the third dimension in housing. In the study of human reproduction we must therefore consider our sexuality as the major force towards human breeding, mitigated by cultural prohibitions, by contraceptive practices, and by other internal (cultural) regulators to a doubling of the population only every half-century or thereabouts (*Figure 2.6*)! Whether external controls will correct our reproductive arithmetic is doubtful, because our technology is too successful and too committed for its maintenance to human increase. Like the angiosperms or the teleost fishes, we have succeeded in colonising our new niche to the point where our major competition is from other organisms of the same kind. They diversified genetically, and we diversify culturally, both of which increase the 'carrying capacity' of the habitat. In a real sense, the development of urban populations resembles the development of parasites like dodder or mistletoe, or predators like pike or *Hydracyon*. Like angiosperms and teleosts, too, the interesting reproductive arithmetic is small-scale, comparing similar or otherwise competing species or cultures. The large-scale expanding arithmetic of *all* angiosperms, *all* teleosts or *all* humans may be trivial even in an evolutionary sense. It is the tactics of competition within and between groups which concerns us most in the study of human reproduction (pp. 65 and 289).

REFERENCES

1. Darwin, 1859
2. Malthus, 1803
3. Herskowitz, 1967
4. Spector, 1956
5. Altman and Dittmer, 1962, 1972
6. Engelmann, 1970
7. Porter, 1972
8. Haldane, 1954
9. Bannister, Harding and Lockwood, 1974
10. Galtsoff, 1964
11. Wallace, 1970
12. Sadoglu, 1967
13. Jones and Wilkins, 1971
14. Whyte, 1965
15. Mather, 1964
16. Salisbury, 1930
17. Fisher, 1930a
18. Fisher, 1930b
19. Marinkovic, 1967
20. Gordon and Gordon, 1957
21. Smith, 1958 (p.130)
22. Manwell and Baker, 1970
23. Allison, 1955
24. Ohno, 1970
25. Smith, 1975 (p.150 *et seq.*)
26. Fisher, 1936
27. Haldane, 1957
28. Lewontin and Hubby, 1966
29. Williams, Koehn and Mitton, 1973
30. Selander, Hunt and Young, 1969
31. Harris, 1971
32. Cohen, 1967
33. Rothschild, 1956
34. Austin, 1965 (p.64)
35. Cohen, 1973
36. Cohen and McNaughton, 1974
37. Baker, 1972
38. Roosen–Runge, 1962
39. Austin, 1972
40. Saunders, 1970
41. Sadleir, 1973
42. Asdell, 1964
43. Kimura and Crow, 1964
44. Lewontin, 1974
45. Darlington, 1969
46. Fremlin, 1972

2.8 Questions

1. Which kinds of animal populations are covered by the Axioms of Reproduction, and which are not?

2. From the Hardy–Weinberg ratios, can you work out other proportions of zygote loss which result in the coincidence of *Figure 2.3* (in this example a balanced selection mimics the repeating gamete frequencies which would be found without any selection)?

3. What is heterosis, and what is its contribution to a 'balanced' heredity in animal populations?

4. Compare the concepts of 'redundancy' and 'relative fitness' in relation to the composition of successive generations of sexual animals.

5. Does the multiplication of human numbers invalidate general ideas about the constancy of animal populations?

6. In what general ways may human population increase be prevented? Attempt to classify the methods you invent as contributing via factors r (difference between individual zygotes), m (accident) and n (reproductive strategy).

3

TACTICS OF SEXUAL CONGRESS

3.1 The Contradictions Between Vegetative and Sexual Needs

In the last chapter we saw that only a small proportion of sexually-produced organisms are privileged to mate. Thus there is an élite which, having survived by virtue of its adaptable genetic endowment and a modicum of sheer luck, now expresses those of its functions which, in its ancestors, allowed them finally to become ancestors and not just survivors. Very few organisms simply emit sperms or eggs when they become adult. The emission is at least modified by season, often by the presence of the opposite sex, and usually by visual, olfactory (chemical), and tactile stimuli as well as internal changes mediated by hormones. The sexes nearly always have some form of 'sexual congress' so that, somehow, eggs and sperms may meet.

Many organisms, for example some stick insects and the cockroach *Pycnoscelis*[1], are always parthenogenetic and are all female, not having any form of sexual congress. They are females who reproduce (females) without fertilisation. These creatures simply commence breeding when they are old enough or big enough; but many organisms, notably greenflies (aphids) and the water flea *Daphnia*, have parthenogenic reproductive cycles for most of the year and then interpolate a sexual cycle associated with the production of overwintering eggs or other resistant stages. The 'Amazon' mollies, *Mollienisia formosa*, are all females which produce diploid eggs, but these still require activation by sperms of related species of molly before development begins[2]. So these Amazon **(gynogenetic)** females, too, are dependent upon sexual congress for successful breeding even though the male does not contribute genetically to the progeny.

The need for sexual congress imposes a complete change of life style upon most organisms. Each survivor has usually grown up in effective isolation, selected for the single skills of feeding, fleeing, hiding, adapting to physiological variation in the environment, and perhaps especially avoiding parasitic infection by direct transfer. But for sexual activities, individuals must come together, albeit briefly. Not surprisingly, the adaptations of life style resulting from this contradiction between vegetative and sexual needs are dramatic. They are best seen as tactics within the overall reproductive strategy of the species, and one species may use several of these tactics. Rather than the anecdotal listing adopted by such authors as Wendt[3] or de Ropp[4] ('Love at arm's length', 'The long squeeze'), we will organise these tactics under separate headings[5] as tactics of incorporation, tactics of separation, and tactics of alternation, followed by a discussion of copulation as a major tactic. Then the special problems of sexual congress in social organisms will be considered. Finally we will consider human sexual congress similarly.

3.2 Tactics of Incorporation

There are a few organisms which manage to incorporate sexual needs into their vegetative ways of life. They change their vegetative habits little, if at all, whilst fulfilling their sexual needs. We may consider three distinct tactics of this kind. First we will consider shoaling, then the adoption of a parasitic male, and thirdly self-fertilising bisexuals. It should also perhaps be mentioned that there are a few oddities such as *Crepidula*, whose individuals can change their sexual pattern according to vegetative circumstances, so incorporating sexual phenomena into the vegetative life style.

3.2.1 SHOALS AND OTHER AGGREGATES

Many fish shoal as juveniles, and it seems that there is selection for this life style because only the 'edge' fishes are preyed upon; predators may also see the shoal as a large 'organism', too large to attack. Some species, for example herring and the zebra fish *Brachydanio*, maintain the shoal until the fish are adult. Thus the shoal serves as a sort of multi-individual bisexual organism. Usually male and female zebra fish spawn in shoals just before dawn whether or not the other sex is present, though some smaller male-female associations may also occur, particularly if there is a variety of 'geography' in the breeding tank. Such shoals may be all of an age, and with a high proportion of siblings and half-sibs so that inbreeding is fostered, as in the zebra fish and some midge swarms. This is unlike shoals of herring or flocks of some migratory birds, in which many ages may participate in the breeding.

Aggregates of mussels, polychaete worms and other effectively sedentary marine forms change their life style little if at all in the first and successive breeding seasons. Gametes are released either in response to environmental stimulus or to substances (pheromones or gamones, pp. 100 and 135) emitted into the sea-water with the gametes of other individuals. The depleted individuals continue feeding throughout. Echinoderms such as starfishes and sea urchins do not form such obvious aggregates for vegetative purposes, but here again breeding causes little interruption of such behaviour as these creatures can show; they even seem not to cluster together when gametes are emitted.

3.2.2 THE PARASITIC MALE

This tactic has been adopted by several unrelated creatures. *Bonellia* is a very aberrant sessile annelid in which the female has a dumpy body but a long proboscis. It hatches as a trochophore larva which soon elongates into a minute miracidium-like larva. If this fails to find a female, it simply becomes a female; but if it finds one, it takes up parasitic residence in the proboscis with other males, remaining tiny and producing sperms which are released when the host female releases her eggs. A vertebrate, the ceratoid angler fish, behaves in a similar fashion (*Plate 2*, p. 15). What is normally brought up from the abyssal depths is the female, but she has several tiny males attached to her head or to protuberances on her body. It is not certain, but probable, that the young become males only if they find a female to attach themselves to.

3.2.3 SELF-FERTILISING BISEXUALS

Again, a few examples of this tactic are found widely scattered in the animal kingdom. Some ciliates, instead of conjugating, demonstrate **apomixis**, in which a pattern of nuclear division and degeneration is followed by fusion as in conjugation; but all nuclei come from the one organism. A few species (e.g. *Paramecium bursaria*) do not show degeneration of the meganucleus at the time, so the sexual and vegetative forms co-exist. Tapeworms are the paradigm of the self-fertilising bisexual, but also show a change of hosts to initiate sex; most or all of the vegetative growth may have occurred in an intermediate host, as in *Schistocephalus*, (*Plate 2*, p. 15 and *Figure 4.11*, p. 90). Nevertheless, sexual congress does not usually complicate vegetative life even for those forms in which several individuals may co-exist in the vertebrate gut. Tapeworm proglottides are initially male then change to female (**protandrous**); it is uncertain whether the proglottides always mate, but the presence of elaborate intromittent and recipient organs (*Figure 4.11*) suggests that they do, probably across the coils of the same individual. The fish *Rivulus marmoratus* is a bisexual vertebrate which, although normally protandrous, can self-fertilise; it probably does upset its vegetative routine to do so, however, so should not be considered as using a tactic of incorporation.

3.2.4 *CREPIDULA* AND OTHER ODDITIES

Crepidula fornicata, the slipper limpet, is a very aptly named protandrous gastropod which is a sessile filter feeder. Young individuals settle, by preference on settled *Crepidula* shells, and become functional males with a long penis which they use to inseminate any female that is under them. If a male happens to be settled upon, it in turn becomes female, and receives the attention of any male above it in the pile of shells. Piles are usually only one shell 'wide' with old productive females at the bottom and newly settled males at the top. All individuals feed all the time. Another oddity is the ectoparasitic monogenean flatworm *Gyrodactylus* (*Plate 2*, p. 15). This feeds on the epidermis of fishes, and moves over the surface like a tiny leech, attached by the posterior haptor with claw-like hooks. Copulation occurs rarely, and the bisexual organism is 'hyper-viviparous'. Most adults have a juvenile in the uterus, which in turn has

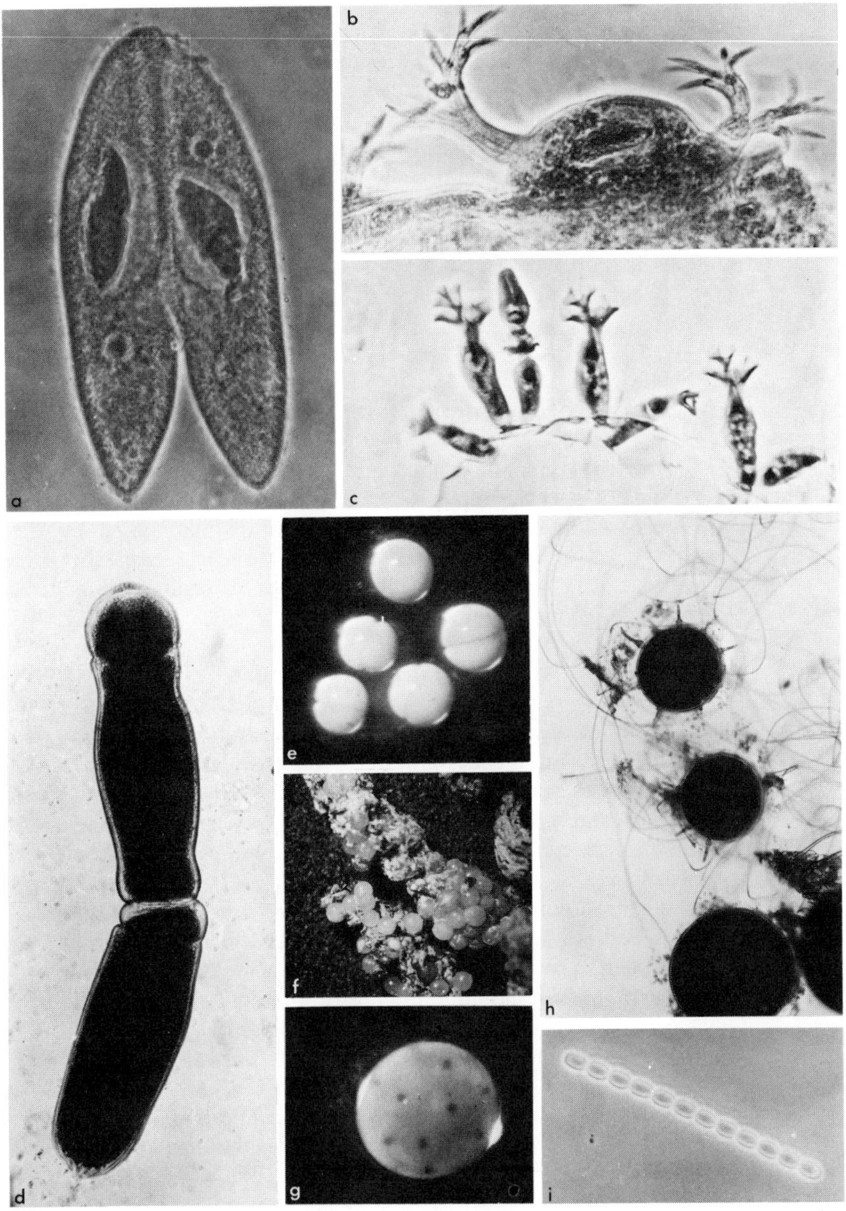

Plate 3 REPRODUCTION IN PROTOZOA (a) Conjugation in Paramecium *(compare* Figure 1.2*). (b) The suctorian* Dendrocommetes *on the edge of a* Gammarus *gill; a ciliated 'swarmer' can be seen in the brood pouch (compare* Figure 10.10*). (c) The ciliate* Spirochona *also lives on the edge of* Gammarus *gills. The spiral frill is used for feeding; conjugation is seen here, where two individuals have their frills apposed. (d) Two gamonts of* Gregarina *have left the gut wall and met in the lumen (*Figure 1.2*) (all gregarine photos by O.G. Harry). (e) The gamonts come to lie side by side and form an 'association cyst'. (f) Cysts on the faeces of locusts. (g) Inside the cyst, gametes have been formed and have fused; 'basal discs' of spore ducts are visible as dark areas. (h) Long ducts release chains of spores. (i) A chain of spores, each with eight sporozoites inside*

another embryo or juvenile in *its* uterus. Three 'Chinese boxes' nested inside one another is usual. Presumably insemination of all the generations results from one copulatory act, so that after birth, an individual might not need to acquire sperms for sexual reproduction to occur, because it has received sperms while *in utero*. The guppy, *Poecilia* (*Lebistes*) *reticulatus*, occasionally shows the same phenomenon, young females producing a few babies at about three months old without ever having 'seen' a male. This has been described as parthenogenesis[6], but is more likely to be carry-over of sperms from mother's hollow sperm-and-baby-containing ovary, since both sexes are to be found in such a litter.

3.3 Tactics of Separation

Many organisms separate sexual and vegetative life styles completely, so that complete specialisation is possible. The separation may be spatial, as in many colonies where a few breeders are maintained by the vegetative efforts of other individuals, or it may be temporal, so that the vegetative life is left behind when the organism is ready to breed.

3.3.1 COLONIES WITH SEXUAL UNITS

The familiar *Obelia* (*Plate 1*, p. 3) has a ramifying coenosarc from which feeding **polyps** and medusa-producing **blastostyles** branch (*Figure 1.1*, p. 6). The **medusae** escape into the sea, and after a period of maturation form ripe gonads and produce sperms or eggs; they are therefore the adults. But the polyps multiply, albeit asexually, and some polyps (e.g. *Hydra*) are clearly adults too. There is a case for regarding medusae as mobile adults of the colony, specialised for sexual function, but this presents a problem, because the common jelly fish (*Aurelia*) would then have to be regarded as the sexual analogue of its tiny asexually reproducing scyphistoma. Perhaps we should only regard those medusae which do not feed themselves as sexually specialised; those that do feed are adults, asexually-produced unspecialised organisms which are capable of both vegetative and sexual function.

Many coelenterates, notably the hydrozoan *Siphonophora* like the Portuguese man-of-war *Physalia*, are very complex colonies of polyp-like and medusa-like units (called **'persons'** in some older books) adapted for feeding, locomotion and protection, and gonophores (styles) upon which gonads are borne directly, or which bear sessile medusae which have gonads (*see Plate 1*, p. 3).

A *Volvox* colony, some thousands of photosynthesising *Chlamydomonas*-like flagellated cells up to 2 mm across, is a beautiful sight as it rolls gloriously on its way. There is said to be a group of cells without chlorophyll at the 'posterior' end, which are specialised for reproduction. In such an integrated colony, perhaps we should consider these not as separate organisms but as a gonad.

The true colonies of social insects relegate all sexual activity to very few animals. Termites (*Isoptera*) have kings and queens which copulate frequently, but hymenopterous social insects such as the honey bee may have permanent **queens**, the **drones** mating once and then dying. Perhaps this is because hymenopterous males are haploid, so their sperms would be free of 'chiasma-associated unacceptability'. The five million or so sperms inseminated by the honey bee drone are enough to fertilise up to half a million worker eggs, so there is a low sperm redundancy. On the other hand, in termites the male is chiasmate and diploid, so it might be surmised that only a small proportion of the sperms will be 'acceptable'. The spermatheca of the female simply would not have the capacity to accept enough high-redundancy sperms for the production of millions of progeny unless the sperms could be provided over a long period. Unfortunately, we do not know the chiasma-frequency and sperm redundancy of any termite, so at present this can only be an attractive explanation for a phenomenon which has puzzled entomologists since the last century.

Some other reproductive strategies, notably the harems of seals, many wild antelopes and other mammals, also restrict breeding to a few males; but the breeding males do their own feeding and territorial protection, so this is not true relegation of exclusively sexual function to a few individuals only, as in coelenterates and social insects.

The palolo worm and some other polychaetes, on the other hand, could be said very aptly to show a tactic of separation. Asexual reproduction by addition of segments is common in these worms (*Plate 2*, p. 15), and a new head appears when the worm has enough added length, i.e. is adult. However, during May–July, the asexually produced product of the palolo worm of the Pacific reef is the sexual form, a variant adult. During July, in a night of a new moon, these sexual forms are released and swarm at the surface, releasing gametes. They then die, so they may be considered to employ the next tactic too, terminal breeding.

3.3.2 TERMINAL BREEDING

Although many organisms restrict their breeding to a relatively short period towards the end of their lives, we are concerned here only with those which change their life style totally and, for example, stop feeding. There are three rather different examples which illumine this tactic; protozoa, mayflies and fishes like salmon and eels. These are organisms which have most clearly found the demands of food and sex to be antithetic.

The sexual protozoa are nearly all terminal breeders, if only because the whole vegetative organism actually becomes the gametes. Sometimes, even where some residual cytoplasm remains, as in the eugregarines after the synthesis of the association cyst around the gamonts (*see Figure 1.2*, p. 8 and *Plate 3*, p. 43), it is used to make complex eversible spore ducts and other associated gadgetry for spore protection and distribution and does not persist as a parent organism. Some purists would insist that if there is an organism remaining after the gametes are formed then we didn't have a protozoan to start with; i.e. that a *Volvox* is a multicellular organism. However, we could counter this; there are clear cases of *non*-terminal breeding in the protozoa, occurring in the *Ciliophora.* The conjugation of *Euciliata*, like *Paramecium* for example, is effectively terminal in that ex-conjugants rarely feed before division into daughter cells. But the situation in *Suctoria* is much more like the metazoa. *Dendrocommetes*[7] lives epizoically on the gills of *Gammarus* the freshwater shrimp and looks, when viewed from above, like a desert island, with palm trees which are the tentacles (*Plate 3*, p. 43). The eccentric hillock is the nucleus, which, prior to sexual reproduction, divides; but the cytoplasm does not cleave normally. Instead it cuts off an 'internal bud' into the 'brood pouch', which grows into another hillock (*Figure 10.10*, p. 195). This is the ciliated larva. Male gamonts are probably cut off, like polar bodies, by unequal divisions from the edge. So much for the protozoan counter-example, which surprisingly does *not* show terminal breeding

The mayflies (*Ephemoptera*) live for one, two or three years as highly specialised larvae in streams and rivers. They frequently produce a subimago, then the **imago** (adult) mayfly emerges. Its gut is air-filled and usually blind-ending. It cannot feed, but mates, lays eggs and dies. Other insects also, of course, restrict sexual activity to the adult, but this usually feeds too. However, some of the beautiful large silk moth imagos are also entirely sexual; they are also short-lived – mating, laying eggs and dying within one or two days.

Many fishes, but especially salmon and eels, show terminal breeding, and often change their habitat in order to breed. Both these examples will be described, as each shows several reproductive adaptations of wider interest.

Eels are commonly catadromous, that is, they mature in rivers and migrate to the sea to breed. Many fishes do this, including many races of the ten-spined stickleback and a few of the three-spined stickleback. Eels invade European and American eastern seaboard river mouths in vast numbers as young elvers in springtime, and feed as they swim upstream, finally achieving territories in slow-running waters, or even in ponds some distance from the river (eels certainly can travel miles across dewy grassland). The eels grow fairly slowly for about 10 years, achieving a weight of about 1 kg and a length of some 80 cm or more. At this time they are bronze in colour, and dissection reveals only 'streak' gonads; there is little morphological differentiation of the sexes. The liver is very large, and it and the muscles are very densely packed with glycogen. This 'brown eel' may apparently stay in fresh water for up to 40 years, but some begin metamorphosis at 10 years (*Figure 3.1*). The 'face' sharpens, the eyes enlarge, the colour tends toward the silvery, especially on the flanks and gill covers, and the behaviour changes. Brown eels are crepuscular, but the silver eels move during day and night, creeping across fields and following waters downstream until the estuary is reached. Some food is taken early in the metamorphosis, but as the change progresses feeding ceases. Males, who are slenderer and sharper-nosed, can now be easily distinguished from the bulkier females, and dissection reveals a body

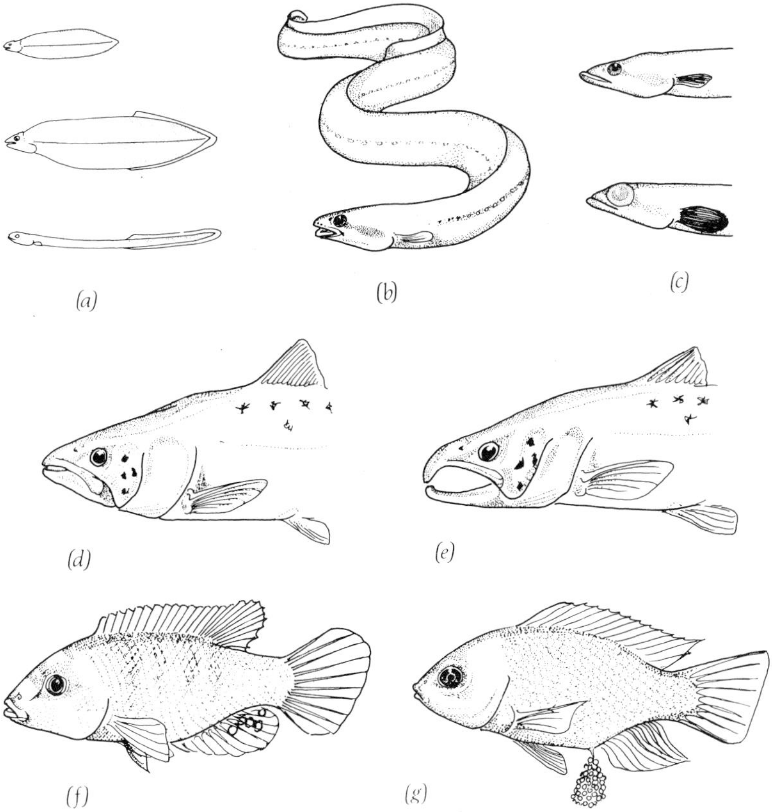

Figure 3.1 SEXUAL TACTICS OF SOME FISHES *(a) A young leptocephalus, an older one (compare* Plate 2*) and a young elver. (b) A 'brown' (river) eel. (c) Upper, 'silver' eel anterior end; note the underslung jaw. Lower, a transformed (sexual) silver eel; note the large eye and pectoral fin (from Meske, 1973)*
(d) Mature salmon, female. (e) Male sexual form showing peculiar jaw structure.
(f) A male Haplochromis *(e.g.* wingatii*) with 'egg mimics' on the anal fin*
(g) A male Tilapia *(unknowr species) showing a mass of egg mimics as an 'anal tuft'*

cavity full of a developing gonad, which is still rather diffuse, and also a large liver. The gut is pale, often transparent and difficult to find, and in several places it may lack a lumen (there is disagreement among authors[8,10] about this).

These reproductive organisms now swim out into the ocean, and some achieve the sea-bed under the Sargasso sea off Bermuda. Here all the sex products are shed, and the rest of the eel decays.

There is some evidence that in the last stages of the journey breakdown of many tissues occurs, perhaps to add more nutrients to eggs. From the egg hatches a strange little leaf-shaped fish, totally transparent, called a **leptocephalus** larva (*Figure 3.1* and *Plate 15*, p. 197). These are common in mid-Atlantic plankton over winter, and achieve the American river mouths in January–February and the European river mouths in late spring, at about three years old. Tucker[8] has added a new and important idea. Because mean vertebral numbers are somewhat different in the European and American eels, they have for a long while been regarded as different species (*Anguilla anguilla* and *A. rostrata*), possibly with neighbouring spawning grounds, but with different developmental programmes so that they returned to the waters from which their parents had come, i.e. American or European streams. Tucker, however, suggested that the differences in the time spent in sea-water, and the differences in the temperatures of the water, could amply account for the disparity in vertebral numbers, which increase during the leptocephalus stage in all eels. There need, therefore, be only one species whose vertebral number differs as a result of the various routes the fish take into fresh waters. Further, he made the dramatic assertion that European silver eels might never breed at all. Bertin had shown them to lack the reserves necessary for the required journey of more than 2500 miles without food, and he suggested that there might be favourable currents; nevertheless, the puzzle remained. Tucker pointed out that all the observations could be accounted for if European eels all died off the Continental Shelf. The population of American eels would then provide larvae which would 'diffuse' outwards from the Sargasso and invade rivers as they found them. The failure of most larvae to breed, even if they survive to metamorphosis in European rivers, makes no difference to the basic American population, of course. This suggestion has received adverse comment from many biologists, notably from population geneticists, none of whose models have included waste on such a grand scale. Some biologists have collected evidence, mostly from biochemical polymorphisms[9], which tends to confirm Tucker's beliefs; but there is still a lot of doubt[10] about the truth of this admittedly elegant theory. We would now interpret the phenomenon, if it exists, as a necessary evolutionary consequence of the interaction of a catadromous species with Continental Drift; its breeding ground has moved too far from part of the home range.

Salmon apparently behave in a converse manner; in fact, the differences are more real than the similarities. The smolts move out of estuaries and, after some delay while their osmotic adaptation is made, they move onto the Continental Shelf, which is a very rich feeding ground. After about four years, they attain sexual maturity and return to the same estuary. The tale of their relentless pressing on toward the head-waters of streams has been told too often to need repeating. But a discovery and a suggestion about the biology of their breeding have not received the same publicity. The discovery concerns the paternal role of the 'cock' salmon, with his imperiously hooked jaw (*Figure 3.1*), after the 'nest' has been constructed and the great pair of fish have fanned it clear of sand and mud. Jones[11] has clearly shown for British salmon that the eggs are *not* usually fertilised by this male, but by a precociously mature parr who hovers under the female as she lays; this is not an occasional but a regular occurrence in British salmon reaches. Does the great male serve any purpose or is he as reproductively irrelevant as the European eel appears to be?

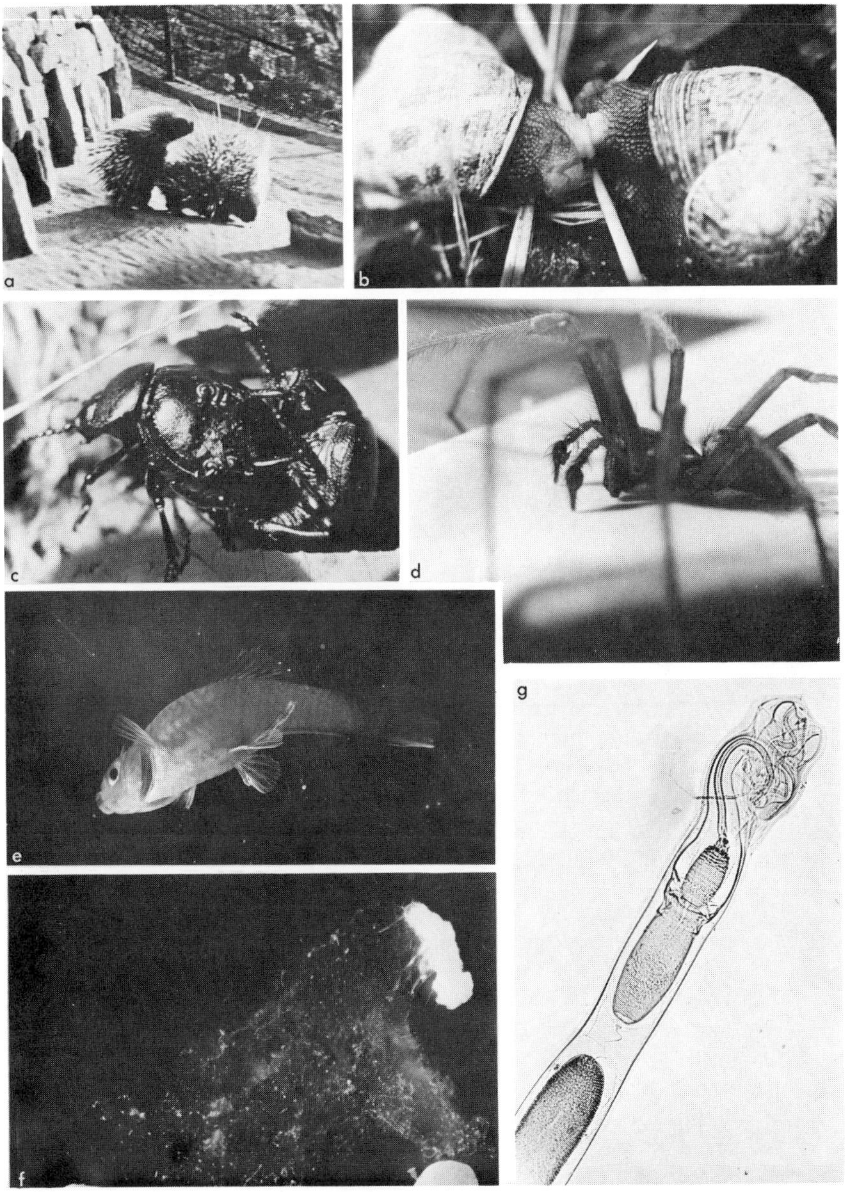

*Plate 4 THE TRANSFER OF SPERMS (a) Porcupines copulating: a dramatic incompatibility between vegetative and sexual needs. (b) Garden snails (*Helix*) transferring sperms to each other (*Figure 3.2 *and p. 56); a dart can be seen emerging from the head of the right-hand snail, having passed right through. (c) Beetles 'copulating'; note the 'suckers' on the front feet of the male but not the female. This may be a prolonged copulation. (d) The male house spider (*Tegenaria*) on his way to find a female; the pedipalps have special organs containing the sperms. (e) A male* platy *(*Platypoecilus maculatus*) in pre-copulatory position. Note the anal fin modified for spermatophore transit, the gonopodium. (f) An axolotl spermatophore: transparent jelly mass, about 1 cm high with the sperms at top right. (g) A spermatophore of* Sepia *the cuttlefish. Only the syringe-like expulsive mechanism is shown, about 5 mm long. At bottom left is the apex of the sperm sac, about 2 cm long and containing about 10^7 sperms.*

The following theory, held by many American salmon biologists, applies mostly to Pacific Alaskan salmon but may have relevance to the British cock salmon too. The head-waters of these Alaskan rivers are fed mostly by melting snow, and are very close to pure water. There is effectively no mineral content in this water, therefore the web of life therein is a very tenuous one; that is, until the salmon arrive in the spring. The salmon lay eggs and spray semen over them, contributing a little to the water. But then the animals die and decay, and the bacterial bloom supplies an organic boost to the new life that the mineral release has made possible. The emergence of many insect imagos from their pupae is timed to coincide with the arrival of the salmon, and their eggs hatch in time for the insect larvae to catch the bacteria and form a growing population by the time the young salmon hatch. The hatchlings, therefore, find a fair amount of food, much of it transformed from their parents' bodies. That this general pattern occurs is not in doubt; but the theory goes on to suggest that such 'priming' of the head-waters by the dead parents is necessary to provide food for the young, and until more is known of the quantitative ecology of the system this must remain speculative.

3.4 Tactics of Alternation

Most organisms neither assimilate their sexual and vegetative life styles, nor separate them in as dramatic a way as the bee or the eel. They alternate between sexual and vegetative periods, usually on an annual basis, so that breeding seasons alternate with feeding seasons. Strictly, this category should include only those organisms which show no sexual behaviour during vegetative interludes, and do not feed during the sexual periods. But in practice lines are difficult to draw because, for example in many mammals like deer, the sexes show rather different behaviour even out of season, and the animals do graze occasionally, and certainly drink, in the mating season. But there is certainly a dramatic alternation nevertheless. Perhaps we might conclude that unlike many of the tactics considered so far, that of alternation might be used more or less rigorously according to species or demand. The vegetative life may take second place during courtship, during the actual mating phenomenon and during brood care. Examples of all these are so familiar that we need spend only a short time discussing them.

3.4.1 COURTSHIP

Many people must now be familiar, if only via the television screen, with the stamping grounds of deer and antelope, on which dominant stags bicker for the best territories, and then mate with a succession of females, to the virtual exclusion of grazing behaviour. The most dramatic display is probably that of the Kob antelope. Dominant elephant seals, **harem masters**, have also received considerable publicity as they patrol their harems, often mating but not, apparently, feeding. Birds of many kinds, too, often direct all their activities towards attracting a mate; it is usually the male who directs all his energies towards this end, as in cock-of-the-rock, bower birds, and many others. Male frogs, croaking back at the pond before the female arrives, male fiddler crabs, and even the male house spiders (*Tegenaria*), familiar because they are sometimes found trapped in the bath on their way to find a female, are other examples.

3.4.2 AMPLEXUS

This is the maintained close physical association between the sexes prior to the release of gametes, and is universal among frogs and toads (*Plate 5*, p. 61), common in the insects (e.g. locusts), and is found in the ciliates as conjugation (*Figure 1.2*, p. 8). The pair of organisms is often held together by specially developed organs, for example the glandular warty thumbs of most male frogs and toads. Frogs and locusts may remain in amplexus for days prior to gamete release, and it is easily observable that the pair of organisms retains some locomotory abilities. Paired frogs can leap and swim, locusts can jump, and paired beetles can walk (*Plate 4*, p. 49) but not fly. Tree frog females have extra large sucker toes to carry the double weight.

The tactic is in many ways a very different tactic from copulation and not, as many authors have suggested, a step on the way to it. Copulation absolves the pair from the necessity of exact coincidence of timing of egg and sperm release; but amplexus is a device to keep the pair in association while lengthy ripening procedures go on which culminate in a complex synchronised gamete release. The eggs of female anurans (frogs and toads) must pass from the body cavity into the oviducts and receive jelly coats while the male is waiting, so that he can emit sperms at exactly the time when the eggs are laid.

Organisms like lampreys or Siamese fighting fish, in which the embrace of the male is required to assist ovulation, show some features of amplexus, and it is these features which lead to doubt whether such embraces should be classed as true copulation. The synchronous release of gametes into the water, or the production of eggs, is the climax of amplexus; in contrast, after true copulation the female leaves the brief association as an effectively bisexual organism, able to consummate the embrace at her convenience and carrying both sets of gametes protected within her; the climax of copulation, the achievement of fertilised eggs, always occurs long after the gamete transfer.

3.4.3 BROOD CARE

There are many organisms which stop feeding during brood care. The most notable are the various **mouth-brooding** fishes, especially cichlids of the supergenera *Tilapia* and *Haplochromis*, and the emperor penguin.

Birds, nearly all of which feed their young, direct the major part of their activity to brood care for this period. Some mammals, too, like the polar bear, the seal and the walrus, do not feed while the young are small. But most organisms which care for the young contrive to feed as well during this period, so do not employ this tactic of alternation.

3.5 Copulation

There seems little doubt that the first metazoan gametes met in the external medium. If Hadzi[12] is right about primeval ciliates partitioning their cytoplasm to produce our first cellular ancestors, however, this stage may have been preceded by gametic nuclei passing directly from one conjugant's cytoplasm into the cytoplasm of the partner as in all modern ciliates (*Figure 1.2*, p. 8). But the association of external fertilisation with other apparently primitive characters, and the variety of modifications found in different evolutionary lines, force us to the belief that this was the primitive system. It suffers from several great disadvantages, however. First, there is the trauma to the gametes from release into the medium, and the dilution problems attendant on this; although this is not serious for most primitive marine organisms whose cells are isosmotic with sea-water anyway, and seem to manage fertilisation with very simple-looking sperms and eggs. Secondly, both selection of sperms and brood care of the eggs require special adaptations if they are shed, and meet away from both parents' bodies. Thirdly and most important, the timing of release must be exact to the minute, and this makes demands upon the behaviour of external fertilisers which are only solved by the use of external zeitgebers (timing stimuli) such as those that operate in the palolo worm or by chemical stimuli or by amplexus. These methods all allow other species to anticipate the release of gametes. Polynesians, who make palolo soup, take advantage of the lunar information used by the worms.

Copulation is a major tactic employed in the solving of these three problems, and especially in relation to the problem of timing. 'Partial' copulations solve only some of the problems, and we will deal with these first, following on with a discussion of the varieties of copulation.

3.5.1 THE SPERMATOPHORE

Very many organisms package the spermatozoa, so that instead of a diffuse cloud of sperms the parents can deal with discrete parcels. Many animals with true copulation also package their sperms, perhaps retaining an ancestral habit for the regulation of sperm dosage; but many organisms in which the sperms are packaged copulate only 'incompletely', and it is these we are concerned with here.

Fertilisation in most crustaceans, for example, is effectively external but the eggs are usually held on appendages of the female as in decapods such as the crayfish and the crabs (*Plate 13*, p. 187), or in a brood pouch as in *Daphnia* (*Plate 13*). Males produce a large number of small spermatophores and some appendages frequently 'mould' the mass and give it to the female as a number of parcels (*Daphnia*) or as a more or less stringy bundle. The female's appendages draw these over the eggs, and the scarcely motile (crayfish) or immotile (crab) sperms penetrate the eggs. These very successful animals are notable in that their tactics don't really solve the first, 'external fertilisation', problem but do seem very concerned with the second two; that is to say, there may be trauma to the sperms, although there may also be protection by the cases in which they are packed (*Plate 4*, p. 49). But brood care occurs, and the spermatophore transfer to the female can occur several hours before or after ovulation and can be very rapid. In the crayfish *Astacus* egg laying and transfer to the swimmerets may take several hours, but spermatophore transit lasts only about 20 seconds. Therefore, the animals need only be vulnerable, that is to say they need only stop their normal vegetative activity, for about 20 seconds.

Newts and salamanders show a step nearer true copulation. There is a brief 'mating dance', with much tail-waving by the males directed to females with swollen cloacae. During and after this dance the males deposit surprisingly large transparent gelatinous spermatophores, each with a white blob of sperms in the middle: that of the axolotl (*Amblystoma*) is about a 1 cm hemisphere (*Plate 4*, p. 49). The female then picks these up with her cloacal lips, and the purpose of sexual congress has been achieved without the vulnerability associated with amplexus or even with intromission; the female is now effectively bisexual. Some hours or even days later, she lays strings of eggs on water plants, apparently taking care to distribute them widely, and often camouflaging them by pressing the water plant around them while they are still sticky. Contrast this with the anuran amphibia, frogs and toads, in which amplexus requires that the animals cease feeding for days on end while synchronising gamete release, the production of spawn.

Cephalopod molluscs come nearer still to true copulation. The male produces enormously complicated spermatophores, usually tubular and with syringe-like devices that are actuated by water absorption (*Plate 4*, p. 49). Such devices are used to expel the sperm mass forcefully into the female genital tract. The spermatophores of the squid *Loligo* are about 2 to 4 cm long, while that of *Octopus dofleini* is about 1 metre in length[13]! After a mating dance, which is often highly ritualised with much arm-twining and colour changing, the spermatophore, or a bundle of them, is transferred to the mantle cavity of the female by a modified arm (**hectocotylus**) of the male. The opening of the spermatophore may come to engage with the oviduct opening before it discharges the sperms. In some genera, e.g. *Argonauta* the paper nautilus, part of the arm itself acts as a spermatophore, crawling into the mantle cavity of the female by its own muscular movements! Again, the female is now effectively bisexual, and can lay the egg mass at her convenience. A number of terrestrial arthropods, such as mites and scorpions, use watertight spermatophores to transfer the gametes, instead of copulatory appendages. Some mites are very casual about the transfer; males just leave them about and females may pick them up when they find them. But scorpions have a very dramatic dance for the transfer. The pairing partners face each other and rush back and forth. After some minutes of this oscillation, the male contorts and deposits a very complex spermatophore with hooks and springs on it. He then entices the female over it, or seizes her chelae and draws her over it. The lips of the genital aperture face forward, catch the hooks and the whole device pole-vaults into the female opening! She then runs back into her territory, and lays eggs hours or days later, guarding them solicitously and carrying the hatched young on her back when making food and water sorties.

The spermatophore, then, may protect sperms from the environment. But its great advantage is that it allows the female to divorce the processes of sexual congress and egg-laying. The transfer of sperms can occur in an environment of open water, or flat sand. This environment may be totally unsuitable for egg-laying, but allows predators to be seen at great distances.

3.5.2 EMBRACES AND OTHER QUASI-COPULATIONS

There are many complex embraces and behaviour patterns which seem to be 'attempts' at copulation. But it must be emphasised again that the virtue of the copulatory tactic is its simplicity. Two sexual congresses of fish will be described to illustrate this point. Siamese fighting fish have a complex embrace system, rather like intermittent amplexus, but mouth-breeding cichlids achieve the functions of copulation with no embrace at all.

Siamese fighting fish (*Betta spendens*) males, when well-fed, and at a temperature of about 28 °C, choose a territory and build a nest of bubbles at the surface. Females showing the 'prepared' coloration, pale vertical bars on a dark ground (*Figure 1.1*, p. 6) are displayed to laterally with much tail-flapping by the male. After some hours of this, when the nest is about 10 cm across, the male begins to attack the female, and to threaten her with frontal gill-spread displays. She hides within, or at the edge of, his territory, and rises vertically to the surface to breathe, providing him with least provocation (this family of fishes are all obligate atmospheric air-breathers).

On the morning of the next day chases continue, but in the afternoon (with some pairs, the following morning) lateral displays begin again with some chivvying of the female towards the nest. She usually obliges by moving towards the nest, and the pair embrace, the male's body clasping the female across the pelvic region, and the fish sinking in the water, rolling over and separating (*Plate 5*, p. 61). This is repeated about every five minutes for some two hours, during which one presumes that eggs are being prepared for expression. Embraces at about two hours produce one or two eggs, which are fertilised as they emerge, caught in the male's mouth as they sink and placed in the nest (*Plate 5*). Embraces continue into the third hour, producing 10 to 15 eggs on each occasion, and usually become less frequent and associated with more chases at about 150 to 250 minutes. The male stays with the nest until the eggs hatch (after about 40 hours) and he guards them for some days thereafter. Excellent as this species is for demonstrating sexual congress to student classes, because it takes so long, it is clearly *not* really a copulation. It is much more like a staccato amplexus.

Related species, the gouramis and paradise fishes, only have a relatively small number (usually less than 20) embraces, and their eggs float up into the bubble-nest because they contain oil droplets. This is convenient for the male, but not for the developmental biologist because they rest blastodisc *down* in a watch-glass!

The cichlid fishes are a very large family, and mouth-brooding has arisen independently several times, in different genera[14]. This is clearly related to the way in which all cichlids carry the young from nest to nest in their mouths, and derives evolutionarily from the common pathological egg-eating behaviour that is to be found in many fish. When the male picks up the fertilised eggs in his mouth, after the female has laid them and he has shed semen on them, this seems a neat arrangement of quasi-viviparity. But in those species whose females brood the eggs, a problem arises because the male cleans a stone or digs a sand 'nest', the female lays the eggs on the prepared site and then she immediately takes them into her mouth, unfertilised. Clearly this is not very satisfactory, and somehow the problem of fertilisation has to be solved. This is achieved in two genera of cichlid fishes in quite different ways (*Figure 3.1*). In some species of the *Tilapia* group, the male has one or more 'egg mimics' pendant from his vent. The female picks at these after picking up the real eggs, and he releases sperms which fertilise the eggs that are already in her mouth. In the group *Haplochromis*, however, there are no egg mimics of this kind. Instead the male has a remarkable picture of a group of eggs on his anal fin (*Figure 3.1f*). He 'fertilises' the place where the female originally laid her eggs before removing them, and always spreads his anal fin toward her. She then attempts to pick up those eggs that she seems to have missed, and so fertilises the rest.

These two remarkably parallel tactics for solving the same problem were discovered by Wickler[15], who has used such behavioural traits for re-grouping several cichlid genera in a more convincingly phylogenetic way[14]. Our lesson from them is that it is not necessary to have intromission for eggs to hatch within the parent. The theory that copulation is a pre-adaptation for viviparity is well-known[16], but is made less tenable by this example, as by others like the carrying of eggs on swimmerets of female crustaceans. But all the purposes that I have suggested for a copulatory tactic are well served by this behaviour. The transfer of gametes is quick, no territory is needed, and the female is effectively bisexual afterwards. This behaviour, where there is no embrace, is more akin to true copulation than the mating activity of fighting fish which, as we saw, is more of an amplexus. Synchrony of timing of gamete release is still fairly critical, however, so this is not a true copulation.

3.5.3 TRUE COPULATION

Copulation in mammals, with which most of us are familiar, is fairly typical, but there are many variations which are not seen in mammals and we must go further afield for our examples, particularly in three parameters. The extent of penetration of the intromittent organ of the male varies greatly; the duration of the act lengthens from the original rapid sperm transfer, as other behaviour becomes integrated with it; and the relationship of the copulatory act to ovulation varies from casual to causal.

3.5.3.1 Extent of penetration

Most birds do not have intromittent organs but a few do, including ducks, geese and ostriches. The posterior part of the abdomen is rotated (twisted, not flexed) through 90° as the male 'treads' the female, and both cloacae are apposed momentarily. The process takes only seconds and many birds copulate in flight.

Many reptiles, however, and even a few amphibians like the 'tailed' frog *Ascaphus*, have intromittent organs of great complexity, often with complex hooks and scrolls to anchor them in the cloaca of the female. Snakes and lizards have double penes which are often of remarkable size and shape when erect, but they only penetrate the cloaca of the partner and usually do not have tubes but grooves for semen passage. Many insects have an **aedegus** or sclerotised penis, but others have only an arrangement of plates and hooks to anchor the abdomen while gametes are passed, often in a spermatophore tube as in locusts. Other insects, notably the beetles, do have a very complex aedegus for depositing semen far inside the female and this is often asymmetrical. That of the flea is notable in that it must negotiate two bends, and miss a blind alley, in the recipient tube; the intromittent organ is in two parts, a hard one for penetration and a thin, soft one for passage of sperms (*Figure 3.3*).

Gastropod molluscs have true copulation, and the penis, as in the whelk *Buccinum*, is often the largest organ of the animal. *Helix*, the garden snail, has a very complex copulation. After a stately courtship dance the bisexual partners press calcareous darts deep into each other's foot or head, then they curl until their reproductive apertures appose (*Plate 4*). Each has its own sperms stored in its 'flagellum', and this connects to the base of the large erectile penis (*Figure 3.2*). Both penes pass, in opposite directions, and each ejaculates into the other partner. The semen is received into the base of a long tube whose tip is the spermatheca. The snails then part; both now have a spermatheca filled with foreign sperms with which to fertilise their own eggs. It is important to move sperms into the flagellum for storage, because the foreign sperms take nearly the same route up towards the hermaphrodite gland as the snail's own sperms take downward, and they should not be mixed. The hermaphrodite gland produces *both* eggs and sperms, but the animals are effectively protandrous each year, only producing ripe eggs when their spermathecae are full (*see Figure 3.2*).

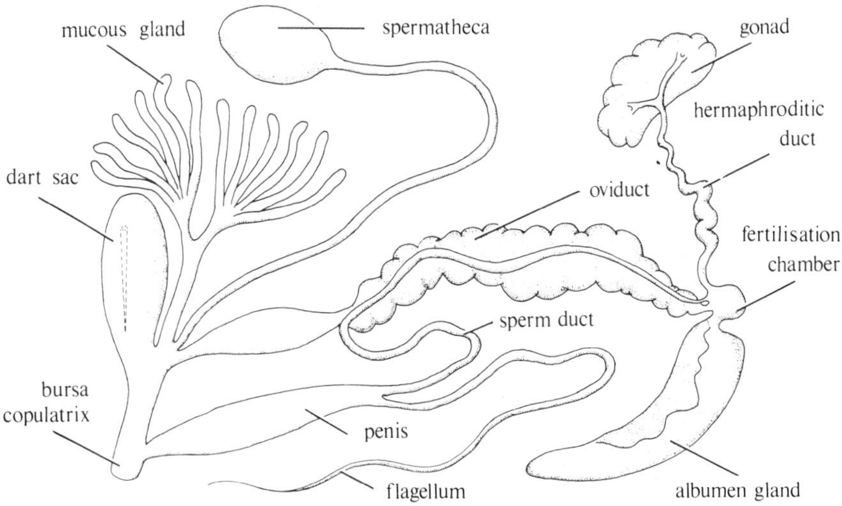

Figure 3.2 GENITAL SYSTEM OF THE SNAIL HELIX *Sperms mature first, and pass to the flagellum, where they are packed into a spermatophore. At copulation, after darts have been exchanged (*Plate 4*), both partners extrude their penes into the other's bursa copulatrix and pass a spermatophore into the spermatheca. The snails part, ova then mature and are passed to the fertilisation chamber, where they meet foreign sperms which have ascended the oviduct from the spermatheca. After fertilisation, the eggs pass down the oviduct, acquiring an albumen coat, and shells and other investments are added from the wall of the oviduct base and the 'mucous gland'. They are then laid in moist soil or under layers of leaves*

This strategy of bisexual copulation is found also in most platyhelminths (*Figure 3.3*). It not only ensures that every meeting can be a mating, but also that *both* partners are fecund when they part; it is thus four times as productive as the more common male/female sexual strategies.

There are many organisms with true copulation, but which use an appendage of other function as an intromittent organ. Mention has already been made of cephalopod molluscs with their hectocotylus arms; many spiders use a similar system. Copulation in spiders is reputably a risky business, and although there has been exaggeration of the dangers to the smaller male from his carnivorous but undiscriminating mate, there is no doubt that he is occasionally caught and eaten. Perhaps because of this extra need for speed and precision, nearly all male spiders have specially designed pedipalps (sensory appendages on the pronotum) with 'spongy' cuticle, a system of folds and spaces, or more usually a complex syringe-like apparatus for the storage of semen (*Plate 4*, p. 61). The male usually spins a small web with a small platform or purse onto which he emits a drop of semen, which he then takes up into his pedipalpal organ before setting off to find his lady-love. Some web-building spiders delay the building of this 'male web' till the female is found, when the web is often a satellite of the female's. But most hunting spiders and some web-builders like *Tegenaria*, the house spider commonly trapped in the bath on his amorous pilgrimage, fill their pedipalpal organs in their own home territory. After the male spider has found the female, a great variety of

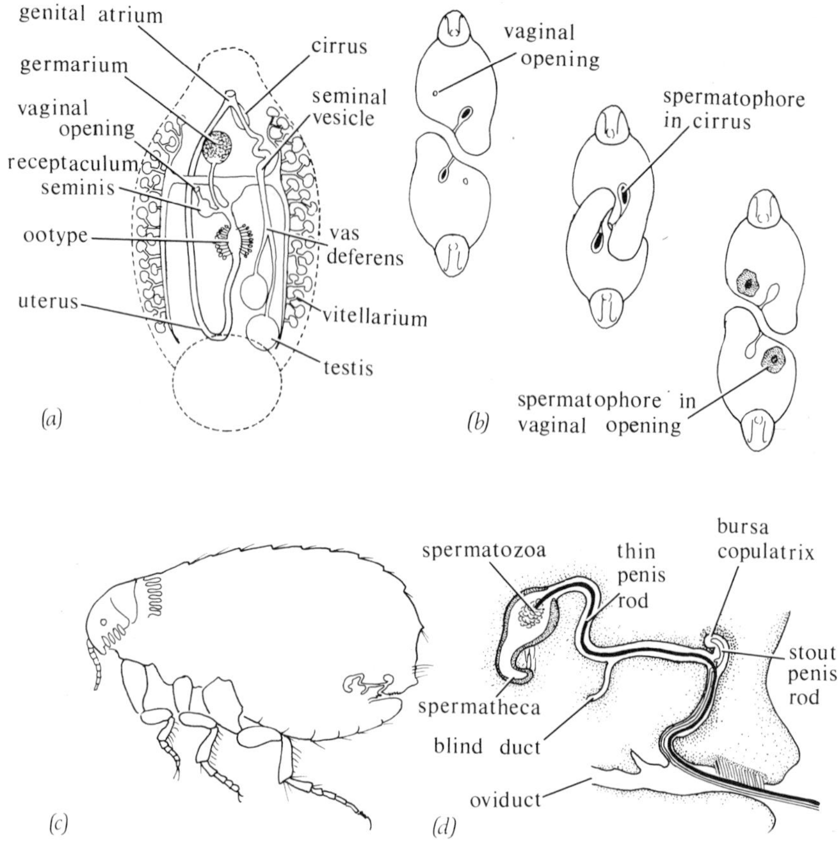

Figure 3.3 GENITAL TRACTS (a) The genitalia of a monogenean. Eggs, foreign sperms from the receptaculum seminis, and vitelline cells mix in the ootype, where a shell is moulded around all of them from some of the vitelline cell contents. (b) Reciprocal spermatophore exchange in the monogenean Endobdella *(after Kearn, 1970). (c) A female flea, showing the genital tract. (d) The two penis rods in this tract; the spermatozoa are deposited actually in the spermatheca (after Miriam Rothschild, 1965)*

morphological and behavioural strategies are brought into play, of which we can mention only three.

Many spiders, especially the lycosids, use special flapping or waving movements of the pedipalps, which seem to turn off the female's predatory behaviour. Surprisingly, the actual process of semen transfer may take quite a long time, up to about 20 minutes; during this time the male may change sides five or six times, inserting the pedipalps alternately into the female's reproductive apertures.

Several spiders, as well as some empid flies, use an 'appeasement' policy, the male bringing a prey animal to the female and mating while she is busy eating it. This has become extremely ritualised in most cases; for example, the female often holds the prey animal in a stereotyped way instead of attacking or eating it. The prey animal is often very carefully wrapped in a silk parcel, and males will sometimes make their parcels part of their competition rituals for females[17]. There is one species of fly in which the male presents the female with an empty parcel, a silk balloon, and then proceeds with copulation while she holds it[18].

In at least one genus, *Gastrotheca*, the male is so tiny that he is beneath the female's notice as food, and can inseminate her without risk. He could actually fit right into the female's reproductive bursa, but I can find no record of the actual copulatory mechanism!

The mechanism of sperm injection is of course very varied. The problem of delivery of the sperms while not infecting the female genital tract with bacteria is sometimes solved by storage of sperms in a special sac or **spermatheca**, remote from the sensitive egg-making organs themselves, which are usually rich in nutrients. Alternatively, a complex penis may be inserted very deeply into the female, as in the flea (*Figure 3.3*).

Most mammals have penes which become erect by blood pressure in **corpora cavernosa**, but there are surprising and illuminating variations. In most primates, like man, penetration is only into the vagina and semen is deposited at the cranial end; the same occurs in the rabbit. But in some rodents, e.g. the mouse and hamster, the penis has keratinised spines on the glans which engage the cervix, so ejaculation of whole semen is into the uterus, and only the coagulum is left as a vaginal plug during withdrawal. The pig, too, engages the cervix with his penis, but here the glans has a corkscrew tip which is gripped by the muscles around the cervical canal; 200 to 500 ml is ejaculated into the uterus. The rhinoceros has a very complex penis[19], which probably penetrates at least to the internal cervical os before ejaculation. At the other extreme, the hyena has very 'primitive' external genitalia, surprisingly alike in the two sexes[20], and it may be that penetration is only into the vestibule in this mammal. There are still many puzzles, as with the apparently unerectile small penis of the male spider monkey, whose female has a very large erectile, indeed mobile, clitoris. Some of the small carnivores, too, have strange sex organs and it is unclear how they work. The name of the palm civet *Paradoxurus hermaphroditus* speaks for itself.

3.5.3.2 Duration

Unless it is complicated by factors involving egg transport as in amplexus, true copulation is usually short. Animals which one would expect to be vulnerable at this time are often spectacularly brief, as in the rabbit, 5 to 15 seconds; some species of deer may complete intromission and ejaculation within two seconds. The bull, whose fibroelastic penis is normally ∽-shaped in the sheath, can straighten it by muscular action and complete copulation in 20 to 30 seconds. Rams take rather less time, but boars, presumably less vulnerable in the wild, take 15 to 20 minutes. Horses take about five minutes in which they ejaculate about 150 ml. In hamsters, the oestrus female is courted briefly by the male, who mounts and achieves intromission many times, though ejaculation may occur only a few times. In the golden hamster 150 very brief intromissions in two hours is normal, with other kinds of behaviour interpolated; maximum sperm donation occurs during about six ejaculations[21] from intromissions 80 to 110. Even here, total time of intromission is probably less than one minute.

Additionally, many mammals have made copulatory activity serve as part of the complex behaviour associated with courtship and in these cases copulation may be extended in time. This is most notable in those creatures with a complex social and behavioural repertoire, especially the carnivores and primates, but some rodents show it too[22]. In dogs there is a complex ritual of recognition between the sexes, with the male presenting laterally, and a receptive bitch allowing investigation of her genitalia by his nose[23]. He erects, mounts and achieves intromission into the swollen vulva. The bitch then rapidly backs against his legs as he thrusts, resulting in very deep penetration into the long vagina. Muscles at the vaginal entry then clamp tight, and the bulbar caudal end of the penis **(corona glandis)** swells to fill virtually the whole pelvic bowl of the female. This 'lock' mechanism, called the **'tie'**, is presumably a device to prevent separation during social interaction in the pack, and remains until ejaculation at least 5 to 15 minutes later, when detumescence occurs and the pair are released. Usually, before this happens, the male dismounts and the dogs stand back to back during the period of the 'tie'. Generally, this 'tie' is necessary to ensure fertilisation, and many breeders even consider that it is related to the size of the litter, although there is no sound evidence for this.

Cats, too, have a form of 'tie' mechanism, in that the penis has very short sharp keratinised 'barbs' on it rather like tiny whiskers, and these probably anchor it in the vagina until detumescence occurs. The 'tie' in these animals, again, might well prevent them being parted prematurely by social interaction.

Dolphins have two kinds of copulation[24]. There is a brief (20 seconds to 2 minutes) intromission which probably serves a social function and is not related to breeding, and a much longer (30 minutes to 6 hours) copulation which may be! During this latter period the vagina, which has a series of interlocking valves, may be flushed with urine by the female, whose cloacal lips are clamped right around the penis. The lengthy procedure may help to avoid contact of the semen with the hypertonic sea-water, and the 'urine' is probably a special isosmotic secretion for the occasion. Breathing is alternate, about every 30 seconds to 2 minutes, during the act, but it is difficult to know when ejaculation occurs. Other marine mammals may mate on shore, as do seals. But sea otters, marine turtles and some penguins seem, perversely, to mate in the sea and apparently have little adaptation to prevent dilution with sea-water. An alternative to the patient approach of the dolphins is shown by two rather different species of whales, the humpback and sperm. Individuals of both are said[25] to dive deeply, press belly-to-belly in the depths, and then leap together out of the sea, achieving intromission in the seconds above water! This unlikely story nevertheless has good evidence for it, but we do not, of course, know if this is the regular copulatory behaviour or the rare

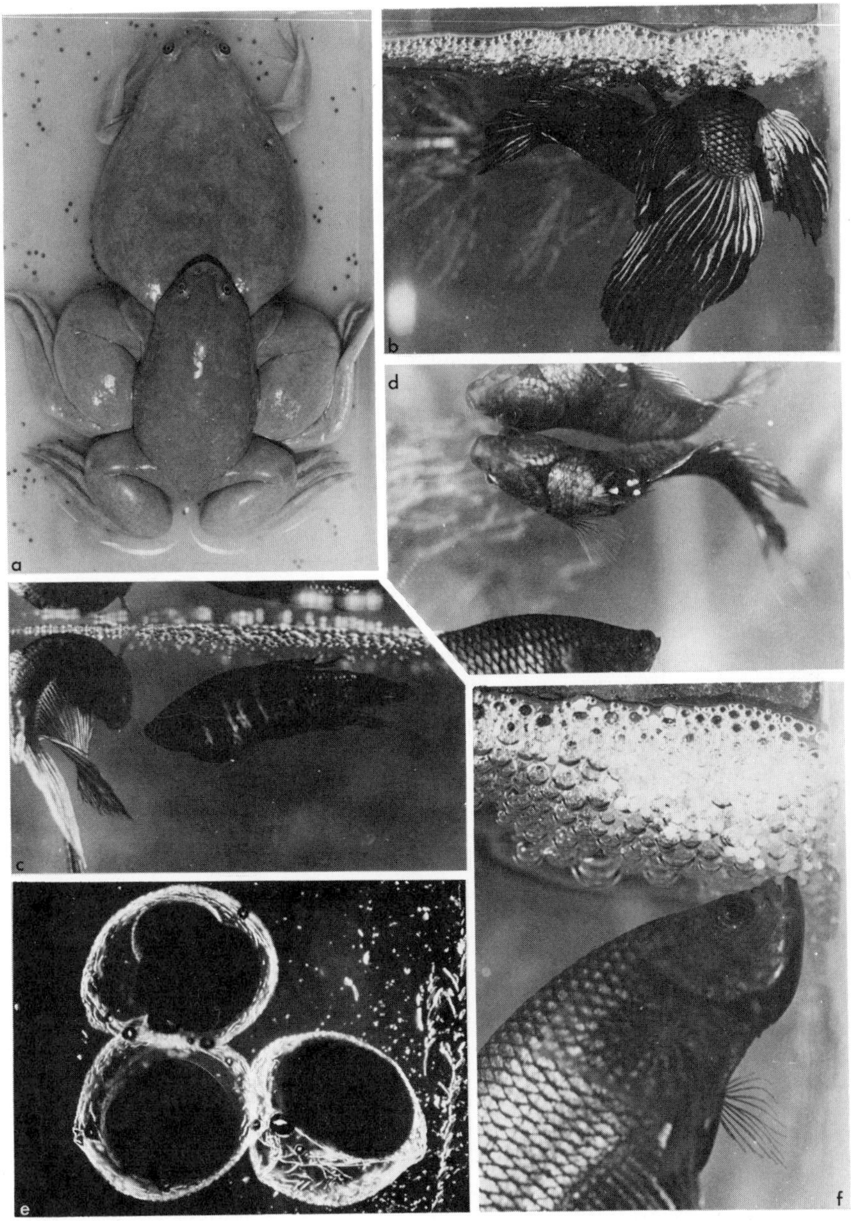

*Plate 5 AMPLEXUS AND COUPLING (a) A pair of toads (*Xenopus*) at the climax of amplexus. Eggs and sperms have been released synchronously. (b) A pair of Siamese fighting fish (*Betta*) coupling under the bubble nest. There are already eggs in this nest (Photo by K. Tyler). (c) One of the unsuccessful couplings, some two hours before (b), the male remains curled as if around the female. (d) After a successful coupling like (b), the female holds a group of fertilising eggs on her anal fin, from where they drift down to the male (Photo by K. Tyler). (e) Three eggs within five minutes of laying. The upper one has already separated its yolk (below) from cytoplasm (upper bump with light crescent) and the chorions are expanding. (f) The male tends the eggs among the bubbles (Photo by K. Tyler)*

exuberant experiments of occasional pairs. If it can occur at all it emphasises the rapidity of sperm transfer made possible by mammalian copulatory mechanisms.

3.5.3.3 Copulation and ovulation

There are a few organisms, including man, some birds and snakes, and many spiders, in which copulation is not related to ovulation but occurs more frequently, leaving the female effectively bisexual, i.e. carrying sperms, all the time. Bats, turtles and seals all copulate a long time before the associated ovulation. English bats, for instance, mate in autumn and sperms are stored in the uterus of the hibernating female till ovulation in spring[26]. Turtles mate at the time of egg-laying, but there is good evidence that these sperms fertilise the next egg layings. Seal females ovulate during lactation, and copulation also occurs irregularly during this time, so it is *next* year's offspring which are fertilised, as in bats and turtles; the major delay in seals occurs during embryonic life, by delayed implantation[20] (p. 202).

Very often, seasonal factors determine both the male's sex drive and the ovulatory mechanism, so that copulation and ovulation are coupled because both are environmentally determined. This is so in most temperate birds, and some reptiles like the slow-worm.

In the mammals there is usually a physiological mechanism which couples copulation and ovulation, instead of both sexes depending on outside factors. This coupling may occur in either direction; the female may time the male, or copulation by the male may elicit ovulation.

Most female mammals are **spontaneous ovulators**, which only elicit complete copulatory behaviour in the male at the appropriate time of their cycle (oestrus); although brief intromissions may occur as part of social interaction at other times. Examples include all rodents, nearly all primates, and most artiodactyls. Male mammals are usually ready for an oestrous female at any time but in some antelopes, e.g. Kob and many temperate deer, the stags sometimes only have a short fertile season (**rut**) and females only cycle during this season; this is an exaggeration of the usual case, when only a part of the year is available for mating because females are pregnant at other times.

The females of some species, however, are induced to ovulate by copulation. This is common among carnivores, especially mink, marten and sable where biting by the male, not penetration, elicits ovulation[27]. It occurs in the rabbit, too, but here there is a complex set of factors which elicits ovulation about 12 hours after copulation. We will deal with these in more detail in Chapters 11 and 16.

The necessary association between ovulation and copulation is not a narrow constraint, as in amplexus; instead it is a broad time bracket which allows considerable variations in timing and still ensures successful fertilisation. Especially, it allows sperm transfer and egg-laying to be separated in time, and usually in space too. The best environment for sperm transfer is not normally the best for egg-laying, birth or care of the young.

3.6 Sociality and the Restriction of Sexual Congress

Most of the organisms which we have considered make their own way in life, and whether or not they survive to breed is an individual matter of successive competences and of luck. A few, like seals and social insects, find their places in a social order as breeders or non-reproductives either by confrontation (bull seals) or by accident apparently unrelated to the genetics of the chosen (bees). The development of reproductive strategies in which some organisms are committed to non-breeding roles poses an evolutionary problem, for it appears to restrict the genetic contribution of such individuals to future generations, and so would seem certain to be lost from the breeding pool. There are three issues here. Firstly, some biologists have proposed a 'group selection': those *groups* better at producing offspring outbreed the *groups* who produce less, and group strategies may be selected even if individuals in the groups do not contribute genetically; most evolutionary geneticists oppose this view. Secondly, the non-reproductives may be regarded as surviving competitors prevented from breeding but contributing vegetative assistance to the breeders by more than the absence of their competition: worker bees and subordinate wild dogs help bring food for the reproductives.

Thirdly, and most important, the arithmetic of 'kin selection' shows that, because 'carrying capacity' of an ecosystem for a particular way of life is limited, marginal organisms (who would not quite survive to breed) can foster the passage of *some* of their genes, those which they share with the reproductives, by raising the likelihood of the reproductives breeding successfully. This is a subtle thought, suggested by Hamilton[28] and Smith[29] and very fully discussed in Brown[30]. Briefly, the selective advantage, in 'inclusive fitness' (passage of genes one shares, as well as possesses) depends upon the extent of shared genes: half with siblings, parents or offspring, a quarter with uncle-nephew, an eighth between cousins and so on. So the advantage *to the individual's* fitness of helping his kin breed must be less for more distant relatives, but is always better than being excluded from breeding by their competition. So non-reproductive positions in a familial society like bees or wolves do, contrary to one's first thoughts, give greater genetic contribution to the future for the genes of organisms filling them. It is better to assist at the sexual congress of kin than to compete with them unsuccessfully. Brown gives an even more dramatic example. He shows how kin selection may lead to sociality in *Hymenoptera*, in which females are diploid and males are haploid. All daughters receive the same genes from the father, but share half their maternal set with the mother or sisters. If a limited breeding budget is allowed, a female may spend this on her own offspring, or on helping to rear her sisters (later eggs from the same mother). In the first case, the proportion of her genes in the young is 50%, but in the second it is 75%!

So genes favouring the care of siblings would tend to increase compared with genes for own-progeny rearing. This is a special case, haplo-diploidy in *Hymenoptera*, and is associated with at least 10 independent origins of sociality among hymenopterans compared with one in all the rest of the insects. There is an assumption in this kin-selection view, that genetic recombinations and assortments of the gene set are being selected, and not mutations; that is to say that the population is adapting (*k*-selection, p. 313) and not colonising (*r*-selection, p. 316).

There is another restriction of sexual congress in most social groups, and this is less easy to understand without some kind of 'group selection'. In non-social animals earlier sexual congress is fostered by the resultant longer breeding period, juveniles which are 'first in the queue' and larger, and by these juveniles maturing earlier and so contributing still more to the gene pool; this effect is well seen in the domestication history of many small tropical aquarium fishes which now breed much earlier and smaller than wild stocks. In the wild this tendency is opposed by seasonal constraints on breeding, by the requirement for metamorphosis or gonad maturation time, and by the size requirement of adults in gaining and holding territories. In social organisms these constraints take different forms, and often seem 'artificial': the ritualised fighting of bull elephant seals over harems, the apparent competition of young queen bees or drones, the status challenges in primate colonies or canivore packs; so much energy seems wasted in these social activities that we feel could usefully be redirected into progeny production. In these social groups, too, effective sexual congress is restricted to *adult* animals, which have usually passed a **ritual test** in competition with their peers. While we can see that this selects for ability in passing such tests, it is less clear that the progeny of the selected reproductives are more competent at the vegetative functions which most of them will perform. Nevertheless, some such competition among peers is usual in social organisms. We may wonder whether the select from *any* test are more efficient organisms, whatever that means; or it may be that these colonies are selecting against deviance from a selected genome by behavioural tests – behaviour is probably a function dependent upon most of the genome. This point is discussed further on p. 224 where we consider the zygote genome and its effect on reproductive status. Whatever functions are served, it is clear that most social organisms restrict sexual congress, by ritual competition, to only a proportion of adults.

3.7 Sexual Congress in Human Society

Human societies vary greatly in their regulation of sexual attempts by adolescents, and in the ritual requirements for enjoyment of sexual intercourse by adults.

Our cousins the great apes show little exaggeration of sexual practice, which is well integrated into their vegetative life style; indeed the gorilla is a very unsexual primate indeed, not only in zoos but also in the wild.

Months may pass between acts of sexual intercourse between apparently compatible adults, just as with some human couples.

It is customary, when considering human sexual congress, to draw titillating examples from 'primitive' cultures, but the comparative sexual literature has a great tendency to generalise[27]: 'Hopi Indians copulate three or four times a week'; 'Among the Siriono, Dusun, Plains Cree, and Trobrianders, men and women often groom one another for some time before beginning to copulate'; 'For Chagga men it is reported that intercourse 10 times in a single night is not unusual'. There are two possible problems with such generalisations. Firstly, human beings in any culture have a vast range of individual variation in their behaviour related, among other factors, to their status in that culture; and their sexual behaviour is certainly not the least variable kind of behaviour. Secondly, word-of-mouth information about sexual practices, although it may be all that can be obtained, should be regarded with great caution: even in our own society, clinical histories suggest a rather different picture from that of the anthropologist. For example, it is apparently common in English and American 'middle-class' morality for there to be a great gap between the social myths and the practice. The social myths about our sexual practices disagree greatly among themselves in a manner from which we may draw instructions about possible variety in practices. The advertising, glossy-magazine myth portrays our young adults as sexually athletic, competent, sensual and avoiding marital sex roles in the search for variety. Co-existent with these conventions is a considerable literature of romantic love. Masturbation is denied yet is apparently widely practised; soft pornography is widely condemned yet even more widely sold; prostitution is a recognised part of all human societies.

Other societies, perhaps without a recent history of Victorian morality, have less of a gap between sexual myth and reality, according to the anthropologists quoted by Ford and Beach[27]: 'On Jaluit in the Marshall islands men make advances by rolling the eyes and pronouncing the name of the sexual organs' and the Crow Indians, although they believe that it weakens them, 'find it difficult to have intercourse less frequently than once per night'. The reliability of these reports should not be questioned, but one might wonder which of the social myths was being tapped by the anthropologist.

It is particularly difficult to disengage the reality from the various myths about selection of sex partners, even in our own culture. It is obvious that courting behaviour, or even 'flirting' has many individual variations but many common elements. Both visual stimuli and speech are usually important, and eye signals are common. Preparedness to reduce 'personal space' is signalled by several routes, and ritual eating or drinking is often involved. Unlike the Siriono, Trobriander, and other human groups where mutual grooming is a common preliminary pattern, in western society this has been reduced to hair-stroking or even to compliments on well-groomed appearance.

Kissing is a general human ritual which, perhaps because of our mobile and sensitive lips and tongues, has both been demoted to a ceremonial greeting and promoted to an element of **coital foreplay**. There are many societies in which it does not occur, and there are certainly many English people to whom the practice is abhorrent. The breasts are involved in sexual foreplay only by the human. Casual caresses may be seen in other species including chimpanzees and elephants; the mammary gland has been given an additional function as an accessory organ of foreplay in the human.

Such apparently unique attributes of human sexual congress find no lack of legendary histories to explain them. Breasts are seen as buttock mimics[3], with a bow towards the Gelada baboon with his (*not* her) bottom ornament repeated on his chest (or vice versa?). Face-to-face copulation has been seen by Hardy[32], and more recently by Elaine Morgan[33], as a relic of our sea-shore life, comparable with the copulation of 'other' sea mammals like whales and sea otters (who have an enormous tail to get around).

Religious regulation of sexual congress by involving sin and guilt is not unusual, nor is the mutilation of sexual anatomy for sexual or anti-sexual ends. Circumcision in Jews and clitoridectomy in Kikuyu both have origins in social philosophy rather than reproductive or even sexual mores. The rituals and myths associated with such practices have had a better press in the Hebrew case, and the practice of male circumcision may indeed have hygienic advantages. The Kikuyu practice seems to have been part of a general subjection of women, i.e. to have a social/political rather than a theological function.

Proscription, or prescription, of sexual congress for ritual or mythical reasons is usual in human societies. The regulation of sexual intercourse to legal relationships, in which responsibility for the care and upbringing of children is clear, is attempted with considerable success by most human groups. Usually behaviour is limited by concepts of evil, perversion, bad luck, or resultant disease or disability rather than by social or legal sanctions. Nearly all societies only explicitly condone many other forms of sexual relationship provided that children are not produced. Casual 'flirting' to 'droit de seigneur' may commonly occur, but there is in addition always a series of **'perversions'** punishable in nearly all societies, for example **incest**. Most include **bestiality**, **anal intercourse**, and **exhibitionism** as less major crimes. Intercourse during menstruation is specifically forbidden in many codes, especially the Judaeo-Christian. Such prohibitions probably do not affect fecundity adversely, as intercourse is usually prescribed during the fertile phase of the cycle. Many human societies have categories of persons ritually forbidden fertile sexuality for magical or mystical reasons[23,27]. **Berdaches**, **shamans**, **alyhas**, and **temple eunuchs** all are genetic males who, as juveniles, are forced into aberrant sexual roles. Monks and nuns are required to be non-reproductive and asexual, for mystical reasons (*see* p. 289).

Some forms of Hinduism and even of Buddhism pursue theological or philosophical ends via sexual arts and gratification. Denial of sexual or, indeed, reproductive function is also common in eastern religions. The *Kama Sutra*[34] *was* concerned with reproduction and with a universal philosophy as well as with the coital tactics for which it is chiefly prized in western society. *The Perfumed Garden*[35], an Islamic equivalent, has little to say about reproduction. Its chief interest to us here is that it perpetrates a very ancient myth, that women can be sexually excited by the sight of the male organs. This is not generally true, but the converse is. Many cultures prevent sight of the female genitals, even if not the breasts, while permitting male nudity, and Ford and Beach[27] list 14 cultures in which sight of female genitalia is always regarded as enticing, without any with the converse. They go on to compare other mammals in which the male is **'presented to'** by willing females; indeed sight may be a usual mammalian sexual stimulus, and may be primary at least in primates.

The last myth of human sexual congress for which room must be found is the Freudian myth. This[36] is that **'libido'**, the sexual urge, provides the motive energy for nearly all human activities, and that infants pass through **oral, anal** and **phallic stages** which leave psychological pathologies if incompletely experienced. The psychoanalyst then recalls the problem area to the patient, who re-lives the traumatic episode with the wisdom of adulthood **(genital phase)**, integrating his personality thereby. Arguments from this belief, ranging from immature **'clitoral orgasm'** to explanations of perversion and neurosis, have had wide currency in the western world for the last 50 years. While it is undeniable that many people have been helped by psychoanalytical methods, it is of course impossible to argue from this to the validity of the underlying beliefs. It is difficult, as Medawar[37] points out, to dissociate the analyst as a sympathetic listener from his Freudian contribution to any cure, or to discover the 'spontaneous remission rate' in a culture eager to cure by any or all methods. Suspicion is excited by the psychoanalytic 'explanation' of behaviour precisely because there is no anomaly which it cannot, in principle, explain; it is undisprovable and therefore unsatisfying as a system. The same criticism removes it from the realm of 'real' science, according to Popper[38]. The same is, of course, true of those extensions of the theory associated with Adler and, especially, with Wilhelm Reich[39]. This latter, whose followers today certainly number tens of thousands, proposed that **'orgone energy'** was the vital force of the universe, as the libido is of the human psyche; its manifestation can be controlled by magic of contagion, and people can influence clouds and perform all kinds of wonders by guiding the orgone energy of the atmosphere, which is blue and accounts for blue skies!

Sex is very important in every human being's life. The attraction of philosophies which tie this concern to universal issues is obvious. It is probably too simplistic to explain the amorous activities of the Greek gods, the central concern of the *Kama Sutra*, and the popularity of Freudian psychology and Reichian cosmology by reference to the popularity of sex, but it is in many ways a more attractive explanation than their mysticisms.

3.8 The Physiology of Human Sexual Intercourse

There has recently been a spate of books attempting to 'bring human sexuality out into the open', to 'remove shame from a natural human function', and so on[4,31]. This is parallelled by a growing number of fictional accounts[40,41] whose attraction (such as it is) depends upon the very shame

Table 3.1 Human sexual response, according to Masters and Johnson (1966)

MALE

	Excitement phase	*Plateau phase*	*Orgasmic phase*	*Resolution phase*
'Genital reactions'				
Penis	Erection, possibly lost and regained if episode prolonged	Increase in glans circumference	Expulsive contractions of urethra; 0.8 second intervals at first, lengthening	Detumescence in two stages
Testes	Elevation by 'shortening of spermatic cords'	Enlargement of testes (50%) further elevation	?	Loss of vasocongestive size increase, and descent
'Secondary organs' (ejaculatory ducts, vas, etc.)	No change observed	No change observed	Contractions which develop sensation of 'inevitability' and initiate ejaculation	No change observed
'Extragenital reactions'				
Nipples	Inconsistent erection	Erection and turgidity	No change observed	Involution of erection
Myotonia	Voluntary muscle tension	Increase in voluntary and involuntary tension	Loss of voluntary control; involuntary contractions and spasm of muscle groups	Myotonia lost after less than 5 minutes
Hyperventilation	None observed	Some, late in phase	Respiration rates as high as 40 per minute	Resolves
Tachycardia	Heart rate increases	100–175 per minute	110–180 per minute	Return to normal

NOTE: changes in scrotum, rectum, Cowper's glands, blood pressure, 'sex flush' and perspiration have been omitted

Table 3.1 (*continued*)

FEMALE

	Excitement phase	*Plateau phase*	*Orgasmic phase*	*Resolution phase*
'Genital reactions'				
Clitoris	Tumescence; increase in diameter and length of shaft	Withdrawal and retraction against symphysis	No change observed	Return to normal position within 5 to 10 seconds after orgasm; detumescence
Vagina	Lubrication and vasocongestion	Development of orgasmic platform at outer third; increase in width and depth	Contractions of orgasmic platform, 0.8 second intervals initially, 5 to 12 times	Detumescence and relaxation
Uterus	Elevation and irritability	Elevation proceeds, creating 'tenting' effect in vaginal depth	Corpus contractions; fluid exudes, *not* drawn in	Gaping of cervical os; later, descent into normal position
'Extragenital reactions'				
Breasts	Nipple erection; increase of breast size and tumescence of areolae	Turgidity of nipples; more size increase; areolar engorgement	No change observed	Detumescence and size reduction
Myotonia	Voluntary and some involuntary muscle tension	Increase in tension with semi-spastic contractions of facial, intercostal and abdominal muscles	Loss of voluntary control; involuntary contractions and spasm of muscle groups	Mytonia lost in less than 5 minutes
Hyperventilation	None observed	Some, late in phase	Respiration rates as high as 40 per minute	Resolves
Tachycardia	Heart rate increases	100 to 175 per minute	110 to 180+ per minute	Return to normal

NOTE: changes in the labia, rectum, Bartholin's glands, blood pressure, 'sex flush' and perspiration have been omitted and disgust which the sex educators purport to be destroying by a more public attitude than our Victorian ancestors.

A good case can be made that shame, and disgust too, are proper parts of the human attitude to sexuality, emotions which in our ancestors served to keep the actions both private and intimate. Lacking proper tactics of alternation (p. 50), or copulatory locks like many other social mammals (p. 60), the regulation of sexual activity by aversive as well as lusty emotions was probably a necessity. Ignorance or confusion about sexual and excretory function is probably not equivalent to these more ancient aversions.

Descriptions of what should perhaps be called the 'Hygiene of Sex' appeared in the thirties[42] and the post-war years[43], and it was perhaps inevitable that a laboratory would describe human sexual function and ascribe norms and acceptable variations. Masters and Johnson[44] used volunteers, and restricted their sample to those who could attain orgasm both by masturbation and in sexual intercourse, and could behave in the presence of observers. They 'discovered' that the events of sexual excitement could be classified into four phases (*Table 3.1*), and described these in language clearly intended to be abstruse and 'scientific': **excitement phase**, characterised by erection in men and vaginal lubrication in women; **plateau phase**, characterised by a high level of subjective excitement and associated with tachycardia and tachypnoea in both sexes, and with the development of the 'orgasmic platform' from the vasocongested outer third of the vagina; **orgasmic phase**, characterised by a sudden rise in tonus to spasm, with contractions associated with ejaculation in men or with the 'orgasmic platform' contractions in women; **resolution phase**, in which the male returns to a non-excited state, but the female may only dip in excitement prior to a repeated abbreviated plateau and orgasm. These descriptions (*Figure 3.4* and *Table 3.1*) have become the goals, the standards, not only of 'sex therapists' but also of the participants. The 0.8 second rhythms of female contractions, 5 if 'mild', 12 if 'intense' have become measures of 'effective sexual stimulation' by the male. Many 'educated' or sophisticated subcultures have now adopted these apparently objective goals as a measure of marital bliss.

Singer[45] has, however, brought some humanity back to this distressing objectivity. He is a philosopher, and deplores the essentialism which defines such a unitary goal, as if physiology (or taste) were universally the same; he advocates a pluralism in the aesthetic goals of sexual activities. That is to say, he denies that Masters and Johnson's description gives all of the aims of human sexual congress. But further, and much more important in the present context, he distinguishes two forms, two approaches, to human sexuality. One, which he calls the **sensuous** end of the spectrum of human sexual activity, is characterised by 'masturbatory' kinds of caresses and leads to those standard orgasms described by Masters and Johnson as physiologically identical in masturbatory and coital stimulation. There is, according to Singer, less emotional involvement in the sensuous activity, and it contrasts with **passionate** experience, which the laboratory setting of the Masters and Johnson experiments would preclude, and so exclude from consideration.

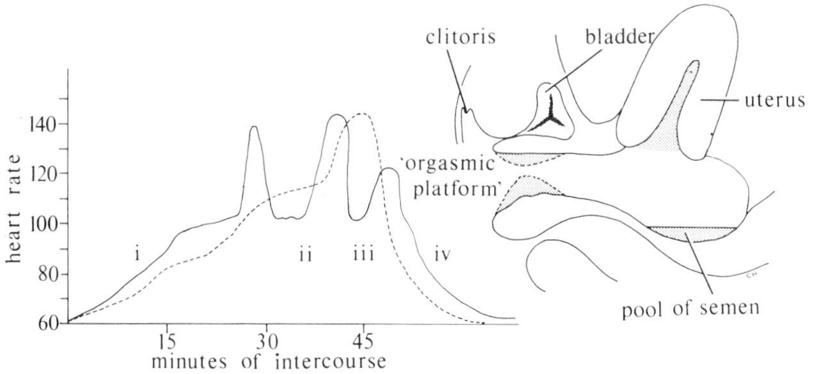

Figure 3.4 HUMAN SEXUAL RESPONSE The graph portrays an 'ideal' sexual intercourse, according to van der Velde or Walker; the man contrives three orgasmic episodes for the woman while he 'holds back' for as long as he can. The drawing is from Masters and Johnson, and shows the female genitalia at her orgasm, with the cervix drawn up out of the pool of semen, and the outer third of the vaginal barrel tumescent. According to these authors, orgasm in women is always accompanied by contractions of the orgasmic platform as well as by the raised cervix; but see Singer, 1973

Passionate sexuality as described by women in novels, by sexually experienced women and by the Foxes[46] (scientists who measured their own *private* sexual activities) may be very different from the public sensuality of the Washington Laboratories of Masters and Johnson. Intromission is often brief, and ejaculation within two minutes is normal, even if not common; female orgasmic response is 'deep', and uterine movement may occur *after* semen is deposited. The cervix does not rise out of the semen pool, and the woman is satisfied with one experience. Unlike the persistent fast breathing of sensual coition, becoming gasps at orgasm, the passionate orgasm is characterised by a glottal closure with contracting diaphragm, and no expiry for some tens of seconds. This difference was not noticed by the Foxes until they examined their instruments, because involvement in the passionate sex act greatly depresses visual, auditory and even proprioceptive sensitivity. The resemblance to coitus in other primates is remarkable, even to some evidence which ties passionate sexuality in at least some women to the preovulation phase of the cycle. Singer argues cogently that this passionate, non-intellectual, sexual intercourse is our basic reproductive activity, while the sensuous use of the same organs is the later acquisition for maintenance of the pair bond.

This duality of human sexuality, a double provenance of our sexual activities, abilities, customs and inhibitions, with reproductive and pleasure-giving facets, would eradicate many of the tortuous arguments implicit in the Masters and Johnson approach. For example, they suggested that orgasmic failure in nulliparous women has been selected because the unresolved orgasmic platform seals semen into the vagina, and so increases fertility. 'Adolescent sterility', especially the remarkable case of the Trobrianders who very rarely, if ever, conceive before marriage, despite a very full sex life, can be explained[45] by the inherently non-reproductive function of sensuous 'play' stimulation. Also, the loss of libido found by some women to be associated with contraceptive pills can be explained by their progestagenic effects.

This dual functioning of human sexual intercourse should not be considered exclusive; most sexual activities may have passionate and sensuous components. Indeed, each human sexual congress must have elements of both, just as it must include many other human activities. Nevertheless, Singer's contribution to the literature of human sexuality has replaced controversy over essentialisms by a recognition that several functions may be served by the same organs, even by the same act. To reduce human sexual congress to the search for one kind of orgasm, which could be achieved more intensely by masturbation in many people, is to simplify beyond reason. Human sexual congress has social constraints, emotional overtones, physiological possibilities and consequences, and possible reproductive sequels which make it one of the most intense, complicated and joyful of human experiences. As we understand more of it, and see it in a more general context, it should gain by association and not lose by abstraction.

REFERENCES

1. Engelmann, 1970
2. Hubbs, 1955
3. Wendt, 1965
4. De Ropp, 1969
5. Cohen, 1971
6. Spurway, 1953
7. Hyman, 1940
8. Tucker, 1959
9. Williams, Koehn and Mitton, 1973
10. D'Ancona, 1960
11. Jones and Orton, 1940
12. Hadzi, 1963
13. Mann, Martin and Thiersch, 1970
14. Fryer and Iles, 1972
15. Wickler, 1962
16. Young, 1957 (p.665)
17. Cloudesley-Thompson, 1968
18. Eibl-Eibesfeldt, 1970
19. Cave, 1964
20. Asdell, 1964
21. Chang and Schaeffer, 1957
22. Dewsbury and Jansen, 1972
23. Bermant and Davidson, 1974
24. McVay, 1964
25. Slijper, 1962
26. Racey, 1975
27. Ford and Beach, 1952
28. Hamilton, 1964
29. Smith, 1964
30. Brown, 1975
31. Morris, 1967
32. Hardy, 1960
33. Morgan, 1974
34. Vatsayana, 1963
35. Nefzawi, 1967
36. Freud, 1905
37. Medawar, 1972
38. Popper, 1963
39. Reich, 1968
40. Roth, 1967
41. Lovelace, 1974
42. Van der Velde, 1926
43. Walker, 1949
44. Masters and Johnson, 1966
45. Singer, 1973
46. Fox and Fox, 1971

3.9 Questions

1. How do adaptations for vegetative needs make sexual congress more complicated? (Start from *Plates 4* and *5*, perhaps.)

2. What adaptations for reproductive purposes make organisms more vulnerable?

3. What are tactics of separation? Think of examples where very different physiologies are required for vegetative and sexual function.

4. Consider the time relations of sperm release and ovulation in a variety of organisms.

5. How does the idea of 'kin selection' relate to the selection of a few breeders in social organisms? (Note: not only in *Hymenoptera*!)

6. How successful are cultural traditions and prohibitions in the regulation of human breeding in *your* society?

7. Suggest functions and antecedents for sexual reticence, gallantry, and coyness in human sexual congress.

8. Pair bonding is a tactic of incorporation, allowing different or alternating functions to the partners; to what extent has human monogamy led to the social subjection of either sex, or to the maintenance of sexual dysfunction?

4

ORGANS OF REPRODUCTION

4.1 The Germ Cells

These cells play a central role in the processes of sexual reproduction, and similar cells may take a comparable part in some forms of sexual or even vegetative reproduction (*see* p. 89). Very early in the development of most metazoans part of the organism is set aside and is not involved in the development of the rest. It is often recognisable by the basophilia of its cytoplasm even before that cytoplasm has become cellular, as in the polar plasm of some insect eggs (e.g. *Diptera*[1], *Figure 4.1*) and the posterior periblast cytoplasm of some teleosts (e.g. *Betta*). In other forms it may be only distinguishable by its later behaviour, although it is commonly found restricted to a small part of the egg, e.g. to the 'grey crescent' in frogs and the red cytoplasm of *Pomatoceros* (*Figure 4.1*). Species which show fragmentation of their chromosomes in all but the germinal tissues, like *Ascaris megacephala* and the gall midges (*Cecidomyidae*) protect those nuclei which, apparently fortuitously, arrive in this germinal cytoplasm (*Figure 10.4*, p. 183). It has been shown experimentally that if nuclei are prevented from entering the pole plasm or if it is irradiated, the resultant animal is normal but sterile[2]; equally, if other nuclei come in contact with this cytoplasm, they are 'protected'. So the peculiarity of the germinal tissue of these insects, like that of frogs, *Pomatoceros*, *Betta*, and probably all other animals, resides in the special cytoplasm and not in the chromosomes.

Germ cells are only rarely (e.g. in nematodes) found in or near the developing gonad during early embryology. They usually wait outside the developmental turmoil of the embryo proper and, when organogenesis is well on its way they migrate in and colonise the forming gonads (*Figure 1.2*, p. 8). In the vertebrates, they commonly wait on the yolk sac before migrating onto the lateral gut walls, up into the dorsal mesentery then laterally into the gonadal ridge hanging into the peritoneum (*Figure 16.2*, p. 265). There is some evidence that a few may be carried via the blood stream[3], especially in birds. In mice there are about 100 of these cells, but a sub-fertile mutant has been shown to have much fewer[4].

The sex of the gonad is not determined by that of the germ cells in it but by the genetic constitution of its own cells. The somatic tissues of the gonads usually 'instruct' the germ cells whether to make eggs or sperms[5]; but in some experimental combinations in some species, sterile gonads result from incompatibility of chromosomal sex between gonadal

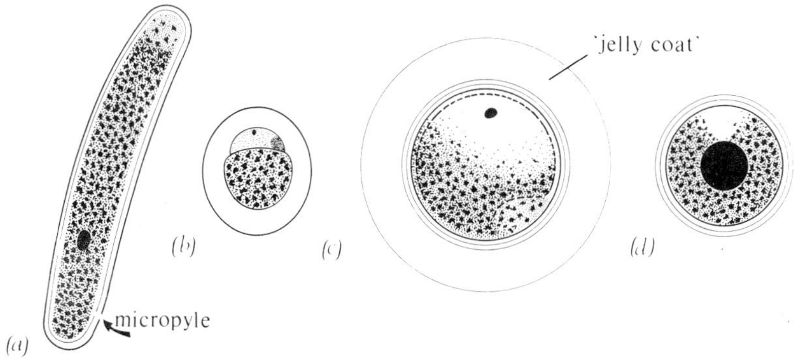

*Figure 4.1 SPECIAL OOPLASMS IN FERTILISED EGGS (a) The egg of an insect (*Cecidomyidae*); the thick chorion is pierced by the micropyle through which the sperm enters. The yolk is central, leaving a thin cortical layer of yolk-free cytoplasm. At the 'vegetal' (anti-micropylar) end is the* ***pole-plasm*** *containing polar granules; cleavage nuclei in this region do not lose chromosomes as the somatic nuclei do (*Figure 10.4*) (after Mahowald, 1972). (b) The egg of a teleost (*Betta*); the cytoplasm has aggregated prior to first cleavage, and a basophilic area of cytoplasm can be seen at one side (after Tin-tin May, unpublished). (c) The egg of a frog, imagined in vertical section with the albumen 'jelly coat'. The pigment layer (shown deep in the cortex for clarity) has dipped further into the yolky cytoplasm opposite the sperm entry point. This cytoplasm, the grey crescent, will become the germ cells. (d) The egg of* Pomatoceros*, viewed from the animal pole. Yolk has cleared from a segment of the egg, and more red pigment can be seen there. This region will contribute to 4d (and 2d) in the larva, and therefore to the adult mesoderm in which the gonads, and probably germ cells, arise*

tissue and its germ cells. That is to say, germ cells may not be completely labile, and in some cases may fail to make gametes at all.

4.2 Ovaries

These are the organs in which eggs are actually produced, and differ greatly in different animals.

Echinoderm ovaries are simple sacs with convoluted walls, some of whose cells are germ cells (*Figure 4.2*). These produce many oogonia by mitosis, which in turn produce many oocytes, which are shed from the wall and come to fill the lumen (*Plate 6a*). They accumulate yolk, grow and mature in this cavity from which they are ejected directly into the sea through the genital pore.

Nematodes have a thin tubular ovary whose tip is the germarium, in which resides the germ cell population. They divide, producing oogonia and, apparently, cells of the ovary wall. The oogonia in the lumen divide several times, resulting in oocytes which meet sperms where the ovary grades insensibly into oviduct (*Plate 10*). Insects have ovaries of three rather different kinds (*Figure 4.3*). All are formed of parallel **ovarioles**, but these differ in structure. **Panoistic** ovaries seem the simplest, and are found in orthopterans and some other more 'primitive' insects. **Polytrophic** ovaries (*Plate 6*, p. 83) are found in *Diptera* (flies) and *Coleoptera* (beetles). The **nurse cells** are cousins of the oocyte,

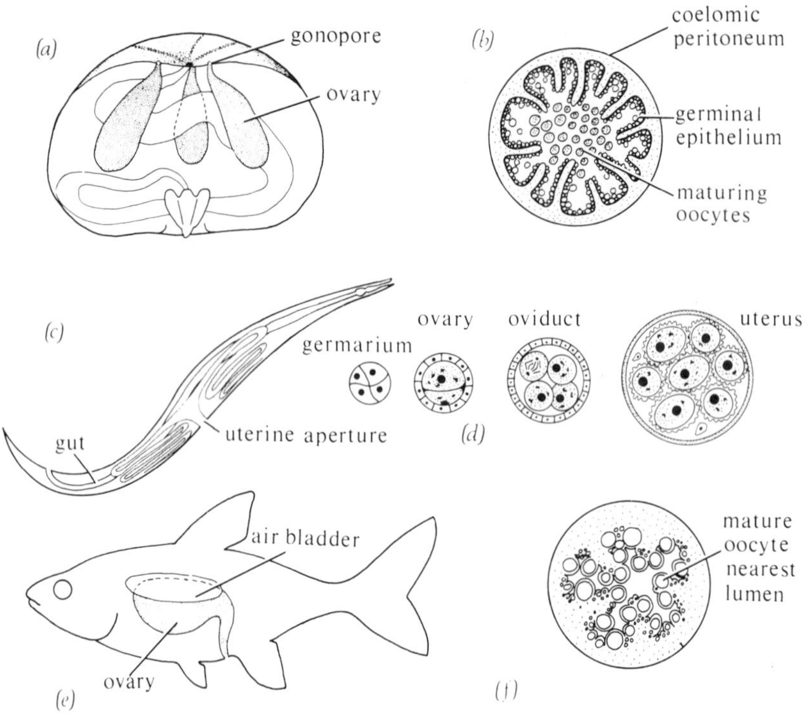

Figure 4.2 OVARIES (a) The ovaries of sea urchins are simple sacs, pendant from the genital pores. Ovaries are stippled. (b) In section, the walls are seen to be folded, with germ cells maturing and being shed into the lumen as oocytes. (c) The ovaries of nematodes are long thin prolongations of the oviduct and uterus, which often opens midway along the female. (d) Sections of ovary-uterus of a nematode. Amoeboid sperms may be found among the fertilised eggs in uterus and lower oviduct. (e) Teleosts usually have a single ovary, commonly hollow and ovulating inwards. In the breeding season it is the largest organ of the body cavity, under the swim-bladder. (f) A section of an immature teleost ovary

from the same oogonium, but behave quite differently. They may commence meiosis (synaptinemal complexes are seen in one sister of the oocyte) but even if so their subsequent development is diverted; the nucleus replicates without division until it becomes 512 to 1024-ploid, and the cytoplasm increases in bulk too[6]. The oocyte nucleus, meanwhile, has taken up the 'lampbrush chromosome' pattern and the oocyte cytoplasm is expanding even more rapidly. When the follicle reaches the posterior end of its ovariole the contents of the nurse cells pour into the oocyte, presumably giving it DNA in a suitable form for use in the rapid cleavage divisions, and extra cytoplasm and yolk. Then the thick chorion (shell) is secreted. Numbers of nurse cells per oocyte vary among the insects, and I have suggested that these other germ cell products might represent 'failed' oocytes comparable with the atretic oocytes of mammals (*see below*). Certainly their number is related to chiasmata in the same way, for *Forficula* with no chasmata has one nurse cell per oocyte, *Drosophila* has about six to seven chiasmata and 15 nurse cells per oocyte,

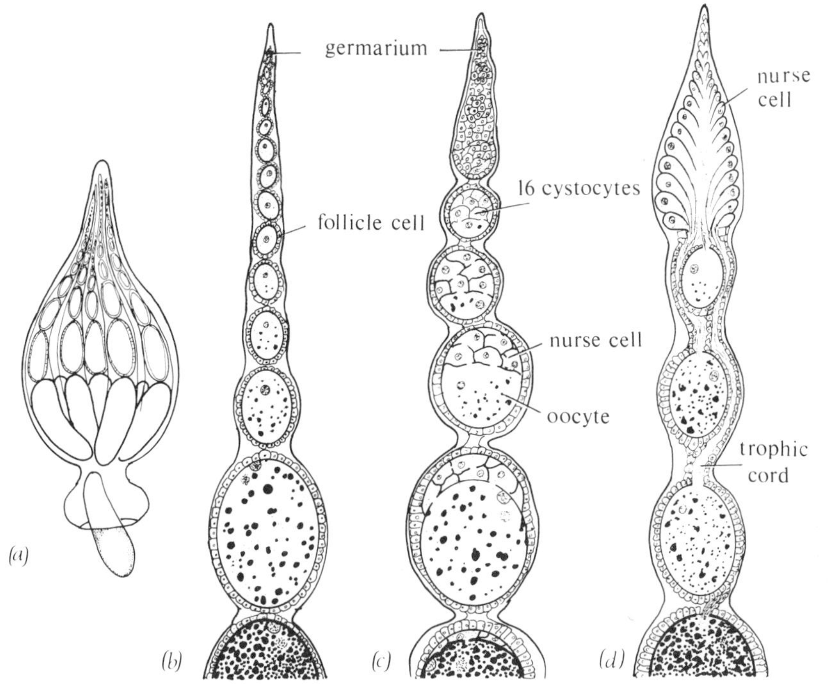

Figure 4.3 INSECT OVARIES (a) The ovary is usually a group of ovarioles, enclosed in a thin membrane. Eggs mature as they pass posteriorly in each ovariole, and are usually ovulated into a common fertilisation chamber, before the chorion is hardened by secretions of several glands. (b) A panoistic ovariole (e.g. of locust). From the anterior germarium groups of cells move backwards and it soon becomes clear that the oocytes are surrounded by follicle cells, which accompany them through to ovulation. (c) A polytrophic ovariole (of a fly or beetle). Cells in the germarium divide, remaining connected by cytoplasmic bridges n *times, where* n *is commonly 1 (earwigs), 2 (many moths), 4 (most flies), up to 7 (some carabid beetles), giving clumps of* 2^n *cells which pass posteriorly. One soon becomes the oocyte and the others, nurse cells, become highly polyploid. Before ovulation the contents of the nurse cells pass into the oocyte (bottom of figure). Compare with* Plate 6. *(d) An acrotrophic (telotrophic) ovary (*Hemiptera, *some beetles). In the germarium a coenocytic mass of nurse cells retains connection to the posteriorly-moving oocytes by trophic cords through which materials pass into the growing oocyte. Some nutrients may also be acquired via the follicle cells*

and *Carabus*, with 127 nurse cells per oocyte[6] has 28 small chromosomes (14 bivalents) and therefore C is probably 14 to 16 (*see Figure 2.4*).

Many insects, however, have **acrotrophic (telotrophic)** ovaries (*Figure 4.3*) and the nutritive cells form a plasmodium at the apical end of the ovariole mixed in with germ cells, oogonia, and young oocytes so that estimation of 'oocyte wastage' is impossible. Nutritive cords extend posteriorly from this mass to add material to the growing oocytes.

The ovaries of vertebrates are all fairly simple organs, with ovulation from the outer surface directly into the peritoneal cavity except in teleost fishes, whose ovaries are often single and hollow, continuous with the oviduct and liberating the eggs into the lumen, not the coelom. Several other vertebrates, including murine rodents, have isolated the ovary from the general peritoneal cavity in a special **ovarian bursa**, into which the oviduct opens, but in these cases the ovary is not remarkable. The morphology of the mammalian ovary is well-known (*Plate 6*, p. 83). The germ cells multiply during embryonic life, producing oogonia in the ovarian cortex in large numbers. Some oocytes become organised within Graafian follicles, and some are finally ovulated with several layers of cells around them (*Figure 4.4*). Most mammalian oocytes, however, are not ovulated but degenerate, in several waves. The first wave of oocyte **'atresia'** occurs just when the oocytes enter the prophase of meiosis, when the female mammal is still embryonic. Its extent can only be estimated in those few mammals, murine rodents and perhaps man, in which oogonial mitosis stops before the oocytes enter meiotic prophase[7]. In

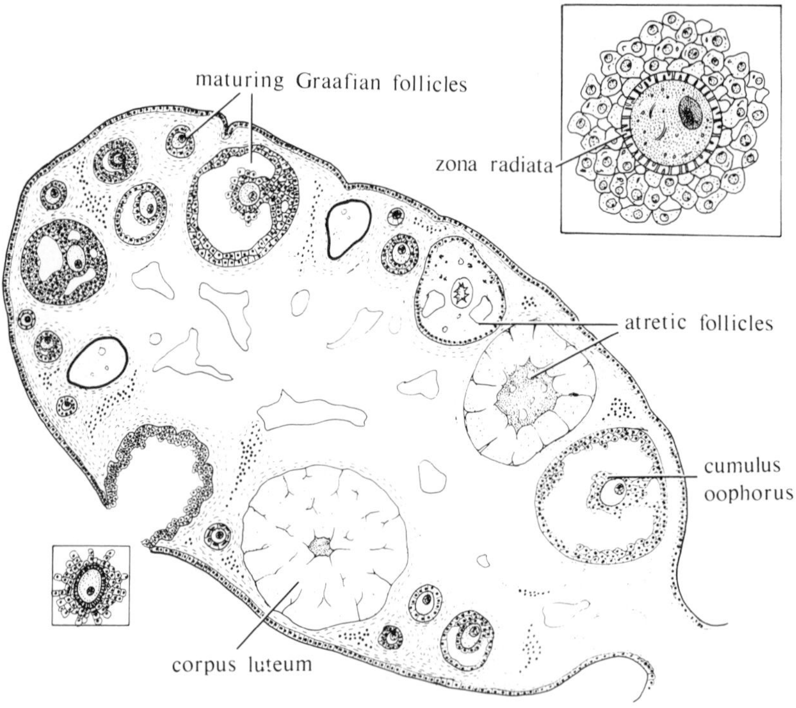

Figure 4.4 MAMMALIAN OVARY A composite diagram of various oocyte stages, as they might be seen in a longitudinal section of the ovary. Upper left, is a 'normal' (i.e. non-atretic) series of follicles, from the youngest oocytes in their follicle-cell coat through the formation of an antrum (space among them) to the mature Graafian follicle with its oocyte in an internal bud, the ***cumulus oophorus****. Follicles may show atresia at any stage, as seen in the series upper right. Nuclear changes are soon followed by the appearance of polymorphonuclear leucocytes, and the destruction of the oocyte itself may occur up to the time of ovulation. A ruptured follicle is shown lower left; its remaining follicle cells fill the cavity to form the progesterone-secreting corpus luteum. This usually regresses unless maintained by a pregnancy.* Inset*: newly-ovulated oocyte. The follicle cells around it had processes through the* ***zona pellucida****, which leave striations; it is often called zona radiata because of this, and should not be confused with* ***corona radiata****, cumulus cells arranged in radiating lines in a jelly-mass. See* Plate 6*, p. 83*

other cases, like the marmoset and the sheep, oogonial mitoses continue into this first phase of oocyte atresia, so it is impossible to estimate what proportion of the mitotic products fail to mature, because dividing and dying cells are present side by side in any section.

In the human, about 6×10^6 oocytes in the three-month embryo reduce to about 5×10^5 at birth. A second wave of atresia cuts down numbers to about 5×10^4 at puberty. In each menstrual cycle up to 40 or more Graafian follicles enlarge, accompanied by more atresia of young oocytes, but only one is normally ovulated. So 6×10^6 oocytes are produced which lead to, at most, about 400 ovulations; in 'natural' circumstances, with the woman usually pregnant or lactating, the number is probably nearer 40. Figures for oocyte wastage in mouse, rat and women compared with their chiasma frequency are given in *Figure 2.4* (p. 28). No mechanism for this association has been suggested; the correlation is even more difficult to explain for the polytrophic insect ovaries, but the association is certainly real. Whether, as I suggest, this is evidence for faults arising from chiasmata the reader must decide.

The egg is often ovulated with a coating of ovarian cells. Not only in mammals but in many other organisms, for example tunicates and lampreys, a layer of follicle cells surrounds the egg. In the lamprey they may function to keep ions from leaking into the fresh water and are not present at the animal pole, where there is a sperm-orienting tuft of mucus (*Figure 6.4*, p. 122). In tunicates they certainly account for the self-sterility of most animals, associated with the presence of ripe eggs and sperms together at the breeding season. But its function in mammals is a puzzle; as suggested on p. 122 there may be a selection of sperms in their passage through this layer.

4.3 Testes

Testes, like ovaries, come in a variety of forms. Most, however, are variations on a theme: several long tubes, blind distally, are twisted together and all connect to a common duct at their proximal end. Spermatozoa are formed from the progeny of cells in the tubule walls, and pass along the tubules. There is usually a proximal segment of each tubule in which no sperm formation occurs but in which sperms may be stored, **vasa efferentia** or **epididymis**, and these tubules fuse to a **vas deferens** or **sperm duct**. Occasionally, as in some polychaete worms, spermatogonia float freely in the coelome, and their products mature there; but this is much rarer than for ova.

The testis of many insects, especially *Orthoptera* like grasshoppers and locusts but also most *Lepidoptera*, forms **'sperm cysts'**, sometimes misleadingly called spermatophores. Spermatogonia are shed from the walls of the tubules, and divide several times mitotically, but without complete cell divisions, so that a group of spermatocytes in close association begins meiosis (*Plate* 7, p. 109). This group forms a clump of 128, 256, etc. spermatids; there may be as many as $4096 = 2^{12}$. These form a bundle of sperms with their heads all together. Hundreds or even thousands of such bundles may be transferred into the female tract; for example in *Locusta* 4–6 × 10^5 sperms are transferred, through a tubular spermatophore which seems to be produced at the male end and 'digested' at the female end during the prolonged amplexus. Guppies, and other livebearing fishes of the family *Cyprinodontidae*, and many crustaceans, also produce such sperm cysts in the testes but in these cases they *are* used as spermatophores, each being enclosed in a membrane which is presumably digested by female secretions.

In mammals the tubules are very long and intertwined. Germ cells produce spermatogonia in the walls, and successive cell divisions produce more cells luminally, so that spermatozoa are finally lost centrally. They spend much of the terminal phase of spermiogenesis with their heads buried in Sertoli cells (*Figure 4.5* and *Plate 7*). The significance of this remains unknown but may be connected with sex hormone production within the Sertoli cell, using cholesterol from the fatty degenerative

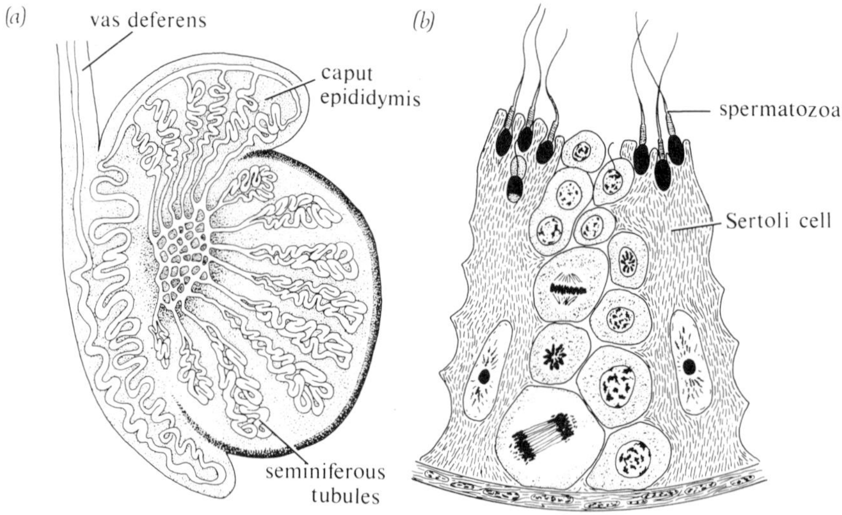

Figure 4.5 MAMMALIAN TESTIS (a) Connections within the system. Seminiferous tubules are linked to the ***rete testis*** *centrally, and anastomose in pairs or trios peripherally. The* ***vasa efferentia*** *(mesonephric tubules) drain the rete into the caput epididymis and the fluid is then resorbed in the epididymis, concentrating the sperm suspension. Sperms mature and interact with epididymal secretions until they reach the cauda epididymis (bottom of diagram) the major storage site in mammals. From this they leave by the vas deferens. (b) Diagram of part of a tubule wall. Sertoli cells form a complete inner layer to the tubule wall, and are in intimate contact with all the germ cells. The spermatids lose their excess cytoplasm in the Sertoli cell (left, lowest spermatid) and are released into the fluid stream in the tubule only when they are of typical shape. They usually cannot fertilise until they have sojourned in the epididymis. In any section like (b), there is a characteristic family of cells. This diagram shows three spermatogonia, one of which is dividing, two stages in meiotic prophase of primary spermatocytes, a first meiotic metaphase of a primary spermatocyte, a secondary spermatocyte and three 'round' spermatids, one with a young flagellum. A telophase secondary spermatocyte and several sperms are shown. Such a conglomeration could never be seen in one tubule section. See* Plate 7b

by-products of sperm production, and lost from the spermatids in the residual bodies[8]. Apparently comparable looking cells, in which maturing sperms lie embedded, are found in earthworms, squids, and butterflies, but perhaps not in some fish, echinoderms, polychaete worms or rotifers[9]. So, if these cells have a common function it must be very general, but is as yet unknown.

Unlike the spermatogenesis of poikilotherms, the various stages of mammalian spermatogenesis have very precise duration, probably because of their constant temperature, usually a little below that of the blood (*see below*). So there are a family of characteristic sequences of cell types from the tubule wall to its lumen for any species (*Figure 4.5* and *Plate 7b*). Much has been made of this 'spermatogenic cycle', and it is certainly a very sensitive indicator of testicular pathology, but it requires great expertise even to recognise such a cycle in a new species. We must be grateful to the French workers, Clermont and his associates[10], who have worked the cycle out for man and the common domestic and laboratory animals.

Nearly all kinds of mammalian testes seem to require a temperature marginally (2 or 3 °C) below that of the rest of the body, and they are usually exposed in a **scrotal sac** with a temperature regulating mechanism. The sac pulls the testes against the body in cool ambient temperature, and relaxes in warm air cooling the testes. There are supplementary behavioural mechanisms too; laboratory rabbits and rodents kept in moderately warm (e.g. 25 °C) animal houses breed better if kept in metal cages, probably because they use the metal to conduct heat from the scrotal sac. Unfortunately, man does not have such a drive, for many infertile couples have been 'cured' by the advice to the man to give up tight underpants[11]. The requirement is not only for spermatogenesis, for there is some evidence that warming the epididymis also causes destruction of the sperms already in it[11].

Yet the sperms will meet a higher temperature in the vas, albeit temporarily, and in the female. Perhaps the higher temperature induces a terminal change of metabolism and so a drastic shortening of life expectancy; but there is no hint of this from *in vitro* studies. Further, there are several mammals, including the whale, the elephant and the hyrax (coney), who contrive to produce sperms in testes retained in the abdominal cavity. Despite suggestions that the testes of elephants are cooled by their proximity to the lungs (!), no suggestion has yet been made to account for the temperature relations of the testes of mammals. Birds, with their higher internal temperature (40 °C+) seem to manage spermatogenesis successfully, after all.

4.4 The Secondary Sexual Organs

These, the ducts and glands associated with the production of eggs and sperms, are of course enormously different in different animals, so no general survey can be attempted even of the vertebrates. Instead, a few non-mammalian species will be mentioned before a very brief consideration of the mammalian ducts.

There are several groups whose ovaries open directly to the outside, with only the barest apology for ducts, for example echinoderms, polychaetes and hagfishes (whose eggs usually develop in the coelome and are released through coelomic pores), bivalve molluscs, nemertines, arrow worms and many crustacea. Many teleost fishes, too, simply continue the ovarian wall to the genital pore. However, eggs of the zebra fish taken directly from the ovary by dissection are not fertilisable, whereas those expressed naturally through this pore are. There may be a subtle change of membrane chemistry as the eggs pass this region; it does not look as impressive as the oviduct of the frog but perhaps it has a function nevertheless.

In live-bearing cyprinodonts (e.g. the guppy) this short tube serves for the passage of newly-hatched fish out of the mother, and for the inward passage of tiny spermatophores. It has rather folded walls, but no dramatic glands.

Nematodes continue the wall of the ovary as a barely broader tube, the oviduct, which itself broadens and becomes the uterus, opening by a pore usually midway along the female. *Rhabditis*, which infests earthworm corpses, is viviparous and transparent; all stages of development can be seen inside large (3 mm) females (*Plate 10*, p. 153), and slight pressure on the coverslip causes 'birth' of young worms.

Many insects possess glands which secrete coats and hardening agents on to the eggs, as in beetles, or the **ootheca** as in the cockroaches. The bombardier beetles (*Brachinus*) have diverted the primarily reproductive function of glands secreting strong oxidising and reducing agents; when threatened, they mix the secretions together with an enzyme, which forms a spontaneously explosive (hypergolic) mixture which the beetle directs and startles any predator. One wonders whether this 'invention' started as an egg-dispersal mechanism.

Selachian fishes (sharks and rays) release their large yolky eggs into the coelome, whence they are engulfed by the funnel-ended oviduct, where many sperms penetrate; they then pass into the 'uterus'. In some forms (e.g. nursehound) development proceeds here (*Figure 4.6*), but in others (dogfish, skate) the horny 'mermaid's purse' is secreted around the egg at this point, and it is then tethered by its filaments to seaweed or rocks.

The oviduct of birds (*Figure 4.6*) works similarly. It has at its posterior end so-called **'infundibular glands'** or **'sperm nests'** where sperms are stored; one copulation fertilises up to 10 eggs in the domestic hen. The egg (yolk) is released into the coelomic cavity and is caught by the oviductal funnel where fertilisation occurs. It is then squeezed down the oviduct, where the **chalazae**, and the denser (thick) albumen are applied. At the **'shell gland'** looser (thin) albumen, shell membranes and shell are all secreted and applied while the contents are rolled, and, mysteriously, the **air sac** is left gas-filled. Then the egg is laid 6 to 12 hours after ovulation. Later, of course, sperms must pass in the opposite direction to fertilise the next egg.

Ducts of three species of female mammals are shown in *Figure 4.7*. The kangaroo, like other marsupial females, has a developmental problem

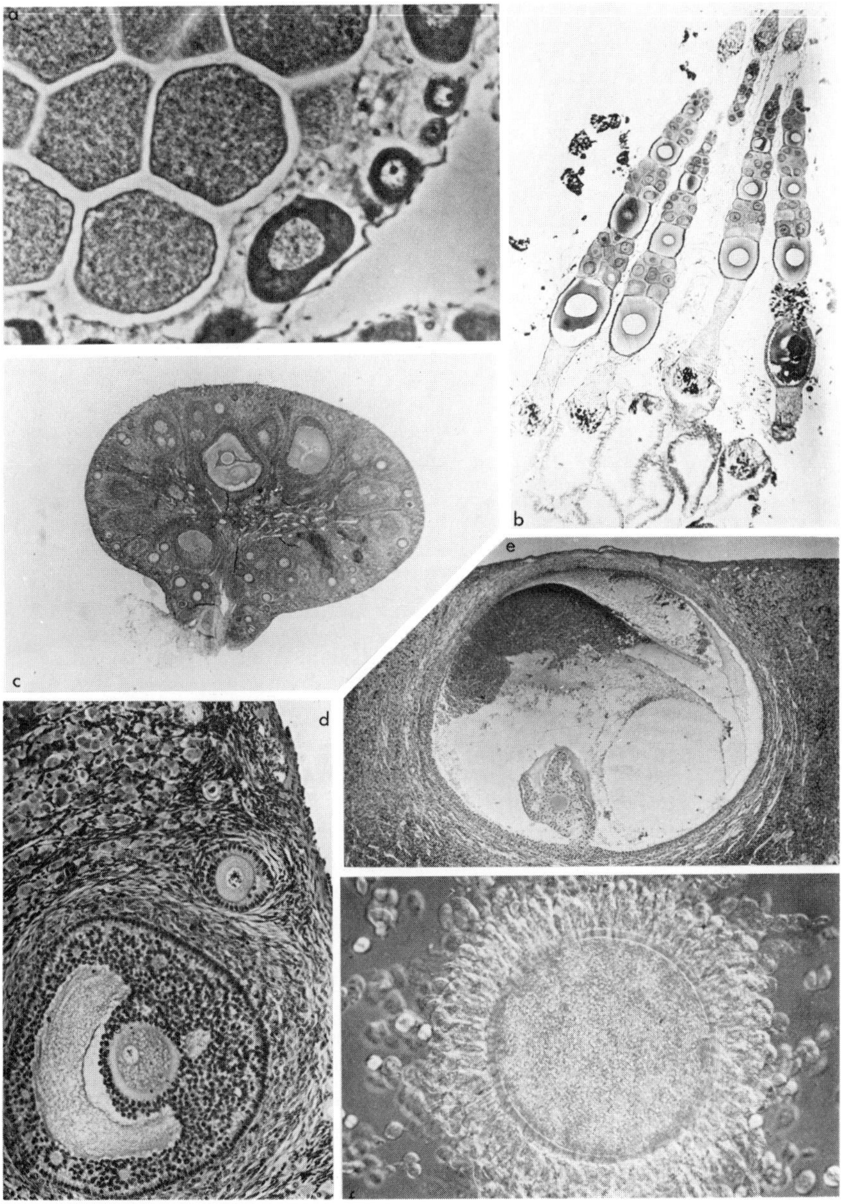

Plate 6 OVARIES (a) Edge of a starfish ovary showing oocytes in the wall, and mature eggs separated by jellies (compare Figure 4.2*). (b) Section of the ovary of* Dytiscus, *a water beetle (compare* Figure 4.3, *p. 77). (c) Section of a whole mouse ovary (compare* Figure 4.4*). (d) Part of a rabbit ovary, with developing Graafian follicles (*Figure 4.4*). (e) Mature Graafian follicle in a rabbit ovary eight hours after copulation. The egg with its 'corona radiata' can be seen on its deep aspect. (f) An ovulated rabbit egg (Photo by K. Tyler)*

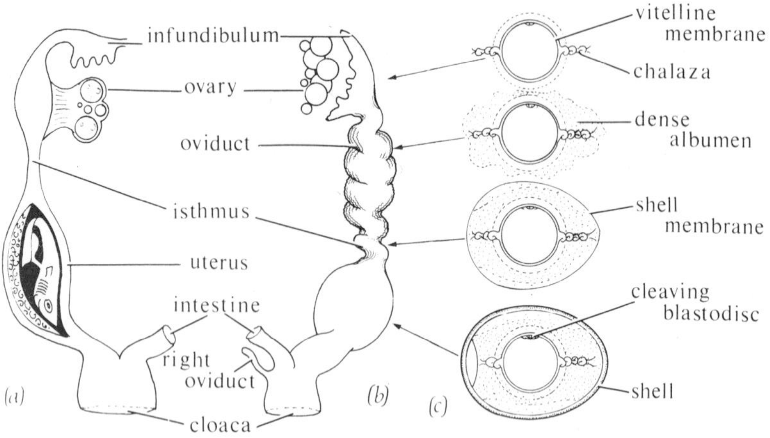

Figure 4.6 FEMALE GENITAL TRACTS (a) Reproduction tract of a viviparous shark (modified after Ballard, 1964). Shark embryos develop a variety of placental adaptations and substitutes, in the 'oviduct' (Amoroso, 1960). (b) Reproductive tract of a bird. Fertilisation occurs in the infundibulum. The right oviduct does not normally develop. (c) The bird's egg in its passage down the oviduct. The chicken may produce more than 10 fertile eggs from one mating

with two sets of tubes, the ureters and Müllerian ducts, rather like that which the male eutherian mammal has with ureters and Wolffian ducts when the testes descend. The solution adopted by the female marsupial precludes a continuous, broad uterus-cervix-vagina; instead two narrow tubes pass either side of the ureters entering the bladder (i.e. the ureters penetrate the uro-rectal shelf). This separation of the anterior vagina into left and right tubes correlates with the bifid tip of the penis of male kangaroos and with the small size of the young at birth. Even though a central canal may open for birth purposes, this can never be as broad as the eutherian can manage; hence the pouch for further development.

The mouse tract (*Figure 4.7*) is specialised in that semen is pumped directly into the uterus, only a 'plug' being left in the vagina; it is also unusual in the separation of ovarian bursa from the general peritoneal cavity. The pig and several other mammals pump the semen directly into the uterus; the boar produces about 250 ml but no vaginal plug.

The human oviduct is long, probably to allow development of the blastocyst to the stage of implantation (but the 1 cm oviduct of the mouse works adequately), but also as a bridge from the lateral ovaries to the central uterus, usual in large mammals where only one young is born at a time. The cervix keeps the flora of the effectively open vagina from the uterine cavity, and also probably selects sperms for passage into the uterus[7,12]. The vagina is kept moderately infection-free by lactic acidity, produced by the Döderleins bacilli from secreted glucose, but this acidity is neutralised instantly by the buffering effect of seminal plasma at ejaculation. Another suggested function of the cervical glands[12], storage of the sperms ascending from the vagina, is dubious; it is difficult to see how sperms could leave them (*see also* p. 113).

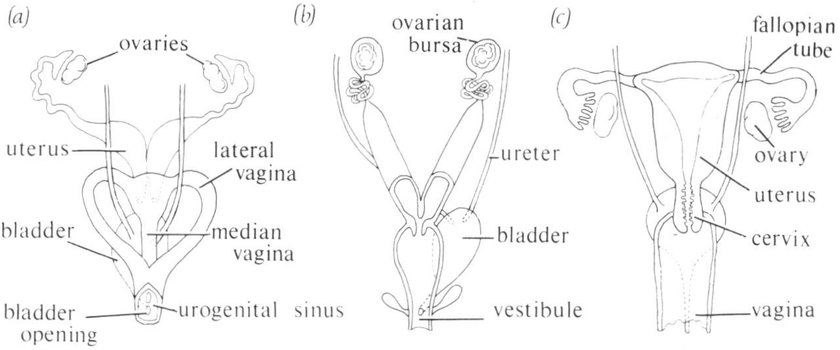

Figure 4.7 FEMALE GENITAL TRACTS OF THREE MAMMALS (a) A macropid marsupial (kangaroo) tract, viewed from dorsal and slightly left of centre. Note the lateral vaginas, avoiding the ureters (metanephric ducts), and the temporary median birth canal (After Nelsen, 1953). (b) Female tract of the rat, viewed from dorsal, slightly right of centre. Note the enclosure of the ovaries in bursae, the double uterus, and the long vestibule, a deep pocket of skin into which the urethra opens and which leads to the vagina proper. Insemination is directly into the uteri. (c) The human female tract, viewed from the dorsal side. Note the single uterus, appropriate for single offspring, and the complex cervix. There is only a shallow vestibule. Note: *In all these genital tracts cavities have been shown as 'gaping' only for clarity. Mostly they are 'virtual spaces', with the walls touching or nearly so; the oestrous rat uterus* is *distended, but this is exceptional*

Male echinoderms, like their females, have gonads opening directly through the body wall at the genital pores. Few other testes open so directly to the exterior, but many have only the simplest of ducts which sometimes seem to function only to 'age' the sperms, which are usually continuously produced in the testes. The ducts are often coiled or convoluted, while retaining their simple structure. Indeed vasa efferentia and vas deferens of unknown animals can often be recognised as convoluted white ducts with a large organ at one end and usually with accessory glands opening to the exterior by the same pore at the other end. Four male genital tracts are drawn in *Figure 4.8*. Mann[13] has tested the secretions of a great variety of mammalian glands and has discovered that the anatomist's name, the homology, and the function are often very different. Many of the substances found commonly in mammalian seminal fluid have not yet been assigned a function, and those functions which have been assigned should be viewed with caution. It seems obvious that fructose or other sugars should serve to provide energy for the sperms – yet Bedford[14] has demonstrated conclusively, by washing the vagina with strong detergent, that enough sperms to fertilise all of a rabbit's eggs have left the vagina, and the pool of seminal fluid, between two and five minutes after ejaculation. Equally, the prostaglandins found in several mammalian ejaculates seem obviously to function in causing uterine contractions – but the amount in one ejaculate is too low to act *per vaginem*, and it is notable that there is *none* in uterine inseminations (except horse). Perhaps it is *excreted* into the vagina, where it does no harm, after having stimulated ejaculation in the *male* tract?

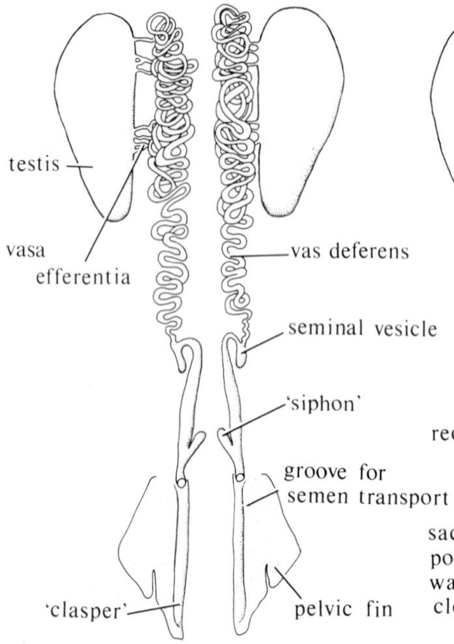

(a) *Squalus* a dogfish

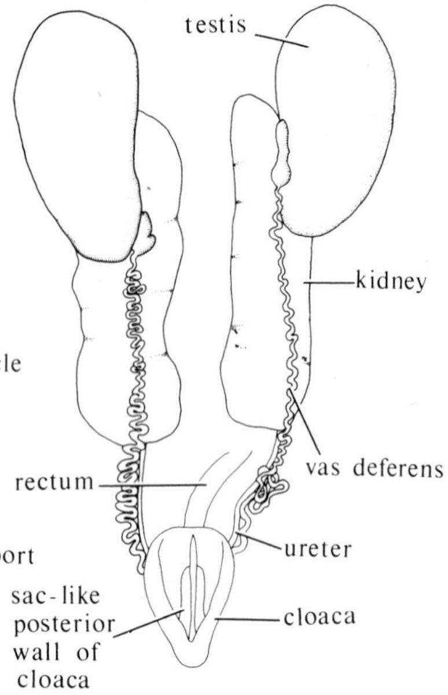

(b) *Gallus* cock

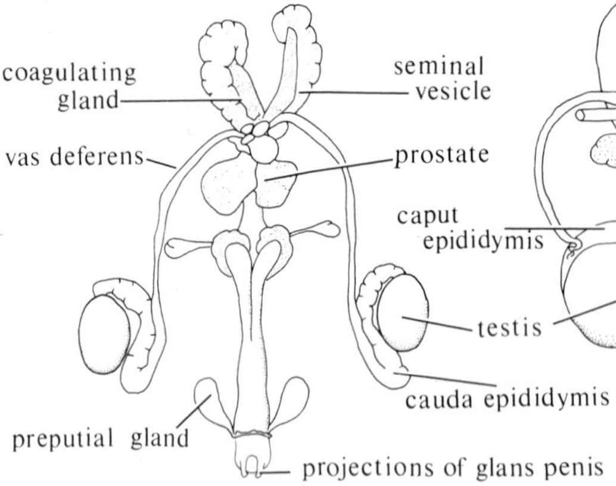

(c) mouse

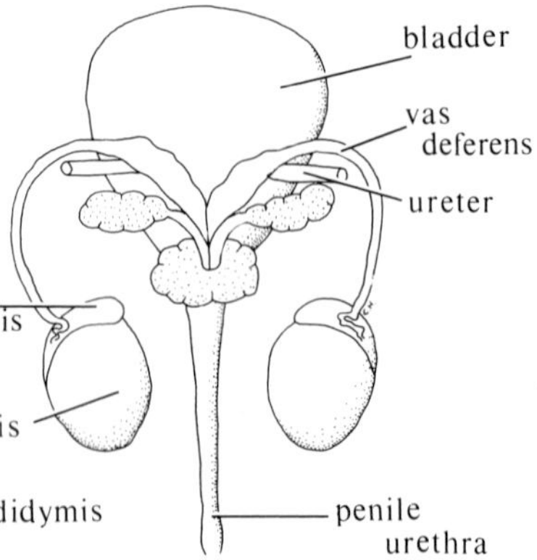

(d) man dorsal view

Male genital tracts commonly have glands whose secretions form solids upon mixture. These may be simple jellies or **coagulum**, as in rabbit or human, whose function is not easy to see, or they may form solid plugs, as in the mouse (*Plate 17*, p. 273), which affect the female psychologically as well as keeping the (uterine) semen inside. If a vaginal plug is not inserted after artificial insemination, the female mouse rarely carries the litter to term. Those males which produce solid spermatophores usually use genital glands to make them, but these are usually considered tertiary sex characters, as are odd transfer mechanisms like syringes on the pedipalps of spiders (*Plate 4*, p. 49).

4.5 Tertiary Sexual Organs

These are the bewildering variety of accessory sexual structures. They range from hectocotylous arms of cephalopods, through penes, suckers on the front feet of some beetles (*Plate 4*, p. 49) and the thumbs of male frogs, through the peacock's tail and the beard of man to sexual dimorphisms of all kinds. No attempt will be made here to list or even classify them; most people are familiar with more examples than I can afford space for here.

4.6 Organs for Asexual Reproduction

Vegetative reproduction, by definition (p. 5) does not require special organs. Branching coenosarc in the coelenterates or ectoprocts may 'die back', isolating parts of the same colony. Buds of *Hydra* may be regarded as special organs, as may the continuing 'embryonic' rear ends of asexually reproducing flatworms or polychaetes (*Plate 2*, p. 15). It may, however, be more profitable to regard these examples as 'healing' processes which have lost those control mechanisms which would stop them when the original animal is complete. Morphogenesis within such buds is better left to the consideration of embryologists, who also find it very puzzling.

---

*Figure 4.8 GENITAL TRACTS OF FOUR MALE VERTEBRATES (a) Dogfish (*Squalus*, modified after Nelsen, 1953, a ventral view). The claspers are anchored in the female's cloaca, and the siphonal contents (including 5-hydroxytryptamine and other neuroactive substances) carry the sperms into the oviducts. (b) Rooster (*Gallus*, ventral view). The posterior part of the cloaca is eversible and at least lodges in the hen's vent. A drop of very concentrated semen is produced, probably directly into the everted base of the oviduct. (c) Mouse (*Mus*, ventral view). The seminal vesicles do not store sperms, but produce a complex additive to the coagulating gland substance, which clots forming the vaginal plug (*Plate 17*). The prostate and other glands around the base of the urethra add substances to semen, but the preputial glands secrete pheromones into the urine which regulate the female's oestrous cycle. The projections on the glans are variable, sometimes resembling a double fish hook or symbolic anchor. They probably stimulate the cervix violently during intromission. (d) Man (dorsal view). This view shows the enlarged bases of the vasa deferentia where they lie against the bladder; these, the* ***ejaculatory ducts****, store sperms prior to the mixing with seminal vesicle and prostate secretion during ejaculation*

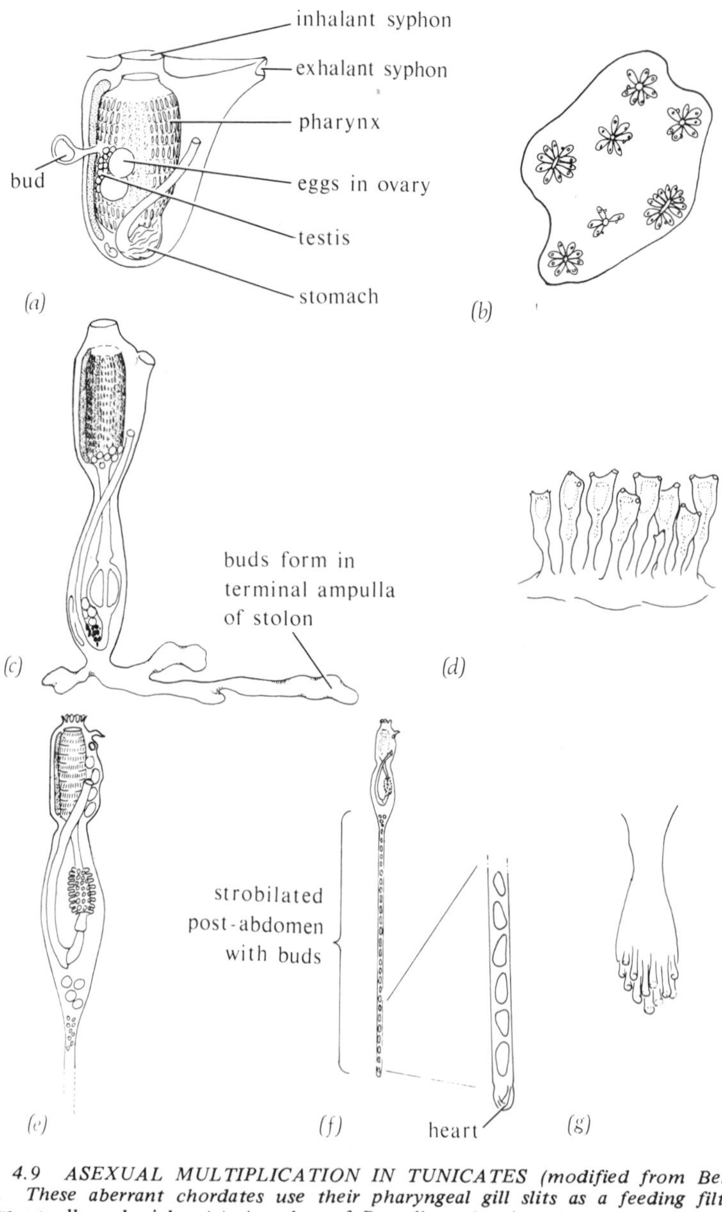

Figure 4.9 ASEXUAL MULTIPLICATION IN TUNICATES (modified from Berrill, 1950). These aberrant chordates use their pharyngeal gill slits as a feeding filter, and are usually colonial. (a) A polyp of Botryllus, *showing the general structure of these strange organisms (see also* Figures 10.10 *and* 11.2*). The bud contains all layers of the body wall. (b) A colony of* Botryllus; *the zooids have separate inhalant openings, but share exhalant pores. The star-shaped clusters of zooids increase their number by lateral budding as in (a). (c) A polyp of* Clavelina. *The stolon is equivalent to the visceral body wall of forms like* Botryllus, *and the tips are swollen. (d) The simple colony of* Clavelina. *(e) The anterior part of a* Morchellium *zooid. The visceral body wall is drawn out 'posteriorly'. (f) A mature* Morchellium *polyp, with the post-abdomen packed with buds. (g) A colony of* Morchellium *(usually pendant as shown; the polyps are reversed); the base of the colony is full of post-abdominal buds, from which next season's zooids will develop*

There are two other major ways in which organisms reproduce asexually; by special spores which usually contain several kinds of cells, or by stolons which also have threads of different tissues along their length. *Figure 1.1* (p. 6) shows statoblasts and gemmules, while *Figure 4.9* shows the reproduction of three genera of tunicates, from essentially stolon-like structures.

One other phenomenon which should be considered here is **larval multiplication**. It is frustrating to admit our ignorance of this as well as the other phenomena of asexual reproduction, but we know little even of the economically important digenean larvae, **sporocysts** and **redia** and their production of thousands of progeny (*Figure 4.10*). There are, however, some interesting examples of larval multiplication which are not yet generally used as examples in textbooks.

Our starting point for consideration of larval multiplication, of course, is twinning. Despite an enormous amount of experimental work, we still have little idea why one blastomere develops into part of an embryo or a whole embryo, depending whether another blastomere is present in the same egg membrane. Natural examples range from *Lumbricus andersoni*, all of whose eggs produce twins, never triplets or quads, through the armadillo *Dasypus novemcinctus* who always produces quads, to the parasitic *Hymenoptera* like *Ageniaspis fuscicollis*. In this species, each egg laid in a host larva breaks up into groups of cells, which each produce

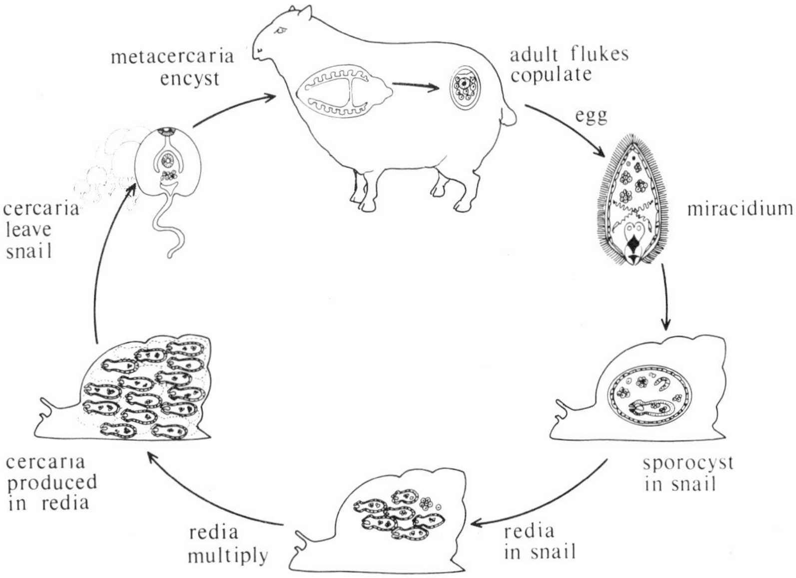

*Figure 4.10 LIFE CYCLE OF THE SHEEP LIVERFLUKE Even on wet ground, few eggs produce miracidia which find snails (*Limnaea truncatula*). Once found, the snail produces many cercaria, all those from one sporocyst having the same genetics. (This is an economical method of multiplying heterozygotes without segregation.) Most metacercaria, encysted on grass (or watercress) fail to find a sheep or other mammal; but this loss comes from the snail's economics, not the adult fluke's.*

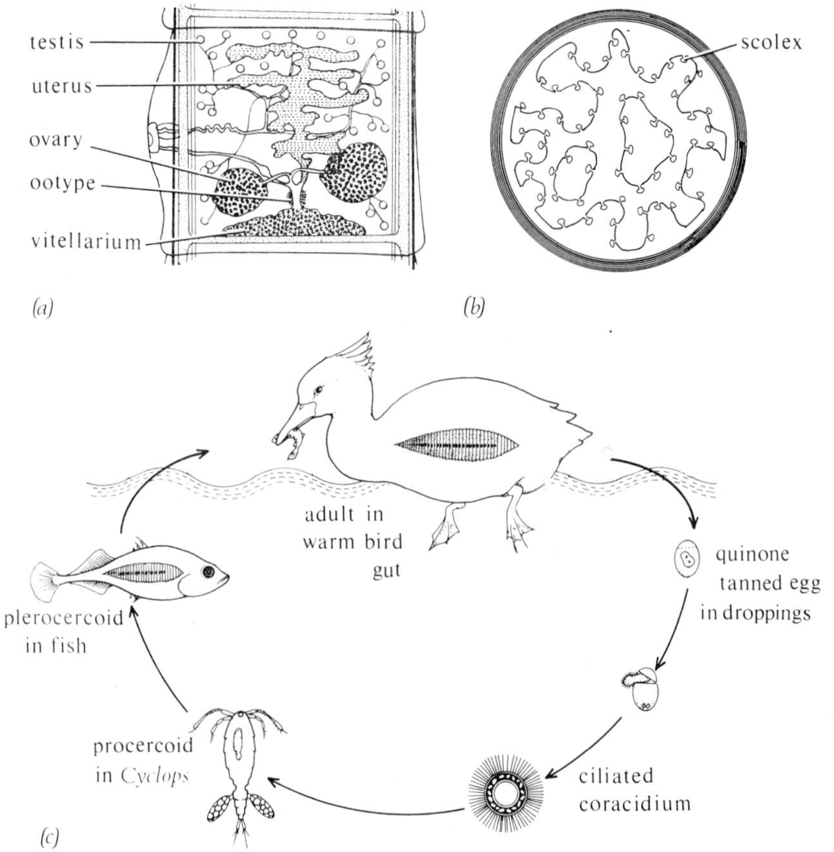

Figure 4.11 *REPRODUCTION IN TAPEWORMS* *(a) One segment of a tapeworm* Taenia solium. *The externally opening genital pore on the left strongly suggests that copulation between segments sometimes occurs; as it is very rare for two tapeworms to share a gut, the function of such copulation is obscure. The repetition of reproductive segments, each with its own ootype, allows parallel production of many eggs, which mature in the uterus as the whole segment is lost in the faeces of the host. It may show active locomotion for hours, even on dry ground.* *(b) The hydatid cyst of* Echinococcus, *a sheep tapeworm. Multiplication of the 'heads' (scolices) of the larva in its penultimate host ensures rapid fixation and metamorphosis in the final host. The adult* Echinococcus *usually has only four proglottides (segments) at a time, and many may co-exist in a gut. Usually, then, any copulation is between genetically identical organisms, and its utility is difficult to see.* *(c) The life cycle of* Schistocephalus. *The plerocercoid is a very common parasite in the coelome of sticklebacks (*Plate 2*). This becomes the adult very rapidly in the gut of fish-eating birds; because of temperature change the gonads mature. The repeated ootypes allow rapid egg production during the four-week adult life. The eggs have a tanned shell and resist digestion during their journey through the bird gut*

a larva[15]. Some pilidium larvae of nemertines produce many baby worms in separate 'amniotic cavities', just like the multiple scolices in cestode 'bladderworms' (**hydatid cysts**, *Figure 4.11*). The next step is the enormous multiplication of redia in snails, and then the limitless multiplication of coelenterate polyps. The polyp is probably best regarded as the asexually multiplying larval stage of the sexual medusa; although if the suggested phylogeny of Hadzi[16] is correct and anthozoans are primitive, the medusa is an added, distributive sexual phase completely exploited in the scyphozoan jelly fishes. The asexual reproduction of corals should properly be considered here, but space does not permit. Nor can we consider the detailed organogeny involved in asexual reproduction in protozoans, although a good case could be made for considering the ciliate cortex an essential organ of reproduction[17].

REFERENCES

1. Engelmann, 1970 (pp.36–7)
2. Gurdon, 1974 (p.87)
3. Baker, 1972a
4. Mintz, 1960
5. Blackler, 1966
6. Chapman, 1969
7. Cohen, 1971
8. Lacy and Pettitt, 1970
9. Roosen-Runge, 1969
10. Clermont, 1972
11. Glover *et al.*, 1975
12. Hafez, 1973
13. Mann, 1964
14. Bedford, 1971
15. Ivanova-Kasas, 1972
16. Hadzi, 1963
17. Tartar, 1960

4.7 Questions

1. Describe the copulation of *Helix* from the point of view of a sperm rather than a participant snail.

2. What are possible functions for glands associated with female and/or male genital tracts?

3. What proportion of various adult organisms' anatomy serves reproductive function?

4. Does asexual reproduction remove or impose constraints on the life style of organisms?

5

SPERMATOZOA

5.1 The 'Primitive' Form

Animal sperms vary enormously in their morphology. The simplest, and probably the ancestral[1], pattern is shown by echinoderms, coelenterates and some annelids and molluscs (*Figure 5.1*). These sperms usually are 'externally' fertilising and may have great distances to swim. Their movement, as observed microscopically, is exceedingly vigorous so that the tails are commonly not visible at all, except by stroboscopy (if the head is stationary) or by high-speed photography. They are also very small compared with the more complex sperms of other forms. Their heads rarely exceed 5 μm in length, and their tails are often less than 20 μm. The tail contains a simple 9 + 2 flagellar fibre pattern. The **midpiece** is represented by only a small number of mitochondria, commonly two, directly at the junction of head and tail. The head is mostly nucleus, but there is commonly an **acrosomal vesicle** anteriorly, with a **perforatorium** between it and the nucleus (*Figure 5.1*). The 'acrosome reaction' of these sperms, commonly elicited by substances in egg-water, usually consists of release of the acrosomal contents by breakdown of the anterior walls, combined with the production of an **acrosomal filament** by expansion of the perforatorium. This carries the original inner acrosomal membrane forward as a projection; this acrosomal filament makes first contact with the egg (*Figure 5.6*).

5.2 Specialised Sperms

There are many organisms in which sperms are utterly different from the basic picture given above (*Figure 5.2*). These differences may be complications of the simple sperm, as in the mammals (*Figure 5.1*). Mammalian sperms have an additional nine fibres in the principal piece of the tail (but not in the terminal segment), a very complex middle piece whose mitochondria are usually spiral, and various complications of head shape associated with rather large acrosomes. Platyhelminth sperms have two flagella, and a very long thin head. Insect sperms have a more complex flagellar apparatus, with a long mitochondrial structure called the **'nebenkern'**. Teleost sperms seem to be even simpler than the basic pattern, for they lack even an acrosome. Sperms of gastropod molluscs often appear as two or more morphologically distinct kinds[2]: **eupyrrhene** sperms are moderately normal-looking, functional sperms and may have

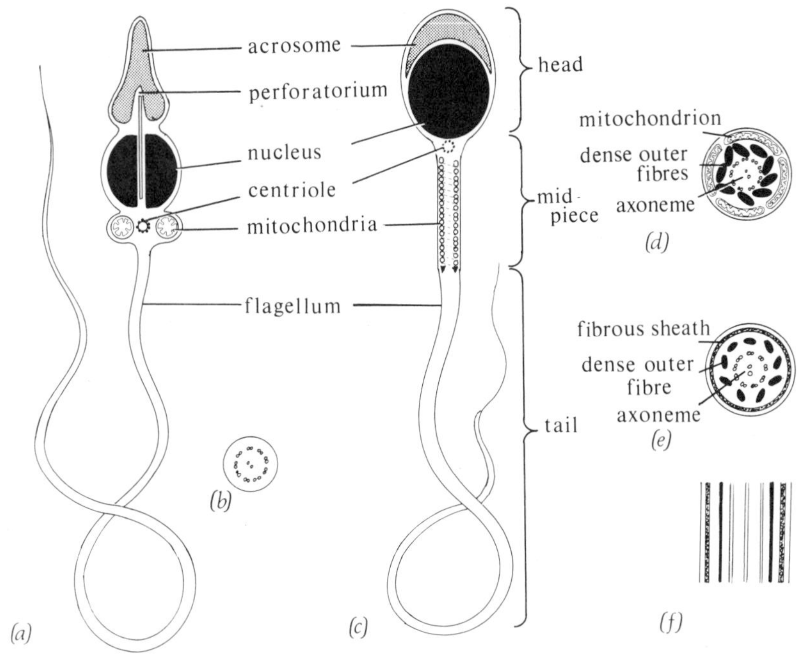

*Figure 5.1 FLAGELLATE SPERM STRUCTURE (a) A 'primitive' sperm, e.g. of marine annelid or mollusc. The midpiece is small, there is a perforatorium which carries the posterior acrosome membrane on its anterior face (*Figure 5.6*), and the tail has simple 9 + 2 fibre structure. (b) Transverse section of tail. Compare (e). (c) A mammal sperm. The midpiece is long with spiral mitochondria, and the tail has a complex structure. Although there is usually no perforatorium, the nuclear material is –S–S-linked for added rigidity to penetrate tue zona. (d) Transverse section of the midpiece. (e) Transverse section of the tail, 9 + 9 + 2. (f) Longitudinal section of the tail*

adequate locomotory abilities even though their tails are very long indeed; but **oligopyrrhene** sperms are not functional gametes. They may serve to carry the eupyrrhene sperms into the female tract, or to support or nourish them. There is some evidence that oligopyrrhene sperms of some genera may 'fertilise' those eggs destined to provide food for the normal larvae, e.g. in *Buccinum*[3].

There are many sperms which do not have functional flagella. They may have prolongations resembling tails but these do not lash (*Figures 5.2* and *5.3*). Mite sperms move without bending, with the tail in front and with their nuclear material in the rear end, by a mechanism still not understood but which may involve surface tension differences[4] (*Figure 5.3*); while crayfish sperms, despite their wheel-like appearances, do not actively move at all. Many other malacostracan crustaceans, like lobsters and crabs, also have immotile sperms but these are of less interesting shapes, and are usually released in clumps as spermatophores. As with many other sperms, including motile ones like those of liverfluke, it is the parents' actions which results in sperm-ovum contact in these crustaceans (*see* p. 136). The apparently immotile sperms of the nematode worms and some other

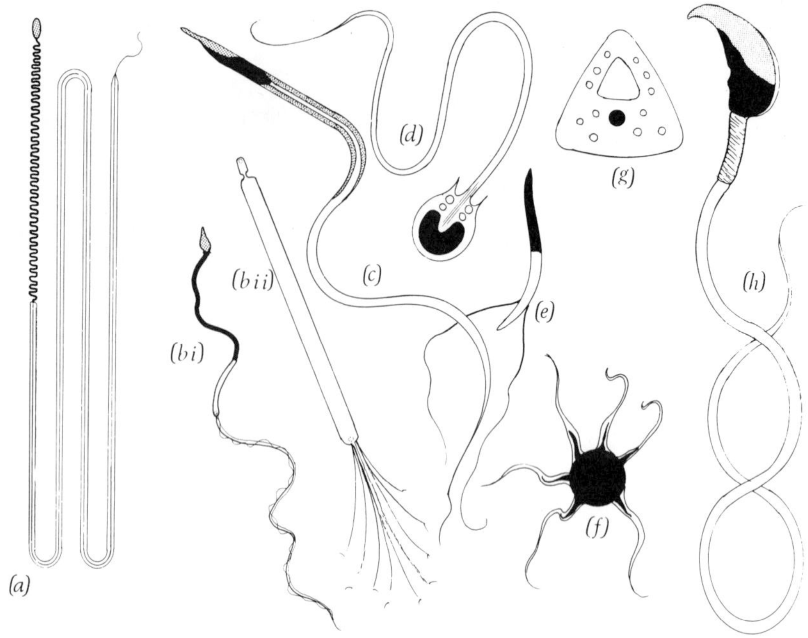

Figure 5.2 A VARIETY OF SPERMS The nuclei are shown solid black, the acrosome stippled. (a) Lithobius *sperm; all centipede sperms have this long head structure. (b)* Paludina *sperms; (i) is the fertilising sperm (eupyrrhene) but (ii) is the oligopyrrhene form also found in the emitted semen of most snails. (c)* Drosophila *sperms, some 2 mm long and with a complex midpiece structure, the nebenkern; many fly sperms are large and complex. (d)* Tilapia *sperm; most teleost fish have tiny sperms which lack acrosomes. (e)* Polycelis *sperm; flatworms often have double-tailed sperms, frequently (e.g.* Fasciola*) with one very long tail. (f)* Procambarus *sperm; many crustaceans show symmetry of an immobile sperm. (g)* Ascaris *sperm; nematode sperms are often triangular when inactive, amoeboid when fertilising. (h) Mouse sperm; most rodents have very hooked sperm heads*

aschelminths are in fact actively amoeboid in the female ducts, and seek out and penetrate ova apparently by their own activity (*Plate 10*).

There is an odd series which can be constructed and which inversely relates the apparent task of the sperm, that is the difficulties of its journey to the egg, to the complexity of structure. Franzèn's primitive sperms, with 'simple' 9 + 2 tails, no proper midpiece but only two or three small mitochondria at the posterior end of the head, and a simple oval head, are those sperms which must travel furthest in the external medium, usually sea-water. Echinoderms, bivalves and marine worms, all with external fertilisation and moderately sedentary adults, have sperms of this kind. Mammal sperms, which must travel perhaps as far (but certainly not in most cases by their own efforts), but with nutrients provided in an osmotically correct solution, are more complex with 9 + 9 + 2 tails, complex head shapes, and very complex midpieces (*Figure 5.1*). Insect sperms, which may have to be stored in spermathecae or ducts but whose path rarely exceeds 10 times the length of the sperm tail (*Drosophila* sperms are about 2 mm long), frequently have a 'nebenkern' and

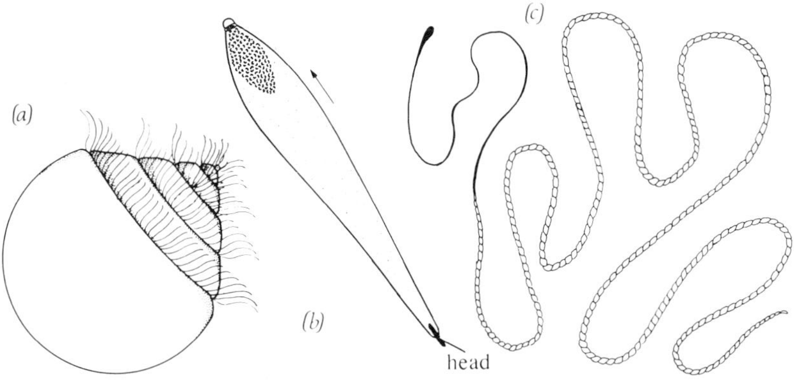

Figure 5.3 THREE PECULIAR SPERMS *(a) Sperm of* Cycas. *These 'tree-ferns' have very complex sperms, which are released from the pollen-tube tip onto the ovum surface. (b) Sperm of the tick* Argas *(after Rothschild, 1965). Locomotion, a gliding movement, is not understood. The nucleus is posterior for the active part of its life, but a peculiar involution occurs during spermiogenesis and another apparently in the female. The sperm may be 7 mm long. (c) Sperm of a* Cypridopsis; *most ostracods have immense sperms (10 mm is not uncommon) and resemble this in the long head and rope-like tail (abbreviated in this diagram)*

very complex swimming movements, mediated by an extraordinarily complex flagellar structure[5]. Octopi[6] and squids have very complex spermatophores (*Plate 4*, p. 149), which pump the seminal fluid into the female; it is uncertain where fertilisation occurs, but the distances involved are not great and the fluid is almost certainly moved by peristalsis of the oviduct. Nevertheless, octopus sperms have a very remarkable locomotor apparatus and even have glycogen granules as a food reserve, as do many other complex sperms, while the simple ones use their own lipids as fuel. For the ultimate in apparent absurdity we must digress to plants. In *Cycas* the pollen tube grows nearly down to the ovum, the tip opens, and from this are released two sperms covered with flagella (*Figure 5.3*); one of these then has to swim a distance amounting to about half its own length.

This series is somewhat contrived and there are, of course, many counter-examples. However, I believe it to be a more honest series than its converse, which would show increased complexity as the task of the sperm appears greater; it persuades me that we still know little of the real functional problems faced by sperms in female tracts.

5.3 Spermatogenesis

This is, of course, a process as various as the sperms themselves. Nearly all the work has concentrated on 'normal' flagellate sperms, especially those of mammals, but there has been much work, too, on the electron microscopy of spermatogenesis and spermiogenesis of arthropods, notably by Nath and his school[7] in the 1950s but with many workers now turning their attention to odd sperms[8].

The nuclear processes nearly always involve meiosis, and result in four haploid sperms from each spermatocyte. The spermatocytes are usually several spermatogonial generations from the original germ cells, and two tactics are common. Either all the products of one early spermatogonium may stay together because they do not separate at mitosis, giving a clump of 2^n spermatozoa where n may be 5 or 6 (*Steatococcus*) 9 or 10 (*Carcinus* and *Locusta, Plate 7*, p. 109) or even 15 or 16 as in cyprinodont fishes. Or spermatogenesis may be 'continuous' as in the mammals (*Plate 7b*), resulting in this case in an observable cycle in each part of each seminiferous tubule (*Figure 4.5*, p. 80 and p. 81 above). Here again all the products of a spermatogonium are connected by cytoplasmic threads, but the spermatids are shed into the lumen of the duct.

In those species where the males are haploid, no reduction division need occur. Such haploid males are found in rotifers, some *Cladocera*, *Hymenoptera*, coccid bugs and many mites, and the spermatogenic nuclear division is often more like mitosis than meiosis; but in some species, including bees, unipolar meiosis occurs. Meiosis may be aberrant even in diploid males; *Drosophila melanogaster* males have a reduction division, but no apparent chiasmata or genetic crossing-over. Some species (e.g. polychaetes and other worms) have been said to have post-meiotic mitosis[7] before spermiogenesis, so that these spermatocytes would each produce more than four sperms. Although such a post-meiotic mitosis is usual among land plants and fungi, descriptions of sporadic occurrence among animals require more validation.

5.4 Spermiogenesis

This is the name given to the final steps of sperm construction, the haploid meiotic product or spermatid becoming the mature sperm. Its final steps are usually concluded outside the testis.

Figure 5.4 shows the development of the sperms of a sea urchin, a crab, a nematode and *Figure 5.5* shows that of a mammal. Shading has been used to represent supposed homologies but this should only emphasise the differences. As would be expected from evolutionary considerations sperms are very different, and the routes from the fairly ordinary-looking spermatid cell to these different sperms are very divergent. Some generalisations seem possible, however.

Most, if not all, true locomotor flagella have a basic 9 + 2 structure, and are derived from one of the pair of centrioles present in the spermatid. The other often, as in the mammals, becomes attached to the posterior end of the head and may, as in many invertebrates, form the 'sperm aster' in egg cytoplasm. This may later contribute to the first division spindle, so that half of the zygotic cells will have pairs of centrioles of paternal origin.

The haploid chromosomes are condensed before they are packaged into the sperm head. Lysine-rich histones are removed and **arginine-rich proteins** appear, associated with the condensing chromosomes. In mammals these are histones, while in non-mammalian species these are usually called

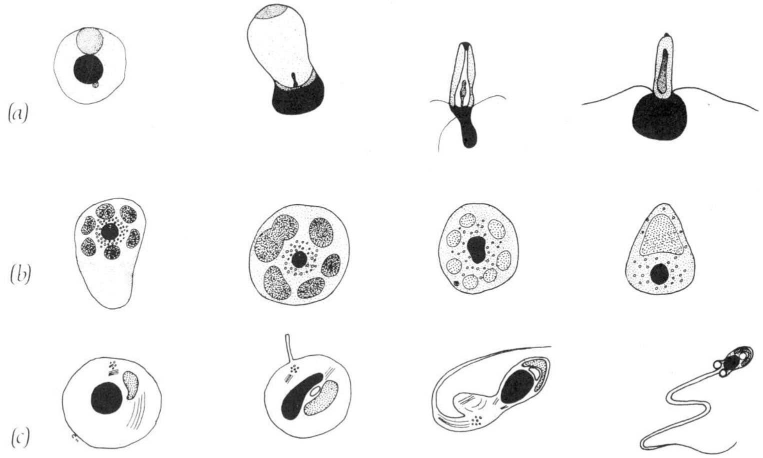

Figure 5.4 SPERM DEVELOPMENT Four stages of development are illustrated, from the round spermatid to the sperm with its characteristic morphology. Nucleus solid black, acrosome or equivalent densely stippled. (a) Spermiogenesis of the hermit crab Clibanarius *(after Nath, 1965). (b) Spermiogenesis of the nematode* Ascaris *(modified from Nath, 1965). (c) Spermiogenesis of a sea urchin (*Echinus*) sperm*

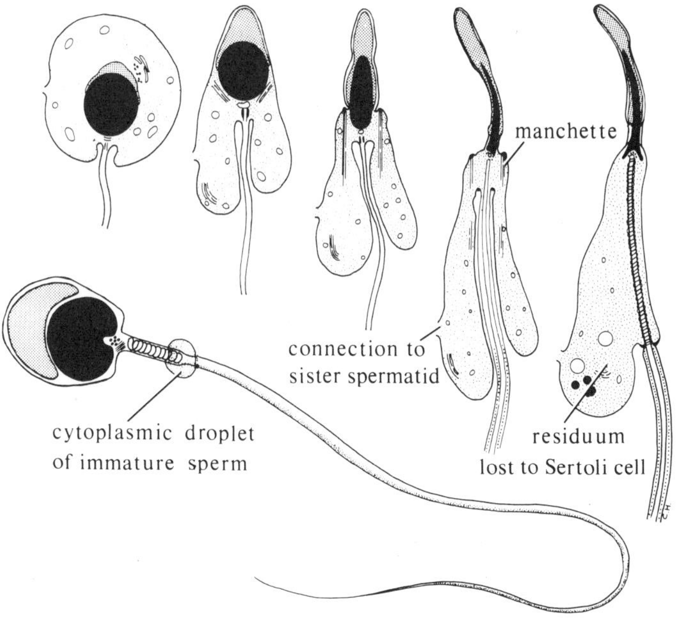

*Figure 5.5 DEVELOPMENT OF A MAMMALIAN SPERM (*CAVIA*) Conventions as in previous figure. Compare with* Figure 5.1. *The cytoplasmic droplet is lost from most sperms during their sojourn in the epididymis. (Modified after Longo and Anderson, 1974)*

protamines, often incorporating the name of the animals, as salmine, clupeine and iridine from salmon, herring and rainbow trout.

Protamines are usually much smaller molecules than histones (25 or so amino acids) but like them seem to come in only four or five kinds in one species. These molecules are thought to couple the nucleic acids so that closer packing results, giving the very condensed structure of most sperm head nuclei. Inoúe showed by sophisticated polarising microscopy[9] that the 11 chromosomes of the locust sperm head lined up serially; but in mammals there seems not to be a regular way of packing the nucleus. In eutherian mammals there seem, uniquely, to be extra S–S cross-links making the sperm head very rigid; Bedford[10] considers this rigidity to be necessary for penetration of the thick zona of the egg without bending and so losing direction during transit of this layer.

The acrosomal space is supposed to always contain enzymes, often in some variety to cope with the various penetration problems to be faced. In mammals at least, perforation of the outer acrosomal membrane and the sperm membrane (*Figure 5.6*) results in leakage of these enzymes and dispersion of the corona cells around the egg. Soluble enzyme fractions from sperm acrosomes of other organisms too can usually be shown to disperse, soften or dissolve egg investments; but the bound enzymes, attached to the outer aspect of the inner acrosomal membrane are usually more important. In marine annelids, molluscs and echinoderms, this inner acrosomal membrane is thrown forward as a long cone by the expansion of an acrosomal granule or 'perforatorium' behind it (*Figure 5.6*), usually in response to sea-water into which eggs have released substances[11]. The resulting new projection is called the 'acrosomal filament'; it should be realised that its origin is *behind* the acrosome proper. This 'acrosome reaction' is probably best regarded as the last step in development of the acrosome.

There is usually much less substance in the final sperm than in the spermatid; much of the excess cytoplasm is lost as **'residual bodies'**, **'cytoplasmic droplets'** and so on. A considerable proportion of the loss or change of cytoplasmic constituents must be invisible but very important chemically. For example, the terminal steps of spermiogenesis (called **spermioteleosis**) usually take place with the sperm heads embedded in special accessory cells. In mammals these are the Sertoli cells, which probably transmute these sperm by-products into essential hormones within the tubule. In earthworm the gregarine parasite *Monocystis* spends its early life as a trophozoite mimicking such a 'sperm mother cell'; one wonders how much trophozoite cytoplasm derives from spermatid cytoplasmic remnants.

5.5 Sperm Transport in the Male

The impression given by most male tracts is that the pressure of actual sperm production forces previously produced sperms into the ducts; these are usually grossly distended, with the white cheesy contents clearly visible within, in the breeding season. This may be a true impression for

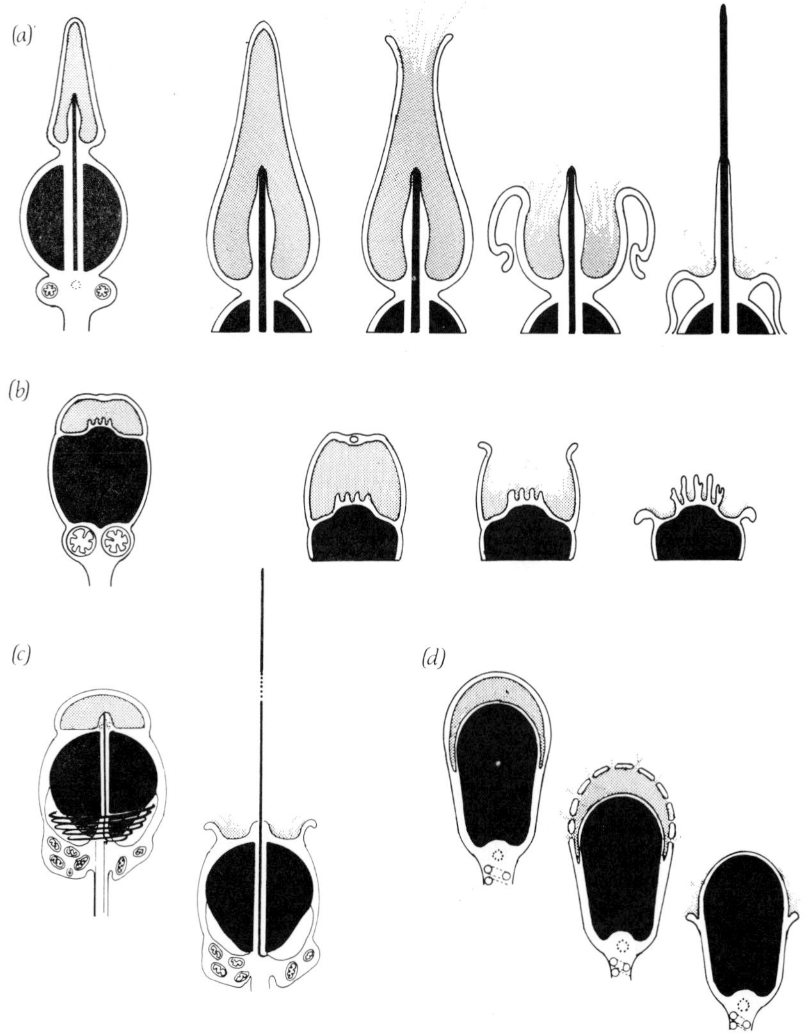

Figure 5.6 THE ACROSOME REACTION. Compare Figure 7.1. *(a) In* Mytilus *sperms (after Dan, 1967), after addition to egg suspension, the acrosome opens, contents pour out, the outer acrosomal membrane and sperm outer membrane fold back, and the perforatorium lengthens, pushing the posterior wall of the acrosome forward as a long rod. (b) In* Hydroides *sperms (after Colwin and Colwin, 1967), there is no rod-like perforatorium, and presumably enzyme molecules are retained on the convoluted rear wall of the acrosome. (c) In the horseshoe crab* Limulus *(*Xiphosura*) there is an immensely long acrosomal filament/perforatorium which is coiled posteriorly to the nucleus prior to extrusion (after Dan, 1967). (d) In mammals the outer sperm membrane and the outer acrosomal membrane fuse at many points and apparently disappear, leaving the acrosome covered with small perforations through which enzymes leak. Later, the outer membrane is lost over the apex of the head, leaving the inner acrosomal membrane, with attached enzymes: the mammal sperm bares its enzymic teeth on approaching the zona pellucida. See* Figure 7.1d *and* Plate 7g, *p. 109*

some organisms, but for mammals at least the situation is much more complicated. Both secretion and absorption occur in the seminiferous tubules, and the net result is the passage of fluid which can be collected from the ductuli efferenti, and which is not simply a serum ultrafiltrate[12]. This fluid at least dilutes the sperm suspension, so that despite appearances it is not simply sperm-to-sperm contact pressure which causes passage into the epididymis. In the epididymis itself the sperms change both physiologically and morphologically until they reach the cauda, where they accumulate as a store. Many mammals normally ejaculate about half to twice the number of sperms found in a vas deferens during a copulatory episode, and the cauda epididymis may contain the equivalent of about 10 to 15 times as much as a vas. About this number again are maturing in the rest of the epididymis and a small multiple of this number will be present in the lumina of seminiferous tubules. It takes about 14 days, in the mouse, from late spermatid to ejaculate; probably this figure also applies to the rabbit and could be doubled for man.

Most male mammals pass sperms continuously through the system even if no copulations occur. Masturbation certainly occurs in wild apes, monkeys and deer and probably in all other wild male mammals. It has been observed in laboratory rats and mice[13] too, and its frequency suggests that an ejaculation about every two or three days is normal in the absence of females. In the presence of females, but without copulation, the rate increases. There is some evidence, too, that regular copulation increases the rate of sperm production, correlated with increased male hormone levels in these animals. In the absence of females, however, the rate does not drop to zero; except in some conditions of pathological overcrowding the rate of sperm production is surprisingly high; usually it is about an ejaculate per two or three days, even for those species with pairing and long periods of non-receptivity by pregnant or otherwise non-oestrous females. In very many mammals, of course, sexuality is seasonal and the males usually only show spermatogenesis before and during each breeding season; so periods of regular and continuous sperm production may alternate with nine months without ejaculation.

5.6 Emission

Occasionally, as in *Hydra* and some marine *Hydrozoa*, mature sperms pass out of the genital pore (in these cases the apical pores of the testes) continually when the organism is in its sexual phase; there is no ejaculation as such, although sperms are released at different rates. More usually, the release of spermatozoa is a process synchronised with the availability of fertilisable ova. Such synchrony is achieved in a variety of ways by externally fertilising forms (*see* Chapter 3). Both sexes may respond to the same external stimulus, or one sex may cue the other in some way. Seminal fluids in the inhalant current of many filter-feeding females (e.g. mussel, oyster, *Pomatoceros*) cause them to release eggs, while egg-water usually has a comparable effect on the males. This gives a positive feed-back situation which leads to all animals in effective aqueous communication releasing their gametes together.

Such an orgy of gamete release has other functions apart from the obvious one that it increases the likelihood of sperm–egg collision. In freshwater species the simultaneous seminal release of several males (females, like *Anodonta*, often brood the eggs) relieves the osmotic strain on the sperms by the release of more ions and organic substances into the water; so a better fertilisation rate is achieved, for example for zebra fish with two or three males chasing each female. All aquatic habitats have many other species avid for eggs or even sperms. By releasing an enormous number all together these organisms are satiated with a proportion of all the emissions, whereas successive smaller emissions might be totally consumed. If eggs or sperms are released successively they might form an attractive food source to which another species could specialise. But synchronous release, only once in a season, removes such a possibility as the predator would starve for the rest of the year.

Because copulation has brevity as one advantage, males adopting this tactic usually store sperms and accessory fluids, and mix and ejaculate them together. It is rarely, if ever, pure spermatozoa which are ejaculated, although a few animals do have very concentrated semen, like the turkey with more spermatozoan volume than the liquid in which they are suspended[14]. More usually many glands contribute to the semen. Some products coagulate to form 'instant spermatophore' or plugs, others stimulate the female or the sperms themselves. But for most seminal additives[15] function is only guessed at, and for some we do not know enough even to guess. The citrate (and other) buffering of human semen changes the pH of the vagina as soon as it enters, so that sperms are never exposed to the normal acidity[16]. That prostaglandins and other pharmacologically active agents, like the 5-hydroxytryptamine found in dogfish ejaculate[15], are required for stimulation of the female tract is an attractive idea; but spermatozoa suspended in simple salines can be artificially inseminated with a good success rate in those mammals with very complex ejaculates, although of course egg yolk or milk does better than simple saline as a diluent. And the attractive, but converse, idea that the prostaglandins in semen may have served to excite the male ejaculatory reflex, simply being 'excreted' into the vagina where they do no harm, is rendered less attractive by the fact that stallion semen contains prostaglandins even though it is a cervical/uterine inseminator; most uterine inseminators have fairly 'simple' ejaculates[15].

5.7 Locomotion

The frenzied swimming activity of flagellate sperms is their most obvious characteristic when ejaculated. In some species, like the bull, this results in a gross swirling movement of the fluid, usually visible as **'wave motion'** under a low-power lens; often indeed it can be seen in the liquid simply held to the light in a test tube. Human and bull sperms are very active,

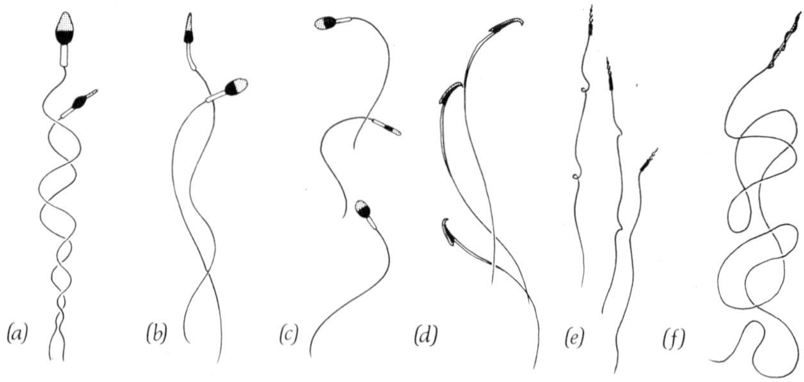

Figure 5.7 SPERM LOCOMOTION (a) Human sperms have several waves along their length. (b) Bull sperms have only 1½ to 2 waves in the principal piece. They turn as they swim, showing the edge of the head. (c) Rabbit sperms also turn, but their movement is less 'fluid' than the bull or human. (d) Rat sperms flex and extend, as if less than a full wave is included. Between cells or in viscous medium, they are very efficient. (e) Octopus sperms show gyres rather than waves, but they do not progress actively on a slide in sea-water. (f) Limnaea *sperms are immensely long (over 0.5 mm) and never seem to swim but only to lash parts of their length. In* Limnaea *ducts it may be another story*

rabbit less so, and mouse and rat sperms swim much less actively; they seem to 'hook' their way along (*Figure 5.7*).

It is moderately certain, however, that the intrinsic locomotion of these sperms is *not* required to bring them to the site of fertilisation. They arrive there so soon after coitus in many species[17] that muscular churning of the organs is usually supposed to account for their distribution, especially as inert particles like Indian ink may also be transported high into the tract. There is evidence that these inert particles, and dead sperms, are washed out of the tract again quickly, probably by ciliary action. Some of these reports of sperms found high in the tract, for example in the oviducts, soon after coitus may be suspect, the result of churning of the tract contents *after* death; we have found[18] that the rabbit uterus sometimes contracts violently as the animal is opened, exposing it to cooling by the air, and in these cases phagocytes from the uterus and abnormally high sperm numbers, presumably from the same source, are found in the oviducts.

Despite such doubts about *very* rapid ascent of sperms, it is still clear that intrinsic flagellar locomotion cannot by itself account for the arrival of sperms at the fertilisation site. It probably serves mainly for location of eggs and for orientation and penetration when an egg has been found. In non-mammalian species with internal fertilisation, sperms sometimes only have to negotiate tubes of a length comparable with their own (*Figure 3.3*, p. 58); nevertheless there is often considerable movement by the sperms. Puzzling also is the continuous movement of some kinds of sperms after ejaculation into a spermatheca; tsetse fly, garden snail and dogfish sperms all show swimming activity when expressed from the female system and examined in its fluids. It is possible, but doubtful, that it is the higher oxygen tension on the microscope slide which allows movement of sperms normally stilled by anoxia. This is a much more attractive explanation for lack of movement in some *male* tracts, for example mouse vas deferens. After expressing the sperm mass onto a slide in saline, movement commences from the edge inwards in a dramatic manner. Although it used to be supposed that a specific substance (a gamone, see p. 135) was involved in sperm immotility during storage, this is now usually attributed to lack of oxygen for many species, especially mammals. (I used to say in lectures on this subject that this gamone, androgamone I, was 'phlogiston, or negative oxygen' but in recent years students seem, alas, not to have heard of phlogiston.)

The mechanisms of sperm motility are very varied. Flagellate sperms are unlike flagellate protozoans in several ways. The flagellum is posterior, and it shows waves which are nearly in one plane[19], not helical gyres. The usual pictures of a sperm swimming like an eel or tadpole is not very misleading, except for oddities like rodent sperms, which have a peculiar stiff looping movement, and the long sperms of some flatworms and molluscs which tie themselves in knots and seem not to be able to show progressive movement (*Figure 5.7*). Amoeboid sperms of nematodes move with only one or, at most, two blunt pseudopodia and closely resemble forms like *Naegleria* (*Dimastigamoeba*) in their locomotion. Some sperms with no intrinsic locomotory ability may have special organs for penetration after the parents' actions have put them on the egg surface. Some arthropods[7], like lobsters, have an explosive mechanism for driving the nuclear material into the egg. *Limulus* has an explosive acrosome reaction which ejects a filament many times the length of the sperm itself (*Figure 5.6*); this draws the sperm to the egg instead of active swimming[20].

Sperms may be carried by other cells, as in the calcareous sponges[21] where a choanocyte is penetrated and carries the sperm to an egg, or in some gastropods where eupyrrhene (gametic) sperms are carried by oligopyrrhene (aberrant, with poorly staining nuclei) germ cell products. This has also been claimed for some species with 'hypodermic' methods of insemination, notably the bed-bugs (*Cimex*)[22] and *Peripatopsis*. In these forms (and in some others like the leeches) insemination is by injection through the body-wall, and the sperms make their way to the ovary via the open blood circulation. Many of them are to be seen inside cells, however, especially in special glands (e.g. Berlese's organ in bed-bugs, also called Ribaga's organ or the **spermalege**) and this has led to the supposition that they are transported by these cells. It is more probable[22] that these sperms are being phagocytosed, as most are when inseminated by a more orthodox route (*Plate 7f*, p. 109).

Much is now known about the biochemical mechanisms involved in sperm swimming. Although the mitochondria of the mammalian midpiece appear normal, an unusual variety of substrates may be utilised; it also seems that, contrary to expectations, changes in the pattern of sperm locomotion induced by changes of medium result in changes of the biochemical tempo. That is to say, the enzymes which produce and degrade cyclic-AMP are themselves subject to sperm demands rather than determining them. The contractile proteins of the sperm flagellum, called sperm myosin (**spermosin**) and sperm actin (flagellar actin or **flactin**) will make isolated tails 'swim' if provided with ATP, and Gibbons[23] has localised the contractile proteins in the side branches of the nine doublet fibres of sea urchin sperms.

5.8 Autonomy of the Sperm Genome

Do the genes carried by a sperm affect its chances of fertilisation? This question is central to any consideration of selection of sperms in female tracts, or heterogeneity as an explanation of high sperm numbers. If sperms cannot be differentiated on the basis of their functional or defective haploid genome, then a strategy of reproduction which required that a large surplus of defective sperms was rejected and not permitted to fertilise cannot operate. Most of the evidence, including of course the existence of Mendelian ratios, shows no discrimination between alleles in their frequency at fertilisation.

There are, however, a number of exceptions, two of which will be mentioned, in which a distortion of the expected gene segregation ratios is found. SD (segregation distorter) in *Drosophila* is a gene locus at which several alleles have been described. All except + are lethal when homozygous, but show an interesting effect in the heterozygote male SD/+; when he mates, the female produces more SD-carrying offspring than +, i.e. sperms carrying SD are more likely to fertilise. Analysis has demonstrated that many spermatids fail to mature, and the presumption is that these are mostly or entirely the + sperms, SD sperms contributing in greater numbers to the ejaculate. But + sperms of identical genotype do not die during spermiogenesis if their sister cells do *not* contain SD. About SD/SD we cannot know, because the male dies in the egg, but it is clear that the presence of SD causes spermatids *not* carrying it to degenerate, a very puzzling case of apparent haploid gene autonomy in the spermatid[24].

Figure 5.8 THE T-LOCUS OF THE HOUSE MOUSE (a) Part of chromosome 17 (= linkage group IX) qk = 'quaking', a membrane deficiency in nerve cell sheaths; 'low' is a non-Mendelian segregator; fu = 'fused', a skeletal defect related to membrane effects; tu = 'tufted', a hair-structure mutant related to interactions between tissues; H2 is a multiple locus specifying tissue-compatibility antigens (whether grafts between mice will be rejected). Tla is the region specifying a characteristic antigen of embryonic cells, probably a protein. At the T-locus are a series of dominant-to-wild type mutant alleles: T, brachyury (see text); T^{hp} *= 'hairpintail', whose effect on embryos varies according to whether they receive it through egg or sperm; the + allele, contributing to a 'normal' mouse with t or if homozygous; and a series of t mutants with no effect if paired with +, but lethal if homozygous or with some other t's (see text), and modifying T to tailless in many cases. (b) Some of these mice. (c) Mendelian segregation of t through female, but not through male. More t sperms achieve fertilisation, which is said to account for its commonness in wild mice: the high efficiency of passage across the generations is thought to compensate for the loss of homozygous embryos. (d) Heterozygotes between t's of different complementation groups are usually viable, female-fertile but male-sterile albeit with normal-looking sperms*

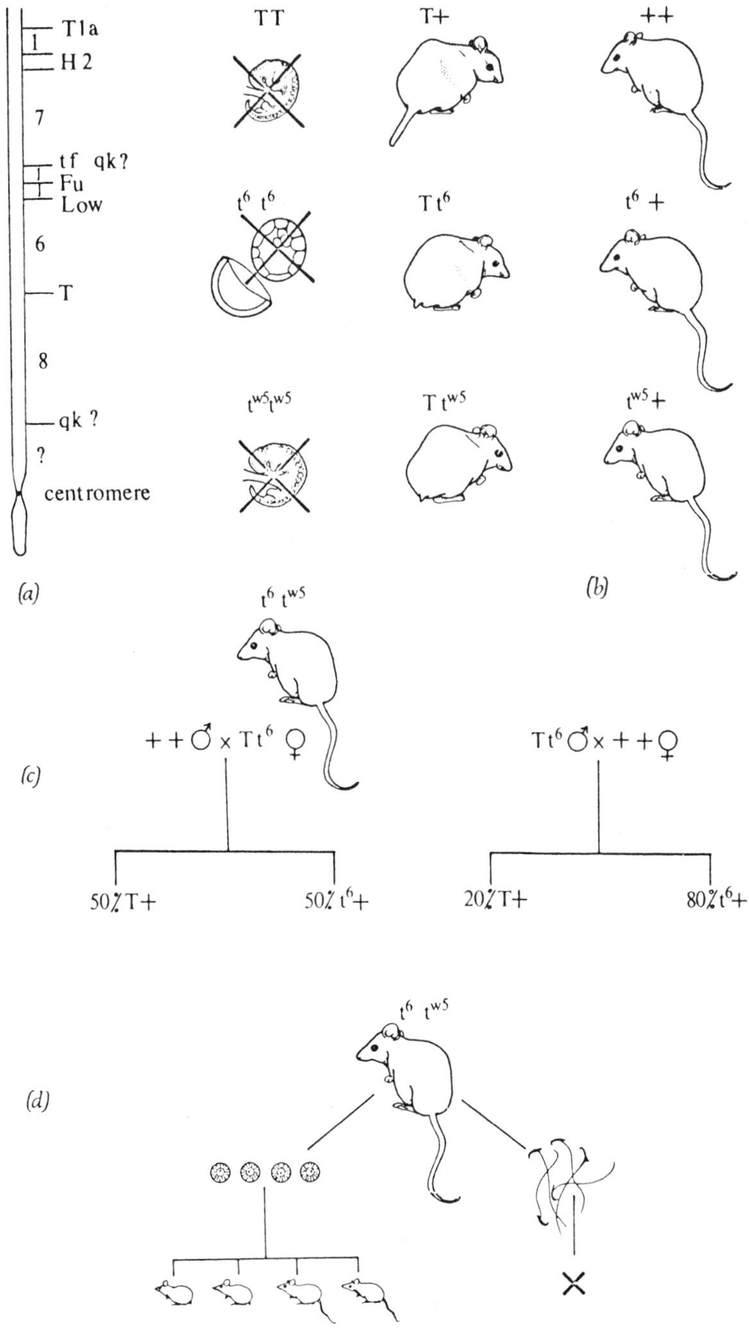
T1a
1
H2
7
tf qk?
Fu
Low
6
T
8
qk ?
?
centromere
TT
T+
++
t6 t6
Tt6
t6 +
tw5tw5
Ttw5
tw5+
(a)
(b)
t6 tw5
++♂ x Tt6 ♀
Tt6 ♂ x ++ ♀
(c)
50% T+
50% t6+
20% T+
80% t6+
t6 tw5
(d)

An even more intriguing case, but with the same puzzle, is found in the mouse at the T locus[25]. This is a complex of genes found in Map Group IX (chromosome 17), between the centromere and a group of genes concerned with immunity and transplantation antigens (*Figure 5.8*). T (brachyury) is dominant to the normal (+) allele and is lethal when homozygous TT; the heterozygote T+ shows a short tail and some pelvic and vertebral anomalies. T segregates in Mendelian fashion. There are other alleles at this locus known generally as t, which give taillessness with T and no apparent morphological effects with +. These recessive alleles (t) fall into several groups (t^0, t^1, t^x, t^y) and nearly all are lethal when homozygous or in combination with another t allele of the same group; with + they show little if any effect except that homozygous lethality reduces offspring number; with T they cause taillessness to varying degree; with another t from a different group (e.g. t^o/t^x) survival is usually possible (*Figure 5.8*). t^6 is one of group t^x and t^{w5} one of group t^y so Tt^6 nd Tt^{w5} are both tailless; t^6/t^6 and t^{w5}/t^{w5}, like T/T, die soon after fertilisation; but t^6/t^{w5} is a viable combination. T/t^6 animals, when mated with +/+, would be expected to give 50% $+/t^6$ normal-tailed progeny and 50% +/T with short tails, and this does indeed happen when +/+ males mate with T/t^6 females. However T/t^6 males mated with +/+ females produce many fewer +/T progeny, and more $+/t^6$ progeny, than expected. Segregation distortion of up to 80:20 is common, very different from the expected 50:50. T/t^{w5} behaves similarly, with t^{w5} progeny favoured. In this T case there is no evidence of selective spermatid failure; and variation of time between mating and ovulation affects the degree of segregation distortion, so it is clear that the variation from Mendelian expectation is caused by events in the female tract. + or T-carrying sperms seem less able to contribute to fertilisation than their t-carrying sister cells. What of t^6/t^{w5} males, in which all sperms carry t alleles? Surprisingly, normal-appearing sperms are produced but none achieve fertilisation; the males are effectively sterile[26]. This example has been given at length to demonstrate that many reproductive situations are not explicable in Mendelian terms, as well as to show a well-documented case of sperm gene autonomy being demonstrated by differential success of the two kinds of sperms in the female tract. Differences in the antigens on the sperm heads of T and + sperms have now been demonstrated convincingly too[27].

Although it was believed for many years that there was little if any transcription of messenger RNA from the DNA of the spermatid, Kierszenbaum and Tres have now produced convincing evidence[28] that RNA *is* being transcribed in mouse spermatids. Is this RNA only transcribed from a few genes like the T-locus? The paucity of evidence for other genes suggests that this is so, but there is a logical hole in this argument. Geneticists can only describe genes for which alternative alleles turn up; if gametes with a particular allele never fertilised because the gene was autonomous, we could never know by breeding experiments, in which nearly all gametes are not used. That is to say, genes which possessed a series of autonomous alleles which denied the sperms carrying them access to the egg would only exhibit this in occasional mutant sperms. No male would be formed by fertilisation from such a sperm, and only in some cases would this condition arise in or through a female, and then it would not be recognised, although her sons might be marginally less fertile as half of the sperms in each ejaculate failed. Even if this phenomenon was widespread it would be very difficult to discover. The ingenuity of geneticists must not be under-rated, however. Helen Spurway[29] discovered a **'grandchildless'** gene in *Drosophila subobscura* (*see* p. 147), and other segregational oddities like 'hairpintail' (T^{hp}) in the mouse[30], which differs in its effect according to which parent it was received from, and t and SD, have indeed been discovered. Perhaps, then, t-locus is the tip of an iceberg whose extent cannot be discovered by breeding tests. Or t-locus might be one of the very few chromosome areas actually to be transcribed in the sperm. It may be significant that the other major claim for autonomy has been for HLA genes[31] in the human which, like H2 in the mouse, are the sites of transplantation antigen variations. On the other hand, these transplantation antigens are found on the surface of many cells, and it has been suggested that heterozygotes may apportion spermatocyte surface to the four resulting sperms in such a way that many sperms have mainly one or other antigen[32]. Differences among the sperms may not then reflect gene autonomy, but only the small number of sites to be distributed at random.

So t-locus and SD may indeed be special. The special oddity of both these situations may, however, be accounted for by the possibility that they are involved very closely with the actual mechanism of sperm selection. On p. 27 and p. 111 space is given to the idea that 'chiasma-associated' error may render sperms unlikely to get to fertilisation. If the t-locus is involved in this mechanism[33] for making sperms vulnerable in female tract, then failure might result in *more* sperms arriving at the fertilisation site – they will not have 'confessed' their 'chiasma failure' and will therefore be over-represented. Other theories to account for the mutant sperms' success have suggested that each mutant produces a toxin to which it, but not other mutants, is immune. There are some 200 described t-alleles so this is unlikely, though possible.

5.9 Numbers of Sperms

Very nearly all animals produce a vast excess of sperms in each ejaculate, compared with the number of eggs offered for fertilisation. This sperm redundancy is small in insects (2–2000) and higher in mammals (10^4–10^{10}). I suggested that it was necessitated by errors[33] which occurred during a proportion of attempted chiasmata, which rendered the sperms receiving the resultant chromatids unacceptable for fertilisation. *Figure 2.4* (p. 28) shows mean chiasma frequency and sperm redundancy for all the organisms for which I have been able to locate both figures. The two lines are those separately computed for 'best fit' for the insects and mammals[34], while the dotted line is statistically less acceptable and made on the assumption that the animals all form one group[35]. The slope of the lines shows the extent of increase of redundancy with chiasma frequency, while the height (intersect with the y axis) shows how many sperms are required even if there is no chiasma-associated error. That is to say, slope tells us chiasma-associated r and the y intersect tells us n (p. 33); m factors probably contribute a little to both[33]. Wallace[34] has criticised this idea on a number of grounds, some of which require consideration. His assertion that the sample is a biased one cannot be countered except by more figures; but the slope has not changed significantly as the number of species has increased from 7 to 25. That only insects and mammals are listed is unfortunate, but the slopes within the groups are consistent, as Mather showed in his appendix to Wallace's paper. Wallace's arguments about each point depend upon separate assumptions, whereas I have used all the figures as published, taking logarithmic means of sperm numbers and arithmetic means of chiasma frequencies where several estimates exist, or using the separate points.

I believe that the idea that most sperms are produced but rejected as gametes for reasons inherent in their manufacture explains the variation in extent of sperm redundancy; also it enables the various factors m, n and r to be calculated separately and referred to physiological, anatomical and biochemical aspects of the reproductive strategy. I have indicated this on p. 30 but some examples should be given here, where we actually consider the sperms and not just the arithmetic.

The sperms of some mites and some ostracod crustaceans are very large indeed compared with the animal itself (*Figure 5.3*). Some mites have haploid or achiasmate males, but I can only find all figures for one species; as few as 16 may be inseminated. Some hymenopterans, in which the males are haploid, have a strategy in which many of the sperms are used. The figures for *Dahlbominus* emphasise this[36]. The male inseminates 200 sperms, then the female lays 120 eggs of which about 70 make females and the other 50 remain unfertilised and produce males. The arithmetic is not so neat in the honey bee, where the queen receives about five million sperms on her mating flight (or when artificially inseminated by the breeder), and may lay more than 100 000 worker (fertile) eggs, over about 10 years, giving a redundancy of about 20 to 50, not bad considering the longevity of the sperms. *Aedes*, the yellow fever mosquito, inseminates about 2000 sperms, and the female then lays about 80 eggs[37]; because three to five chiasmata (C) may be found on the three bivalents, chiasma-associated r may be 10 or so, too little to contribute greatly to the redundancy. But in locusts (*Locusta* and *Schistocerca*) C is 18 to 20, so one might expect chiasma-associated r to be about 1200. Total R is over 10^3, $3.5–5 \times 10^5$ sperms inseminated for some 100 to (a maximum of) 700 eggs. The sperm figure is probably low, because only the contents of the tubular spermatophore can be counted, and sperms are probably passing through into the female tract. The numbers recoverable from the female fall rapidly in the hours after transmission, whether or not eggs are laid, and only about 5×10^4 can be found by our methods 24 hours after the male had dismounted. Even the king crab, *Limulus*, which is an 'external fertiliser' with very odd sperms, needs about 60 000 sperms to be available to each egg for about half the eggs

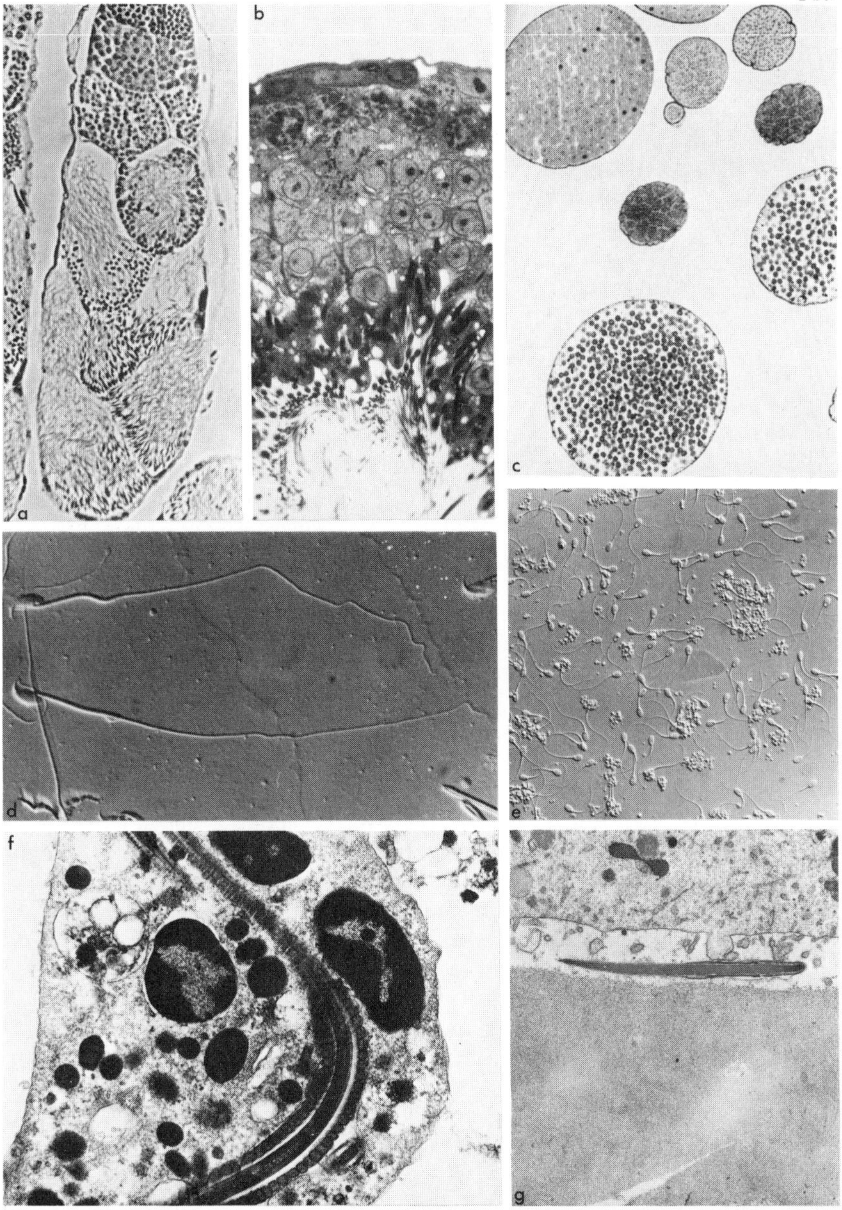

*Plate 7 TESTES AND SPERMS (a) Part of a grasshopper (*Stenobothrus*) testis. Note the separate 'follicles', each with developing sperms at one stage. Mature sperms are at the bottom of the photograph. (b) Part of tubule wall of mouse testis (compare with* Figure 4.5*) (Photo by K. Tyler). (c) Sections of* Ascaris *(roundworm) testis. Mitosis occurs in the thin terminal sections (top right) maturation in the thicker regions (top and bottom left). (d) Mouse sperms (compare* Figure 5.2*) (Photo by K. Tyler). (e) Rabbit sperms, from a c. 100× dilution of an ejaculate (Photo by K. Tyler). (f) Thin section of a rabbit uterine leucocyte containing a sperm in process of digestion (EM and photo by K. Tyler). (g) Thin section of a just-fertilised rabbit egg. Zona pellucida is at the bottom, with a slit from sperm penetration, and egg cytoplasm at the top. A sperm head (possibly a supernumerary sperm) has bared its inner acrosomal membrane and the egg cytoplasm is 'bubbling' (compare* Figure 7.1*) (Photo and preparation by K. Tyler)*

to start development[38]. Unfortunately, C is not known, nor other factors of *r* or indeed *m* or *n*.

The Chinese hamster produces very large sperms and has a very low chiasma frequency for a mammal (15 to 17); not surprisingly, it ejaculates a relatively small number of these sperms. Electro-ejaculation suggests that the number is very low indeed, 10 000; but our experience with stud males mated with oestrus females suggest that about a million are deposited in some tens of intromissions. Female Chinese hamsters are very fierce to males except for the oestrous period of some four hours every five to six days, when they exhibit lordosis and permit him to copulate; at any other time they will tear his scrotum severely and may damage the testes, even if they are put in *his* territory. The wild behaviour may not allow tens of intromissions over three or four hours, of course; nevertheless I feel the higher sperm figure is more likely than the lower, published, one.

The same problem occurs with the golden (Syrian) hamster. Electro-ejaculation gives a figure of 10 000 for this rodent, but males allowed to mate *ad libitum*, achieving intromission some hundreds of times over about two hours and ejaculating 20 or 30 times, produce about 20 million sperms. C is about 27.

The overall number of sperms in an ejaculate, of course, includes *n*, *m*, and *r* and there may well be other than 'chiasma-associated errors' in the *r*. Many ejaculates are much more than enough, and it is possible to extend them using appropriate diluents, often based on milk or egg-yolk to provide large molecules. Semen from a good bull may be extended 500 times in exceptional cases, though 50 to 100 is more usual. It is perhaps significant that animals near my chiasma line, like rabbit or mouse, can only be diluted about 20 or 5 times, respectively. It is said human semen cannot usually be diluted: there is on average about a one-in-ten chance of fertilisation for each copulation at about ovulation time.

Although it may have little to do with *r*, figures for *minimum* sperm number needed to fertilise mammalian eggs are given in *Figure 2.4*, p. 28. Amphibia produce two points which are very difficult to explain in terms of a comparable chiasma-associated redundancy with insects and mammals. The *minimum* number of sperms required for frog (*Rana temporaria*) is shown in *Figure 2.4* from unpublished figures[40]; published figures are much higher, however. Even the low figure is barely within the terms of the theory if one supposes that there may be zygote redundancy among the 500 to 1000 eggs[33]. But axolotl (*Ambystoma*) has some 60 chiasmata[41], and only millions of sperms in each spermatophore. A reasonable estimate is that about 100 eggs can be fertilised from each spermatophore, giving $R = 15 \times 10^4$ at most; more likely 10^4, when 10^9 would be expected by my theory[33]. Axolotls develop despite a surprising amount of aneuploidy[42], but nevertheless have a very low R for their C.

What happens to the excess sperms? The processes associated with fertilisation nearly always look much more like a selection, an 'obstacle race', rather than a strategy to allow sperm and egg to meet easily. A few non-mammalian examples must suffice, with a little more detail for some mammals.

Sea urchin eggs are invested in a jelly coat which, as it dissolves in sea-water, agglutinates nearly all sperms of the same species, i.e. sticks them together head-to-head. This was discovered at the beginning of this century, and as the 'fertilisin' reaction served as the paradigm for sperm-trapping by eggs. But Tyler[43] in 1962 stated that only 1 per 1000 of these sperms *can* fertilise afterwards; presumably the others have had their acrosome reaction prematurely fired, or are simply too stuck together to be effective.

Very few of the sperms of *Hydrozoa* seem to be wasted, according to Miller[44]; but even some of those in the vicinity of eggs fail to swim towards them, and those not in the vicinity are presumably those less sensitive to the attractive substances.

Frog sperms nearly all fail[45] to penetrate the various layers in the jelly around the egg, and can be seen in immotile spherical layers up to gastrula stages and beyond. Only about 1 per million (or even fewer) gets through to the vitellus according to some unpublished results of mine (but 1 per 300 according to Wallace, who is much more familiar with amphibian eggs).

The penis of the male mouse articulates with the cervix and he ejaculates about nine million sperms into the uterus, which is already distended with oestrous fluid containing, among other proteins, a variety of immunoglobulins (antibodies), and complement (the cell-membrane lysing system). As he withdraws, after about two to three seconds' intromission, the male deposits a mixture of secretions in the vagina, which instantly coagulates and becomes the hard vaginal plug. Sperms can be found in the oviducts 15 minutes after copulation and the number increases until some six hours later. Different breeds have different characteristic time relations between oestrus, the willingness and attractiveness of females, and ovulation, which usually occurs within two to six hours of mating. The eggs probably require an hour or so to finish extrusion of first polar body and so become fertilisable, so fertilisation usually occurs well before the maximum number of sperms is found in the oviducts, puzzling for any theory to account for sperm number. Those sperms which achieve the oviduct differ from nearly all those left in the uterus in that their acrosomes are not covered with a layer of immunoglobulin (Ig); nor can they be coated by washing in serum as can nearly all mouse sperms taken from the vas deferens or epididymis of males[46]. Also all these oviduct sperms are complete, whereas in the hours after copulation there is considerable destruction of those sperms left in the uterus – heads and tails usually detach, then the pieces are eaten by phagocytes, as are many whole sperms. Sperms in the anterior end of the oviduct invade the cumulus mass of ova (p. 122) and many simply stop; others reach the zona and halt, often sideways-on to it. One or several penetrate the zona, but only one fuses with the vitellus. What determines which sperms pass through the utero-tubal junction is unknown; the sperms which do get through cannot be coated by antibody, and it is tempting to consider this proof that they have been a different population from the beginning. They may, however, have picked up a protective coat, or lost their antigenic one, in some way unrelated to their spermiogenic history. Again, it is not known what differences allow some sperms through to the egg, but the visible difference is often spectacular – several sperms get stuck in the membranes, then one swims through almost as if the membranes were not there, but these membranes have not disappeared because subsequent sperms again have difficulty.

In the rabbit, non-pregnant does will mate readily nearly all the year round. Males mount anything even approximately rabbit-shaped; they achieve intromission and ejaculation into the vagina in some 5 to 15 seconds, often throwing themselves off the doe as they ejaculate. Some breeds, notably lops, cry out and show quite a prolonged (20 to 30 seconds) spasm *after* the event. 100 to 200 million sperms are deposited, and after five minutes enough have penetrated the cervix to give a normal size litter[47]. By four to six hours post-coitus (pc) there are some six to nine million sperms in the uterine cavity and about 1000 in the oviducts, perhaps moving through into the peritoneal cavity. Nearly all these uterine sperms have antibody (probably IgG) over the acrosomes whereas less than half those in the oviducts can be so coated[46]. At about six hours there is a massive influx of leucocytes into the uterine lumen, and they begin phagocytosis of sperms immediately; despite this destruction the number of sperms in the uterus increases slightly as more are released from the cervix. Even with this recruitment, however, the number falls after about 10 hours until at the time of ovulation, 12 to 13 hours pc, there are less than four million left, mostly without IgG-coated heads. In the oviducts are about 1000 sperms mostly without IgG, of which 5 to 50 enter each egg (*Plate 7g*). As in the mouse, many sperms fail to penetrate cumulus or zona, and again it is not clear why; perhaps these are already 'capacitated' sperms (p. 140) which have prematurely exposed their acrosomes and have no enzymes left. Again, while it is tempting to relate the IgG-coated sperm population to those which may have 'failed' in spermiogenesis because of chiasma-associated error, the connection is tenuous.

In the human, some 350 million sperms are ejaculated. Below about 60 million is considered possible grounds for clinical sterility whereas oligospermia, less than one million sperms per ml, is supposed never to achieve paternity[48]. The cervical mucus changes its character at the time of ovulation and becomes more penetrable by sperms; nevertheless, many more sperms fail to penetrate even this 'ideal' mucus than succeed in swimming within its fibrous matrix. Within the mucus, between the threads (**micelles**) of mucoprotein is a more fluid substance rich in antibodies, some secretory antibody (IgA) as well as the characteristic antibodies of serum. The secretion comes out of the cervical glands, and many sperms follow the pattern of cervical micelles up into these glands, where they accumulate. Hence many workers have seen these glands as a sperm store whose function is to release only a trickle of sperms into the uterus so that sperms may be available for ovulation even several days later. The general suggestion that a major function of the mammalian female genital tract is to attenuate the sudden vast sperm number into a trickle of long duration is an attractive one[49].

Unfortunately the accumulated sperms, in human cervical glands or pig utero-tubal glands, probably are not released but phagocytosed. Mammalian sperms seem not to have a 'reverse gear', and it is common to see leucocytes containing sperm debris in the local connective tissue and even in local blood capillaries.

Few sperms can be seen in human uterine washings, and at most a few thousand sperms can be found at the fertilisation site. *In vitro* penetration of human eggs, followed by early development, suggests that only one sperm usually penetrates, and that a period of capacitation of human sperms might not be required.

It is supposed that some women are effectively sterile because they produce abnormal anti-sperm antibodies. Some prostitutes, for example, have been shown to have a high titre of anti-sperm antibodies of unusual kinds, and it is possible that in such cases *all* sperms are prevented from attaining the site of fertilisation, instead of *nearly all* sperms.

REFERENCES

1. Franzén, 1970
2. Fretter and Graham, 1964
3. Dupouy, 1964
4. Rothschild, 1961
5. Baccetti *et al.*, 1975
6. Mann *et al.*, 1970
7. Nath, 1965
8. Dallai *et al.*, 1975
9. Inoué and Sato, 1962
10. Bedford and Calvin, 1974
11. Dan, 1967
12. Setchell *et al.*, 1969
13. Kihlstrom, 1966
14. Nelsen, 1953
15. Mann, 1964
16. Fox and Fox, 1971
17. Austin, 1975
18. Cohen and McNaughton, 1974
19. Woolley, 1975
20. Dan, 1967
21. Gatenby, 1920
22. Hinton, 1964; Davey, 1965
23. Gibbons, 1975
24. Hartl, 1975
25. Bennett, 1975
26. Dunn and Bennett, 1969
27. Yanagisawa *et al.*, 1974
28. Kirszenbaum and Tres, 1975
29. Spurway, 1948
30. Johnson, 1974
31. Fellous and Dausset, 1970
32. Goldberg *et al.*, 1970
33. Cohen, 1975b
34. Wallace, 1974b
35. Cohen, 1973
36. Wilkes, 1965
37. Jones, 1968
38. Brown and Knouse, 1973
39. Chang and Schaeffer, 1957
40. Wallace, 1974a
41. Wallace, 1974c
42. Fankhauser and Humphrey, 1954
43. Tyler, 1962
44. Miller and O'Rand, 1975
45. Bernstein, 1952
46. Cohen and Werrett, 1975
47. Bedford, 1971
48. Fletcher and Johnson, 1974
49. Braden and Austin, 1954

5.10 Questions

1. Describe four kinds of sperm, and the means by which they achieve the vicinity of the egg.

2. Why do male animals overproduce sperms so vastly? (Consider *all* sperm production, not only the redundancy within an ejaculate, and compare with other cell losses, e.g. from skin or gut.)

3. List possible functions for substances found in seminal fluids, and give examples.

4. If the genes carried by a sperm affected its chances at fertilisation, how would this manifest itself? Design experiments to discover alleles which affect a sperm's chances.

5. Contrast sperm production in a locust, a mammal (e.g. mouse), an eel and a snail.

6

EGGS

6.1 General Structure

Eggs look even more varied than sperms (*Figure 6.1*), but to some extent the variation is misleading because it is associated with the obvious protective coats and food reserves. The cytoplasmic and nuclear structure of eggs, perhaps more similar than their gross differences would suggest, is difficult to discern because of the presence of these membranes and the yolk; so we know less about chiasmata in oocytes than in spermatocytes, for example.

Eggs have some of the properties of individual body cells, and are often and misleadingly called 'egg cells' in elementary textbooks (*but see* p. 9). There are three successive nuclear generations within the same cytoplasmic envelope, however, and these must be distinguished. The early **oocyte**, a multiplication product of **oogonia** in the ovary, is usually the longest stage of development. In most animals the prophase of oocyte meiosis takes a significant fraction of the life span, four years in temperate frogs and 12 to 50 years in women, before the meiosis continues through the **secondary oocyte** stage, producing one **polar body**, and then to the **ootid** losing another polar body and leaving the maternally derived nucleus haploid. During the long meiotic prophase the chromosomes assume an extended stage, seen as **'lampbrush'** chromosomes in newts and some insects (*Figure 7.2*, p. 131). The enormous amount of RNA transcribed during the long oogenesis has several known functions, but these do not account for all this synthesis. Early in oogenesis a small (5S) component of ribosomes, and messenger-RNA (m-RNA), are made. The larger (18S and 28S) ribosomal-RNA (r-RNA) is a major product later in oogenesis, the structural DNA which codes for it usually being associated with nucleoli. Often these genes are 'amplified', that is to say the structural DNA is replicated many hundreds of thousands of times[1], making micronucleoli scattered through the nuclear mass or at its membrane, as in *Xenopus* and teleosts. Sometimes r-RNA is made in associated nurse cells (*Figure 4.3*, p. 77) and passed into the egg cytoplasm, as in some insects and molluscs. Much m-RNA is certainly transcribed too in the late oocyte nucleus, and most of this is later tied up in 'informosomes' to await decoding during early embryology[1]. Some of this m-RNA, however, is translated during oogenesis to provide many enzymes and structural elements; the array of 'housekeeping' enzymes needed by the oocyte simply to metabolise, as well as special enzymes required for the assembly of yolk particles and the egg's primary membranes and cortical granules, are usually acquired from its own nucleus.

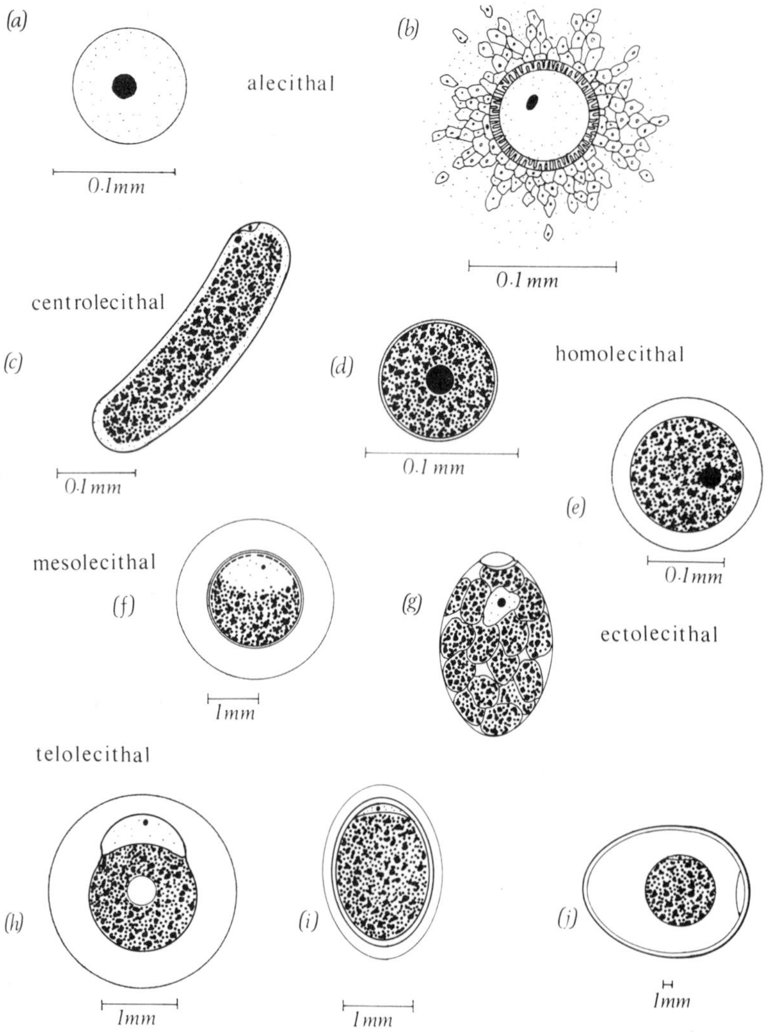

*Figure 6.1 EGGS AND YOLK All eggs are shown as vertical, A/V, sections. (a) The yolkless egg of a slug, suspended in a nutritive albumen in the shell. (b) The yolkless egg of a mammal, surrounded by the zona pellucida and corona radiata (radiating lines of cumulus cells); nutrients are provided from maternal fluids of oviduct and uterus (*Table 12.1*). (c) Oval egg of an insect, like a housefly, with a thin shell of cytoplasm around a central yolk. The egg nucleus waits under the micropyle (by polar body at top of figure) and the zygote nucleus usually moves to the middle of the yolk before cleavage. (d) The egg of* Pomatoceros, *a marine worm. Yolk is evenly distributed before fertilisation. (e)* Echinus *(sea urchin) egg has evenly distributed yolk granules and a jelly coat. (f) Frog's eggs have yolk mostly in the vegetal hemisphere. The pigment layer is broken at the animal pole; and there is usually an albumen coat. (g) Platyhelminths have a variety of cells inside the egg-shell: an ovum (shown clear), vitelline cells with much food reserve, and sperms (not shown). The liverfluke egg has an operculum (lid) shown at the top. (h) Fish (teleosts) separate the massive yolk from the 'active' cytoplasm soon after ovulation (*Plate 9a*). (i) Cephalopod molluscs (e.g. squids) have much yolk, making up most of the egg volume. (j) The yolky bird's egg has albumen as a food supply as well as the massive yolk; indeed much of the yolk is used after hatching, but the albumen serves embryonic needs like yolk of other forms*

Often some egg cytoplasm is contributed from other cells as well as from its own nucleus. Nurse cells, for example in the polytrophic ovaries of some insects (*Figure 4.3*), and follicle cells, as in panoistic insect ovaries or mammals, may contribute proteins and other molecules, ribosomes, mitochondria, cytoplasmic volume and even much DNA. In the polytrophic ovary the whole contents of all the nurse cells, usually with 512 or 1024 times the haploid DNA content, are poured into the mature oocyte (the main function of this multiplied DNA may have been to provide many copies of the nucleolar ribosomal DNA segments so that enough 18S and 28S ribosome bits were made). In locust and the domestic fowl the components of the yolk are elaborated elsewhere, in the fowl liver for example, but picked from the blood by follicle cells and transferred into the egg cytoplasm.

While this massive synthesis and transfer are in process the oocyte nucleus is usually very large, and it is often called the **'germinal vesicle'**. It is usually much less basophilic than the RNA-rich cytoplasm surrounding it, because of the extended state of its chromosomes. There is much **'nuclear sap'**, too, which may have an important morphogenetic effect once it is released into the cytoplasm when the nuclear membrane breaks down as the first meiotic division starts.

The mature oocyte is an assemblage of cytoplasmic and even nuclear contributions from a variety of sources, not a 'cell' except in the trivial sense that it has a nucleus and cytoplasm. In most ways it is quite unlike cells; its nucleus divides twice, and one grand-daughter nucleus shares 'use' of nearly the whole cytoplasmic structure with an alien nucleus; even then it does not multiply as do cells, but divides into portions with different functions, largely determined by events two nuclear divisions and frequently some years before.

Textbooks are often confusing, too, about the haploid state of the egg. Very few animal eggs are ever truly haploid. Most are shed as primary or secondary oocytes and complete the first meiotic telophase, producing the first polar body, outside the ovary (*Figure 6.2*). Usually the sperm enters before completion of this division, and only very rarely (e.g. in the sea urchins) is sperm penetration delayed until telophase of the second meiotic division. Except in these latter, then, the egg has at least two chromosome sets throughout and is *never* actually haploid. This may seem a trivial distinction, but it may be enlightening. For example, there are several reasons for supposing that eggs cannot show **autonomous haploid gene expression**, but the strongest is certainly that most eggs are never haploid, and only very briefly even possess a separate segregated haploid chromosome set. Theories which relate oocyte atresia, for example, to gene expression or chromosome properties of any other kind must not include haploid expression as a necessary event. My suggestion, that 'chiasma-associated errors' cause most oocytes to be wasted and result in use of only a small proportion, must depend upon synthesis of an unusual protein by the complete diploid (crossed-over) chromosome set of the oocyte, or the absence of a necessary substance from this complete set, and cannot posit a particular property of the final haploid segregate. Oocyte atresia in mammals occurs long before any haploidy. I have in fact

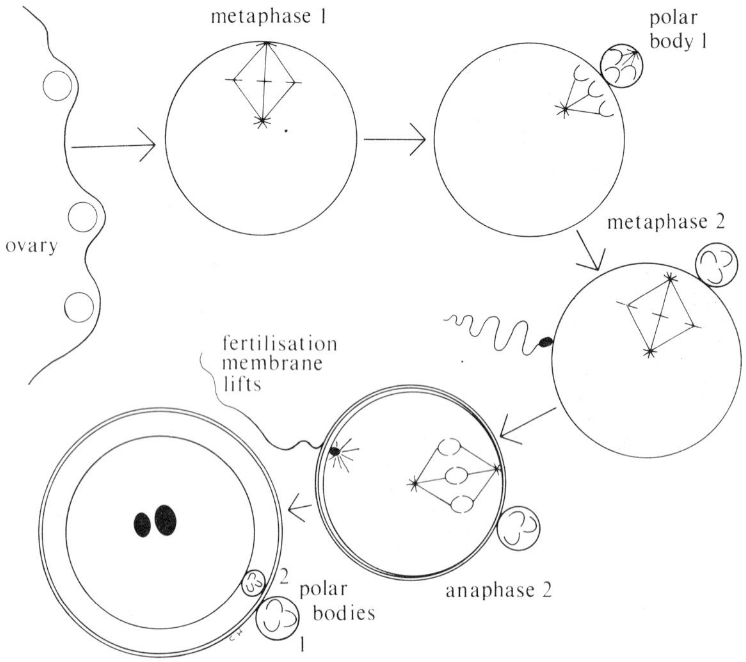

Figure 6.2 THE EGG FROM OVULATION TO FERTILISATION Eggs are usually ovulated after meiosis has resumed: the lampbrush chromosomes have contracted, the nuclear membrane disappears, and the spindle apparatus is formed. This oocyte is shown at first metaphase (M1). This first meiotic division proceeds, and very unequal cleavage of cytoplasm results in a 2° oocyte and the first polar body (PB1); both have n double chromosomes with some exchanged segments. The 2° oocyte usually proceeds directly to the second meiotic division, but as a rule halts at the second metaphase (M2), as shown, to await the sperm. When the sperm enters, the division resumes, anaphase (A2) proceeds and a second polar body is cut off. The sperm, when activating the egg, usually causes changes in the egg membrane, e.g. the lifting of the fertilisation membrane from it. In this case the first polar body is outside, the second inside, the fertilisation membrane. In mammals, however, both are within the thick zona pellucida, which persists from pre-ovulation through the ootid (post-meiotic) stage to late cleavage

proposed that some repetitive DNA (p. 172) might make a characteristic protein only when mis-read, frame-shifted, and not when read properly without interpolation or deletion[2].

The architecture of eggs is considered at more length on p. 116. Usually, accessory cells have caused considerable organisation of the egg by donating material which does not mix, or affecting the cortex beneath them. Here it must simply be emphasised that during oogenesis the egg cytoplasm is organised so that, usually upon interaction with the sperm path, a series of events is set in motion whose result is a complex embryo or even larva, *without* morphogenetic action by the zygote nucleus or its descendants (Chapter 8).

6.2 Yolk

Many different substances are used to transfer energy, material or hard-to-obtain substances from one generation to the next. These are commonly associated with phospholipids or lipoproteins in formed structures called **yolk spheres, platelets, droplets** and so on, and there may be many kinds in one egg. Thus fish eggs often contain much phospholipid yolk, considerable protein, some carbohydrate (glycogen), and the eggs of fresh-water species have a high salt content. Many, like herrings and gouramis, have an oil droplet too which makes them float.

These yolk substances may be **intrinsic**, synthesised and structured by the egg cytoplasm using enzymes coded by the oocyte nucleus. Or they may be **extrinsic** like the phosvitin and proteins of chicken egg yolk, synthesised in the liver and transferred via the blood. Usually some are transported via the circulation, and then a process of synthesis and assembly results in the final formed yolk[3].

Yolk is rarely distributed evenly through the cytoplasm (*Figure 6.1*). Even in almost **alecithal** eggs like those of some gastropods and eutherian mammals, the '**yolk nucleus**', where deposition starts, is excentrically placed, not around the central nucleus, so animal and vegetal poles can be distinguished. The same happens in **homolecithal** eggs like *Pomatoceros* and *Echinus*. In **centrolecithal** eggs, like those of most insects, the yolk fills all the egg but for a thin film of cytoplasm under the chorion; the nuclei, undergoing meiosis, are to be found nearly at one pole (*Figure 6.1c*). **Mesolecithal** eggs, like those of the frog, are said to grade through eggs like lungfish and *Gymnophiona* to **telolecithal** eggs like selachians and birds. But this series does not ring true as an evolutionary ladder; the different kinds of telolecithy have clearly evolved separately, and in fact can be shown to be basically different by the ways the yolk is involved in development[4].

There are several kinds of organisms in which nutrients are not incorporated into the oocyte itself as **entolecithal** yolk, but are supplied from outside after development starts. The mammal is a familiar example, and the general case of viviparity is shown by a series of selachian fishes (*Table 6.1*). When the embryos gain much nutrient from the maternal circulation after development starts in the female tract, this is associated with increased intimacy of placental contact or some other supply mechanism, and with less entolecithy to start with (*see* p. 204). The gastropod molluscs, while not usually viviparous, nevertheless often leave yolk out of the egg; examples include many common water-snails like *Lymnaea* and *Planorbis* (*Figure 6.3*). These eggs float in nutrient albumen, and like mammal eggs they grow in size during cleavage.

Most platyhelminths divert the function of a large part of their complex genitalia, originally part of the ovary but now called **vitellarium**, to the production of special yolky cells. These are enclosed within the egg-shell with an oocyte and sperms. These vitelline cells, or nurse cells, also produce substances which contribute to the formation of the resistant egg-shell. This process occurs in the ootype (*Figure 3.3*, p. 58), and is complex and

Table 6.1 Sources of embryonic material in selachians (from Amoroso, 1960)

Organism	*Scyliorhinus canicula*	*Torpedo ocellata*	*Mustelus vulgaris*	*Trygon violacea*	*Mustelus laevis*
Common name	Dogfish	Electric ray	Common smooth nursehound	Sting ray	Nursehound
Egg status	Oviparous	Ovoviviparous	Ovoviviparous	Ovoviviparous	Viviparous
Gestation (months)	6	4	10	2	5
Wt of ovulated egg	1.31	6.78	3.93	1.9	5.54
Wt of 'completed' embryo	2.69	13.37	60.16	118.0	189.6
% change	+ 105%	+ 97%	+1432%	+6105%	+3326
% change water	+ 213%	+ 251%	+2490%	+10 412%	+5608%
% change ash	+ 292%	+ 157%	+2800%	+10 250%	+7609%
% change organic substances	− 21%	− 23%	+356%	+1628%	+1064%
Placental type	None	Milk from villi on uterine wall, into gut of embryo	Uterine secretion copious	Villi thick, milk-secreting and long into gut of embryo	Intimate yolk-sac

time-consuming; indeed strobilisation of tapeworms has been seen as a tactic to multiply the number of ootypes (*Figure 4.11*, p. 90) so that eggs may be produced faster[5]. Likewise the larval multiplication of digeneans has been considered a compensation for the slow egg production of the adults.

Other eggs are often used as a supplementary food source, either as successive ovulations which feed the young in the oviduct as in some shark larvae, or as **trophic eggs** penetrated and eaten by the first young to emerge as in the whelk *Buccinum*. Such trophic eggs are in fact very common[6], and are found in planarians, oligochaetes, molluscs, spiders and newts as well.

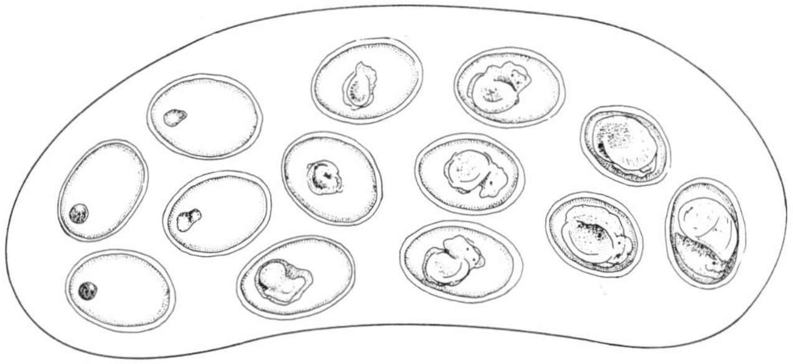

Figure 6.3 AN EGG MASS OF A WATER SNAIL, E.G. LIMNAEA *These are often found on the glass sides of aquaria, or water plants. The capsules are embedded in a jelly, and each has a tiny ovum developing, and growing into a juvenile snail by transforming the albumenous solution which surrounds it. At the left are shown eggs (about ×10 relative to their capsules) developing via trochophores and veligers to snails at the right of the mass. Unfortunately for the biologist, the eggs of a mass are in fact all at one stage*

6.3 Primary Egg Membranes

Very few, if any, eggs have only a cell membrane separating their cytoplasm from the external medium. There is nearly always a thin **vitelline membrane** or a thicker **chorion**. These are formed at the interface between the accessory cells (e.g. follicle cells) and the growing oocyte, and there are often microvilli from both cytoplasms extending across the membrane from opposite directions as it is being formed (*Figure 4.4*, p. 78). When the accessory cells lose contact, for example at ovulation, the membrane frequently has radial lines showing where these fine processes have been withdrawn. The **zona pellucida** of some mammal eggs has been called 'zona radiata' because of these fine striations, but the name is more properly applied to the vitelline membrane of yolky eggs like those of reptiles.

The vitelline membrane can usually be penetrated anywhere by sperms, and at this point **cortical granules** usually release their contents, making a perivitelline space by lifting the membrane from the surface (*Plate 8c*). It is changed in the process, becoming a **fertilisation** membrane and impermeable to sperms. On the other hand eggs like those of insects and many teleosts which start with a thick chorion have **micropyles**, areas or pores where sperms may enter. In some fishes at least, the micropyles secrete a sperm-attractive substance[7] unusual among animals; but coelenterates also have ova or accessory cells which attract sperms chemically[8].

Eggs may have a cellular membrane around them as well as a vitelline membrane or chorion. That of tunicates may prevent self-fertilisation, but we are unsure of the function of **cumulus cells** or **corona radiata** around mammal eggs; these follicle cells are embedded in a jelly or mucoid matrix (*Plate 6f*, p. 83 and *Figure 4.4*, p. 78). The eggs can be fertilised after the matrix has been lost, so its penetration by sperms is not a necessary part of the fertilisation process. The suggestion that it forms a large target for sperms which, when they contact it, are guided centripetally by the cell arrangement is made a little more plausible by the situation in the lamprey[9] (*Figure 6.4*). Here the major part of the oval egg is covered by follicle cells, but at the animal pole they are absent; instead there is a 'tuft' of transparent mucoid fibres extending a surprising distance away from the egg. Any sperms which touch them attach and then slide between the filaments to the egg surface. Presumably this also protects the successful sperms from osmotic problems for the minutes it takes to penetrate the thick membrane beneath the tuft. In mouse, rabbit and rhesus monkey a similar trapping function of cumulus may occur, for it is commonly observed that more sperms can be found among the corona cells of any freshly-ovulated eggs than in the rest of the oviduct, and there is no evidence for chemotaxis. Again, however, it is possible that selection of a certain sperm population occurs here, in which case the sperms observed in the jelly between the cells may be those which have failed a test.

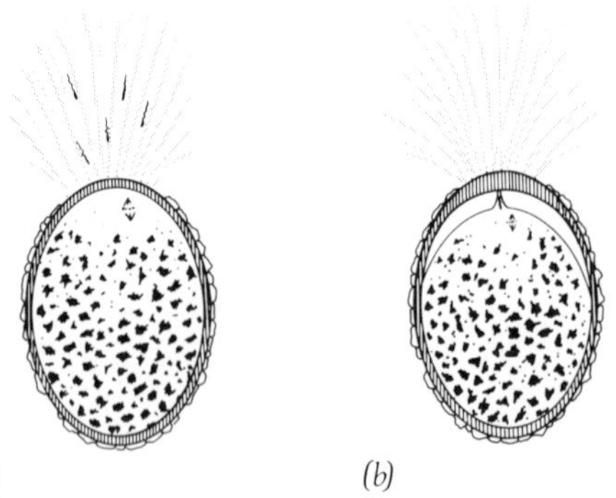

Figure 6.4 LAMPREY FERTILISATION (a) The ovulated egg has follicle cells over most of its surface, but at one end these are absent and a tuft of fibrous jelly protrudes into the medium. Sperms in the jelly move slowly towards the egg, looking like a 'reversed fountain'. (b) 5 to 10 seconds after insemination, the egg contents contract, drawing one sperm head, in a thin thread of cytoplasm, from the thickening chorion to the vitellus

6.4 Ovulation

This is the process whereby eggs leave the ovary, and it is usually associated with sexual congress. In species with copulation, however, it is usual for inseminated females to delay ovulation for considerable periods.

Sometimes, as in many echinoderms and polychaetes, eggs are shed directly into the external medium. Usually they are accompanied by substances which affect sperms, and which elicit sperm release in males. These unprotected eggs are usually released at a synchronised spawning time rather than continuously; the advantages of this have already been considered (p. 101) and are related to possible loss from specialised egg eaters as well as to higher chances of fertilisation. In *Echinus* contraction of the smooth-muscle wall of the sac-like ovary simply squeezes out any eggs which are inside; the effect can be produced by injecting 10% potassium chloride through the test. Some tubicolous polychaetes, like *Pomatoceros*, release their eggs when taken from their tubes; these eggs come from the coelom, where they develop, and not directly from the ovaries which they leave as small oocytes, so ovulation as a process does not occur.

Ovulation is also very difficult to define in those species whose germarium is continuous with the oviduct, like nematodes (*Figure 4.2*, p. 76) and insects (*Figure 4.3* and *Plate 6*, p. 83). It may be defined in relation to the maturation stage of the egg (*Figure 6.5*). Yet other species, like the guppy and other live-bearing cyprinodonts, fertilise eggs still in the ovarian wall, and the contents of the ovary leave as baby fishes. Most vertebrates do have a real ovulation, usually into the peritoneal cavity, from which the eggs are commonly guided into oviducts. They are coated variously in jellies and shells before being released (*Figure 4.6*, p. 84). Sperms must ascend this complicated oviduct in reptiles and birds so that they may penetrate the freshly ovulated eggs (which start development as they pass down the oviduct). In amphibia and many fishes, on the other hand, the sperms must penetrate the egg investments after it has passed down the oviduct. The same is probably true in the octopi and squids, in which sperms are released onto the egg mass as it is laid, from the oviduct glands or from the receptaculum seminis below the mouth.

6.5 Tertiary Membranes

Such egg investments as those in amphibia and cephalopods are called **tertiary membranes**, and are formed by accessory sex organs of the female such as her oviduct and its glands. **Secondary membranes** are those formed by, or consisting of, the follicle cells around the egg itself, while the **primary egg membrane** is the vitelline membrane or its equivalent in mammals, the zona pellucida, secreted at least in part by oocyte cytoplasm.

Familiar tertiary membranes are the albumen, shell membranes and shell of birds' eggs (*Figure 4.6*), the frog egg jelly, the 'mermaid's purse' around dogfish eggs, cocoons of planarians and the oothecae of cockroaches. Their primary purpose is usually protective, against drying, mechanical damage,

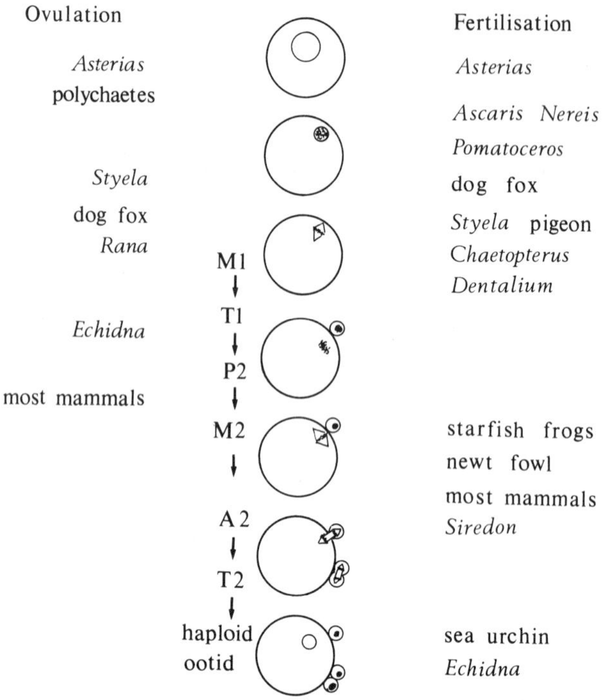

Figure 6.5 THE TIMING OF OVULATION AND FERTILISATION Ovulation usually occurs after meiosis has resumed after the extended nuclear secretory ('lampbrush') phase of oogenesis. Polychaetes, tunicates, and some echinoderms ovulate at the 'germinal vesicle' stage; the amphibian oocyte usually leaves the ovary at M1; most mammals ovulate oocytes after M1 and before M2, at which they await the sperm. Sperms can enter the Asterias *egg at ovulation or at any time during the meiotic divisions, but other starfishes' oocytes are normally penetrated at M2, like most animals.* Echinus *egg is one of very few actually to become haploid; sperms only enter after meiosis is complete. Dog and fox are unusual mammals in that sperms may enter follicular eggs, or at least oocytes which have not even reached M1*

ion loss, and so on, but some obviously contain food materials so that the juvenile can hatch later, and be more mature, than if the yolk were the sole source of nutrients.

6.6 Viviparity and Oviparity

This will be considered at greater length in Chapter 11. Two rather minor points may be made here, however. One concerns the absence of yolk in the eggs of eutherian mammals, with marsupials having very little and the monotremes laying tiny, but yolky, eggs. This kind of series is usually assembled to show that viviparity has achieved its peak in the eutherians. This may be so, but it should be remembered that the mother is involved as well as the offspring; she releases nutrient slowly over the whole gestation, instead of over a short period as a bird must. The overall cost is probably about the same. Indeed, this series emphasises the problems faced by *non*-viviparous forms which must release a potential organism into the environment, not merely move a tiny egg from one body organ to another. Viviparity represents an *escape* from some of the demands of egg-laying, although the price of the escape may be high because of the continuous metabolic demand (p. 203) of the parasitic embryos.

The other minor point also concerns a difficulty of oviparity, the problems posed by the yolk mass as the embryo develops. Cleavage of the cytoplasm is prevented except at the animal pole of most yolky eggs; where it does occur in the vegetal half, as in amphibia and crossopterygian fishes, the 'cells' are so large that they must be engulfed or overgrown by the animal cells. In teleosts, reptiles and birds the yolk is not involved in embryogenesis, and development is considerably modified so that it finally rests in the peritoneal cavity (teleosts) or gut of the hatchling. Where the viviparous mother establishes a continuous nutrient passage or placenta, this usually only modifies later development; the crucial early development can proceed in a nutrient fluid, unhampered by internal yolk. Perhaps this is the explanation for those forms, like freshwater gastropods and platyhelminths, with nutrient supplied outside the embryo (sometimes called **ectolecithal** eggs); development is not hampered by the inert yolk mass as it obviously is in insects, cephalopods and birds (*Figure 12.1*, p. 211).

6.7 Numbers

Table 2.1 (p. 16) lists the egg numbers produced by a variety of species. In Chapter 2 we considered this range as a spectrum of redundancy, showing production of a few (selected?) oocytes full of yolk or otherwise well started in life, contrasted with the production of tens of millions of tiny oyster eggs most of which die as planktonic larvae.

Atresia (resorption) of most mammalian oocytes is well documented (*Table 6.2*); it also occurs in fish, echinoderms and insects[6]. As well as this pre-ovulatory loss, there are various post-ovulatory oddities. The plains viscacha, a large South American rodent, ovulates nearly 1000 oocytes of which only about six implant[10]; elephant shrews show a similar, but lesser, loss between ovulation and implantation.

There is often an explicit mechanism in the egg for rendering many zygotes redundant, as in some molluscs like the whelk *Buccinum*. A large proportion of the fertilised eggs in the egg masses of several gastropods are destined to be food for the other juveniles; these are called **trophic eggs**, and the embryologist Raven[6] sees these, which he calls 'alimentary eggs', as the final term in a series which commences with internal yolk and continues with ectolecithal eggs, whose nutrient is produced by nurse cells of the same lineage as the egg, as in insects. Then he quotes rotifers, in which the ovary has two parts, a **vitellarium** which makes nurse cells only and a small **germarium** of oocytes. Platyhelminths

Table 6.2 Oocyte atresia and related losses in the female germ line

Organism	*Common name*	*Kind of redundancy*	*Stage of life cycle*	*Ratio*	*Source*
Diopatra	Marine worm	Nurse cells	Adult	48–78	Anderson and Huebner, 1968
Spirorbis	Marine worm	Nurse cells	Adult	1 (*no* nurse cells)	P.E. King, 1968
Stylocidaris	Sea urchin	Atresia. 'Rate of production and destruction nearly equal'	Adult	High	Holland, 1967
Forficula	Earwig	Nurse cell	Pre-adult	2	R.C. King, 1964
Drosophila	Fruit fly	Nurse cells	Pre-adult	15	R.C. King, 1964
Carabus	Beetle	Nurse cells	Pre-adult	127	Chapman, 1969
Notopterus	Knife fish	50% (constant) atresia	Juvenile and adult	High	Shrivastava, 1969
Pleuronectes	Plaice	Atresia	Before and after spawning	?	Wheeler, 1924
Mus	Mouse	Atresia	Embryo–adult	28 000?/100? = 280?	Jones and Krohn, 1959
Rattus	Rat	Atresia	Embryo–adult	160 500/150? = 1130?	Thibault, 1969
Ovis	Ewe	Continuous atresia and production	Embryo–adult	>10 000/20 i.e. >3500	Thibault, 1969
Bos	Cow	Atresia	Embryo–adult	2 700 000/10? 270 000	Thibault, 1969
Homo	Woman	Atresia	Embryo–adult	6 100 000/40? 1 500 000?	Thibault, 1969
Elephantulus	Elephant shrew	Wastage of ovulated oocytes	Each ovulation	400/2 = 200	Van der Horst and Gillman, 1941
Lagostomus	Plains viscacha	Wastage of ovulated oocytes	Each ovulation	900/2 = 450	Weir, 1971

NOTE: *Lagostomus* and *Elephantulus* probably have oocyte atresia as well as 'wastage' of ovulated eggs. Lemurs and macaques apparently continue ovogenesis into adulthood, like *Stylocidaris*, *Notopterus* and ewes, so complex kinetic studies would be required to assess oocyte number per ovulation. Ovulations for mouse, rat, woman are in assumed 'natural' state, with pregnancy and lactation alternating, except for anoestrous seasons. References are to be found in Cohen (1971) divide the ovary into several organs of which the vitellaria are very widespread through the tissues (*Figure 3.3*, p. 58) and the germarium (ovary proper) is a discrete organ. He continues with the cases where most eggs are consumed by the others as food, found in some polychaetes, oligochaetes and prosobranchs like *Buccinum*. In my view, this last situation would more easily commence such a series than form its climax; this would make the series show a progressively earlier separation of trophic and developmental function.

This series also raised an important question relating to redundancy. If *any* eggs may become larval or trophic, but 3% regularly become larval, then a real redundancy related to a developmental fault may be considered. If, on the other hand, it is possible to point out in advance which cell will become the oocyte and which cells will become nurse cells, this is a case of specialisation for function and should not be misinterpreted as a redundancy derivable from stochastic processes (e.g. 'chiasma-associated errors') during oogenesis. Oocyte atresia in mammals, and trophic oocytes, seem to come into the first category; but it *is* possible to predict which of 16 cystocytes will become the oocyte in *Drosophila*. Nevertheless, both categories fit the C/R line (*Figure 2.4*, p. 28), and it is puzzling that this should be so[2].

REFERENCES

1. Davidson and Hough, 1971
2. Cohen, 1975b
3. Williams, 1967
4. Nelsen, 1953
5. Llewellyn, 1965 (p.63)
6. Raven, 1961
7. Austin, 1965
8. Miller and O'Rand, 1975
9. Kille, 1960
10. Weir, 1971

6.8 Questions

1. In what respect may eggs be usefully regarded as cells?

2. What are the relationships of yolk to the rest of the egg, in a variety of organisms?

3. Eggs must be protected from the environment, but sperms must be allowed in and there must be at least gaseous exchange. What solutions to this problem are in use by organisms?

4. What relationship is there between size and number of eggs?

7

FERTILISATION

7.1 The Processes

Fertilisation involves a large number of processes. We tend to think of a 'moment of fertilisation', but because some of these processes take longer than others, we cannot really speak of the egg as being fertilised until all of them are complete. This is usually just before the first cleavage. The processes themselves differ also from animal to animal, but there are basically five which should be expected.

7.1.1 PENETRATION TO THE VITELLUS AND THE 'ACROSOME REACTION'

The sperm must penetrate the membranes surrounding the egg, and this is usually achieved by means of sperm enzymes borne on the exposed acrosome; sometimes, as in insects and fish for example, there is a special pore, the **micropyle**, through which the sperms penetrate a tough resistant chorion around the eggs. As we have seen, sperms vary greatly in their morphology, but most of them undergo a change in their acrosomes, or a comparable change in other structures in the vicinity of the egg. The outer membrane of the sperm usually fuses with the outer membrane of the acrosome lying directly under it, and at the points of fusion pores appear, through which the acrosomal contents are released into the surrounding medium; these usually contain egg-membrane-penetrating enzymes. This occurs in mammals and is known as the **'acrosome reaction'**[1]. In the annelids and some echinoderms, an apparently different sort of 'acrosome reaction' occurs (p. 98)[2]; the *internal* wall of the acrosome is pushed right out, through both the outer acrosomal membrane and the sperm plasma membrane, as an 'acrosomal filament' which projects in front of the sperm and contacts the egg surface (*Figure 5.6*, p. 99).

7.1.2 MEMBRANE FUSION

When the outer sperm surface, which is by now the surface of the *inner* acrosomal membrane, perhaps projecting as a filament as in annelids and *Limulus*, touches the cell membrane of the egg the two fuse like two soap bubbles, and the intermediate wall disappears. This leaves the egg cytoplasm in contact with sperm cytoplasm (*Figure 7.1*) over a small area, which increases, allowing the contents of the sperm head access to the egg.

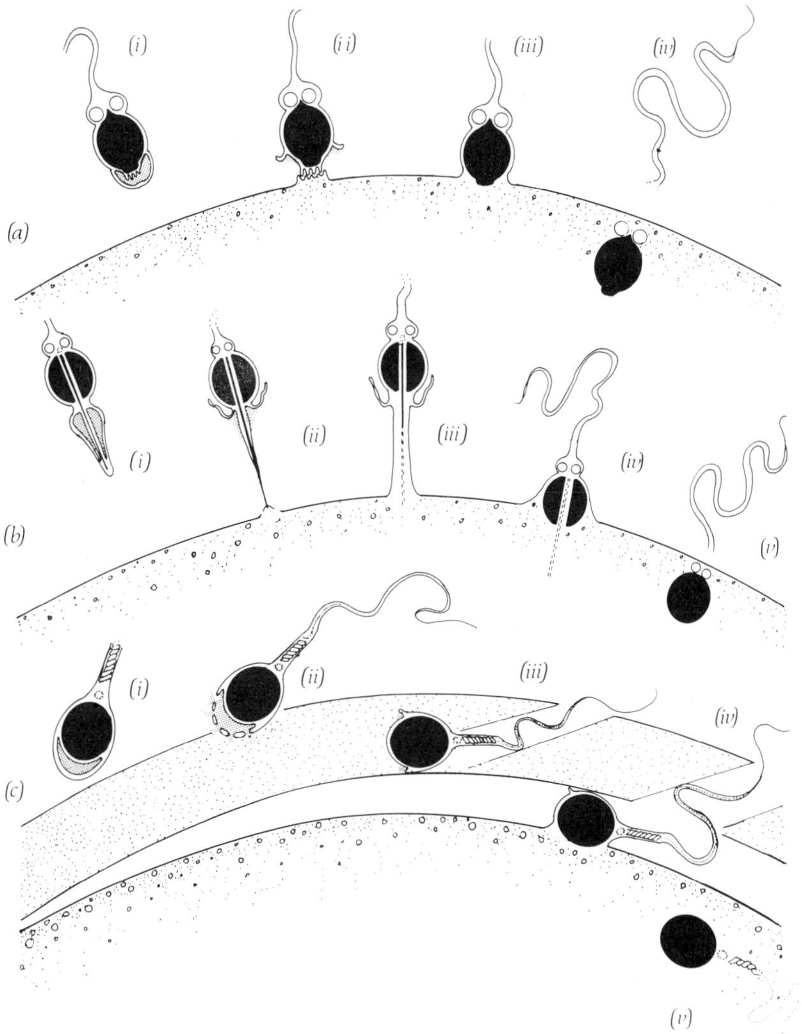

Figure 7.1 EGG/SPERM FUSION (a) In Hydroides, *a polychaete like* Pomatoceros. *i. The sperm approaches; ii. the acrosome reacts, losing most of its anterior covering; iii. the adnuclear acrosomal wall fuses with the egg membrane, allowing nucleus centriole and mitochondria into the egg cytoplasm; iv. the tail is shed, possibly leaving a patch of male-derived membrane as part of the egg surface (modified after Colwin and Colwin, 1967). (b) In* Mytilus. *i. The sperm approaches; ii. the 'perforatorium' pushes the posterior wall of the acrosome forward as an acrosomal filament which contacts the egg. Acrosomal contents (stipple) may have dissolved a path through egg membranes; iii. the membranes fuse. iv and v, as (a) (modified after Colwin and Colwin, 1967). (c) In a mammal (e.g. rabbit). i. The sperm approaches the zona surface; it may not have undergone the acrosome reaction as it penetrates the cumulus mass; ii. If acrosome reaction (appearance of pores) has not occurred previously, it happens now; iii. The sperm probably lyses its way through the zona using the residual enzymes on the adnuclear acrosome membrane, which now constitutes its anterior tip. It leaves a slit in the zona. iv. One sperm (of many penetrants in the rabbit) fuses with flaps of egg membrane by its* posterior *head membrane, and the contact enlarges. v. The whole sperm is commonly engulfed (after Austin).* Note: *Compare this figure with* Figure 5.6

7.1.3 PENETRATION OF THE VITELLUS

The area of contact now grows, and at least the nucleus and a centriole of the sperm are forced into the egg, the sperm membrane becoming a patch in the egg membrane. Most sperms leave the tail outside, and it rarely (for example in the freshwater oligochaete *Tubifex*) remains stuck on this patch. The sperms of mammals usually go in entire; that part posterior to the head retains its membrane and part of the head leaves some of its membrane at the surface (*Figure 7.1*). Most flagellate sperms enter perpendicular to the surface, but the mammalian sperm is the exception and enters tangentially; the first contact looks like the approach of a boat to a dock. The mammalian vitellus virtually engulfs the sperm with a sort of pseudopodial action around the post-nuclear cap: amoeboid nematode sperms also enter entire, and the granular mass of acrosome-like bodies usually precedes the nucleus.

7.1.4 ACTIVATION

The egg responds variably to penetration by a sperm. In many marine invertebrates, especially echinoderms, the contents of **cortical granules** of the egg burst and create a space between the cell membrane and the vitelline membrane. The substance extruded has a high osmotic pressure, and water is drawn through the vitelline membrane, making an overall volume increase and lifting the membrane off the vitellus; it probably also hardens the membrane, since no more sperms penetrate it, and this is now called the **fertilisation membrane**[3].

In mammals, there is usually a less dramatic elevation of the thick **zona pellucida**, and the perivitelline space is often seen to increase somewhat after fertilisation (*Plate 7g*) and even during the first cleavage. However, again there seems to be a **'zona reaction'** which prevents other sperms from gaining entry. The rabbit is peculiar in that many sperms normally penetrate the zona, but only one penetrates the vitellus; up to 50 so-called **supernumary sperms** can be found actively moving in the perivitelline space for some hours after fertilisation. There are usually two or three in the mouse, too, and it is not an unusual phenomenon, despite myths to the contrary, in other mammals including man. The vitelline surface is obviously the main defence against pathological polyspermy in mammals.

The eggs of fishes also usually exhibit a dramatic elevation of the chorion, but this sometimes happens *without* fertilisation, and is merely a response to the external medium. Siamese fighting fish eggs show this, but zebra fish eggs only elevate their membranes if they are fertilised. Gourami eggs seem to have a very impermeable chorion; when they are fertilised, a perivitelline space appears, but not because the membrane elevates. Instead, the vitellus shrinks, presumably because liquids are drawn from it rather than from the medium; this also happens in lamprey eggs.

These visible reactions are only a small part of the processes that are activated by sperm entry. Oxygen uptake also rises for example, and protein synthesis begins, having been turned off in the late oocyte; the **egg meiosis** also proceeds and usually one or both polar bodies are formed (*Figure 6.2*, p. 118, and *6.5*, p. 124). Even if the sperm nucleus has been inactivated by x-rays, egg cleavage will usually still proceed. Frog eggs, and many others, can be activated simply by pricking with a needle (blood, egg fluids or even chicken egg albumen on the tip are more effective than a clean needle). A sudden change of temperature, pH or osmolarity or the addition of detergent, proteases or lipases have all been used to activate eggs. Clearly the sperm function in activation is not a complex message, but simply a trigger; the egg has been 'loaded' during oogenesis, and is 'fired' by the sperm.

Monroy and others have investigated the onset of protein synthesis resulting from activation[4]. Oogenesis has endowed the egg with a very large complement of ribosomes, often, as in *Xenopus* oocytes for instance, by **'amplifying'** the gene for ribosomal-RNA and using some thousands of copies of it to synthesise ribosomes at an enormous rate. During nearly all of oogenesis the chromosomes have been in an extended **'lampbrush'** state (*Figure 7.2*) in which synthesis of messenger-RNA has been proceeding apace. Some of this m-RNA, at least, survives through fertilisation. Ribosomes extracted from unfertilised eggs can make proteins when provided with m-RNA and the requisite transfer-RNAs (t-RNA) and amino acids, *in vitro*. And m-RNA from unfertilised eggs can be

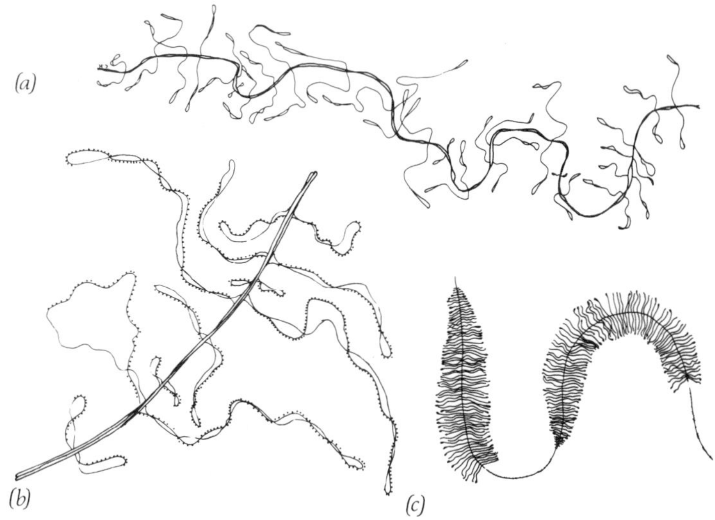

Figure 7.2 LAMPBRUSH CHROMOSOMES (a) Part of the squashed nuclear mass of a newt oocyte; the chromosome length has side branches (seen to be double) and the central axis is composed of the two homologous chromosome regions. (b) A higher power view of a length of such a chromosome, showing the 'beads' on one axis of each side branch. (c) When these spreads are made on electron microscope grids, stained and viewed at high magnifications, the beads are seen to be regions of RNA transcription from DNA, the chains growing as one proceeds along the strand

used to prime foreign, or even homologous, ribosomes. But both these experiments require that the cytoplasmic architecture is destroyed far more violently than with the agents which we know can activate the system and stimulate protein synthesis. If very delicate extraction is used, inactive polysomes (groups of ribosomes attached to m-RNA) can be extracted, on which protein synthesis is not occurring. They may lack one of the t-RNA amino acid components, but it is more likely that they require a simple stimulus or the removal of a simple repressor[4]; poly-adenine may be involved here in some way. By analogy with other systems, this removal could be as simple as K/Na ionic imbalance of the kind which would be expected if the membrane is broached. There is some evidence that these inactive polysomes are inactivated by a protein which is susceptible to removal (digestion) by a wide variety of proteases. As soon as a few of these polyribosomes are activated, of course, m-RNA coding for proteases is likely to be among them and the resultant protease synthesis causes a 'cascade' reaction which releases all the others. Because these inactive polysomes enshrine m-RNA from the oocyte, they are an important source of developmental instruction; indeed they have been called **'informosomes'** by Spirin[5], and we will say more about them in the next chapter.

7.1.5 SYNGAMY

The culmination of the processes of fertilisation is the fusion of the nuclear materials. These usually form **'pronuclei'** which move together and meet in the centre of the cytoplasm, but before this can happen both male and female nuclear material must be organised into pronuclei.

There is a long textbook tradition which speaks of the egg as a haploid gamete, but in fact there are very few eggs in the animal kingdom which complete meiosis before the sperm enters. *Figure 6.5* (p. 124) shows the stages of egg meiosis, and the time at which ovulation and vitellus penetration occur, in different organisms.

By far the most common system is for the oocyte to remain in an extended meiotic prophase for a very long period; but this is odd, and not really comparable with a stage in male meiosis, because the oocyte DNA is so synthetically active. Seasonal, hormonal or traumatic stimuli are all used by some organisms to cause the final ripening and release of these oocytes. As ovulation approaches they commonly proceed through to first meiotic metaphase, with the spindle very close to the surface of the animal pole of the egg and the chromosomes on the spindle equator parallel to the egg surface – then the processes pause. When the egg is released from the ovary into coelomic fluid, or the outside medium, this first meiotic division proceeds through anaphase into telophase, and a tiny **'first polar body'** is cut off, containing the 'north' pole of the spindle and its telophase chromosomes (e.g. *Figure 6.2*, p. 118). Usually without reorganisation of a 'proper' nucleus, the telophase chromosomes at the internal pole of the half-spindle of the egg now form a prophase cluster. This rapidly forms a new metaphase plate as the divided centriole forms the spindle of the second meiotic division. At this stage, with one polar body and the metaphase for another, most eggs again enter a resting phase and await activation by a sperm (e.g. *Figure 6.2*, p. 118).

There are a few eggs, like the tunicate *Styela*, in which the nucleus is still in the swollen prophase secretory state, the **'germinal vesicle'**, when the sperm enters. The sperm enters, then waits for polar body production. Other eggs, like those of some sea urchins, complete meiosis very rapidly and only admit sperms when already haploid.

After completion of the second meiotic division and production of the **second polar body**, the egg nucleus is usually reconstituted and moves towards the centre of the egg cytoplasm (not usually the geometric centre of the vitellus, because of the asymmetric yolk). The first polar body often divides too.

Meanwhile, the sperm nucleus has had a series of problems too. In the terminal stages of spermiogenesis the sperm nucleus has undergone a process of condensation unlike any somatic nucleus. Histones are removed from the DNA and replaced by other basic proteins, which in many species are protamines; these bind tightly to DNA and to the remaining histones. These sperm-characteristic proteins are usually very arginine-rich. This reduces the size of the nucleus and hence of the sperm head, which is usually much smaller than a somatic nucleus and aids in 'streamlining' of the sperm[1]. Especially in eutherian mammals, this arginine-rich basic protein bonding is reinforced by many S–S bonds too, which doubtless lend a certain rigidity to the sperm that is useful in penetration of the egg and its membranes[6].

When the sperm nucleus is released into the egg cytoplasm, it is still, of course, condensed. It immediately starts to swell as the protamines are removed, and it draws in fluid. There is a possibility that these sperm proteins may release inhibition of inactive polysomes in egg cytoplasm, but the evidence is only circumstantial. As the nucleus swells, its attendant centriole usually puts out long fibres into the egg cytoplasm, which resemble one end of a mitotic spindle but may be produced all around, making a 'star' shape, the **'sperm aster'** (*Plate 8e*). It was for a long time thought that this sperm centriole was indispensable for the first division spindle, egg and sperm nuclei each supplying one end. But false 'sperm asters' appear in anucleate bits of sea urchin egg cytoplasm when chilled, or put in Ca^{++}-free sea water and so on, so this contribution by the male is now generally considered to have been over-rated. It may be normal in most species, but not obligatory; in some the egg cytoplasm may actually supply the attendant centriole to the sperm nucleus, the sperm's own centriole being discarded[7].

The two **pronuclei** meet in the centre of the egg cytoplasm. This need not be the geometric centre of the egg, as there may be immense amounts of yolk at the vegetal pole, with a very little cytoplasm; in that case the nuclei will meet near the animal pole. How they meet is not known, but in frogs, sea urchins and mammals the paths curve toward one another as if each 'aimed' at the other's present position – they do not go straight to the rendezvous and wait. Perhaps bundles of microtubules, maybe even those connected with egg centriole and sperm aster, meet and contract. An interesting clue is that the sperm centriole, which followed the nucleus into the cytoplasm, soon swings around it and leads the way. The mutual approach of the pronuclei is called **karyogamy**.

When the actual fusion occurs there may or may not be actual condensation of the mixed **zygote nucleus**. This does occur in a few organisms, but in most the pro-nuclear membranes disappear and the chromosomes, already shortening from a condition resembling mitotic prophase, take up positions on the first cleavage spindle. Each chromosome is two chromatids, one of which will go to each half-egg at first cleavage, so that each half will have a complete diploid complement of chromatids, n from the sperm and n from the egg. This event is known as **syngamy**.

7.2 The Medium

Most fertilisations occur in an isosmotic medium, sea-water or the animal's own body fluids. Sperms do not apparently have osmoregulatory mechanisms, so it is surprising to find *any* which function in fresh water. But a surprising number do, if only for a short time. Lamprey (*Petromyzon planetra*) sperms have a 'half-life' of about a minute, as would be expected, and that of tropical freshwater fish rarely exceeds three minutes except in special cases. Such special cases include cichlid sperms, which are emitted in a fair volume of semen which billows over the eggs, and the bubble nest builders like gouramis and Siamese fighting fishes whose nests may contain motile sperms some hours after embracing has ceased; these may have been protected by the mucous coat of the bubbles. Less easy to explain, however, is the prolonged life of *Anodonta*, *Dreissensia*, *Hydra*, and *Craspedocusta* sperms; even when resuspended in fresh water they may be motile for hours. Perhaps, like the sperms of some bryophytes and pteridophytes, swimming in fresh water with 'naked' nuclei at the top of the spiral, these sperms depend on membrane-associated, already-primed, enzymes for motility and so would not be damaged by loss of electrolytes. Sperm 'models', sperms which have been extracted with various solvents including glycerol/water mixtures, still swim when ATP is added[8], so it is not really surprising that some sperms can swim in fresh water.

Substances may be released into the medium with the gametes or even by the gametes, and these may have effects upon gametes from the other sex. **Fertilisin**, the jelly coat of sea urchin eggs, is the classical example, described by Lillie[9] in 1919. This is a fairly heterogeneous mucopolysaccharide, having molecular weights ranging from 200 to 500 000 and a major component at about 300 000. It has the property of agglutinating sperms head-to-head. This was for many years considered a trapping mechanism so that sperms would congregate in the vicinity of an egg mass. I have recently suggested, however, that this may be a selection process, only the sperms which are not agglutinated being acceptable. Tyler[10] had already discovered that after a sojourn in egg-water only 1 per 1000 of sea urchin sperms could fertilise, but he did not suggest directly that this might be a selection. Evidence for selection is coming in from other quarters, however, and that sperm which enters is now usually considered special[11].

Other substances associated with gamete physiology, in this or in other ways, are generally called **'gamones'**. Those released by the male are called **androgamones** and those released by the female, **gynogamones**. These names have unfortunately also been given to extracts of the gametes themselves, or even to extracts of other cells which can be shown to have effects on the gametes. There are often complex interactions, but organisms differ so much that it is probably unprofitable to give the details here. They can be found in papers by Runnstrom[12] and Metz[13].

Many sperms which normally fertilise in the body fluids of the female, like those of mammals and birds, are very sensitive to lack of large molecules in their environment. Even small quantities of bovine serum albumen, egg yolk or milk aid longevity of mammalian sperms in saline solutions. Polyvinyl alcohol is better than nothing, but perhaps not as useful as proteins. Many kinds of tissue culture cells require this protection in the same way, so perhaps the adoption of internal fertilisation helped solve this problem for many organisms. Natural sea-water, of course, has many organic molecules, some quite large; sperms of *Pomatoceros* remain motile much longer in natural than in artificial sea-water.

Sperms can utilise external energy sources; fructose, glucose and other sugars are used, and in aerobic conditions citrate, pyruvate, lactate and other intermediate respiratory products can be utilised instead[14]. Some sperms, like those of the octopus, contain much glycogen, but it is unlikely that this is used for their energy requirements. The function of this glycogen, like that of many substances in the semen of mammals, is unknown. Inositol and glyceryl phosphorylcholine (GPC) are found in large quantities in mammalian semen[14], and may serve as large protective molecules instead of the male's proteins, which could cause antibody reactions when injected into the female; high in the tract the female's own proteins would be protective.

There is a notable phenomenon which has received little attention, probably because it seems both at first sight, and on further analysis, to be paradoxical. It may have to do with viscosity of the medium, which is why we mention it here. We saw in Chapter 4 that the primitive spermatozoa of sea urchins, externally fertilising polychaetes and teleosts and those of the coelenterates have a very simple structure, including a tiny mitochondrial set, and a simple flagellum. This seems rather curious, because these sperms must deal with the vagaries of an external environment, and often have to swim great distances in order to reach eggs. But the medium they swim in is of low viscosity, and perhaps this is why they do not need the complexity.

There are a few sperms which have lost independent locomotory organelles, and this is associated with their transport by the parents, for example the decapod crustaceans and the nematodes. But usually, in forms with internal fertilisation, the sperms are far more complicated and their locomotory apparatus is enormously sophisticated compared with the simple *Echinus* sperm. It is not too difficult, indeed, to see sperms forming a more and more complicated series as the distance they have to swim becomes less and less! Mammalian sperms have very complex midpieces, with many mitochondria. They have tails with 9 + 9 + 2 fibres and a terminal part (the end-piece) which is usually rather more flexible than the principal piece or main piece. Yet the evidence suggests that the mass movement of spermatozoa in the mammalian female tract mainly results from contraction of the uterine walls. Bird sperms are very complex too, often with spirals on the acrosome and complex tail movements – but they must pass through albuminous secretions on the way up the female tract, and also penetrate a *very* thick egg membrane. Newt sperms are complex and, like all their other cells, are enormous. Yet they are deposited in spermatophores and as they are laid the eggs actually pass the sperm mass. The undulating membrane along most of the tail of the sperm in these species, and the wavy pointed head, may again be related to passage through the albumen egg-coat; but frog sperms make a much longer journey, then penetrate an apparently equivalent albumen, without such specialisation. It is interesting, though, that some toads (*Discoglossus*) have sperms like newts. Insect sperms rarely have to swim more than their own length, but have complex fibre arrays within them and a mitochondrial **'nebenkern'**, and often a gyre-within-wave movement. The sperms of gastropod molluscs are inordinately long, with very attenuated heads (*Figure 5.7,* p. 102) and a very complex locomotion (which usually results in knots on the microscope slide), and those which are not carried to their destination by other (oligopyrrhene) sperms have a system of peristaltic tubes to transport them, as in *Helix* (*Figure 3.2*, p. 57).

Some of the most complex sperms are the peculiarly biflagellate ones of platyhelminths. These have tubes to negotiate which may exceptionally be 10 times the length of the sperm. Non-flagellate sperms of the ticks and mites (*Figure 5.3*, p. 95), with very peculiar locomotor grooves along their length, are always longer than the tubes they negotiate, and may be up to 10 times the length of the whole tick (*Argas*)[1]. Cephalopod sperms, whose spermatophore ejects them into the base of the oviduct, are immensely complex and have glycogen granules. But the sperms do not appear to use the glycogen and at present its function is unknown. This series has already been suggested on p. 94.

Possibly this increased complexity of the sperms is demanded by the complexity of the egg membranes to be penetrated and the various liquid viscosities. But this idea falls down in too many cases to be useful. For example, in platyhelminths sperms are usually wrapped up *with* the egg cell and vitelline cells in the ootype and the shell is formed around them; in gastropods the egg is effectively naked when fertilised. Neither of these demands much sophistication on the part of the sperms, but these sperms are very complicated indeed. This kind of evolutionary paradox is very

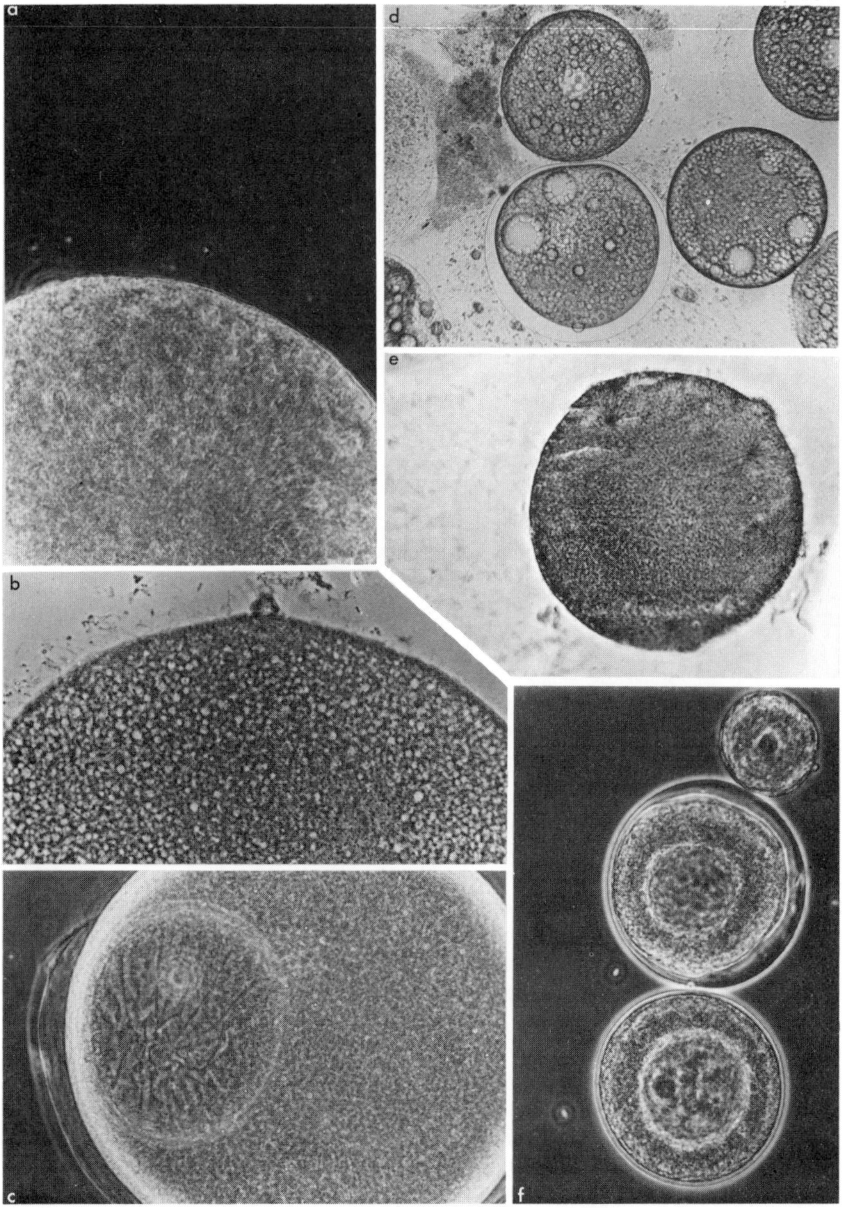

Plate 8 FERTILISATION (a) The egg of a starfish. The dots above it are sperm heads, and the middle one has contacted egg cytoplasm with its acrosomal filament. Compare Figure 7.1 *for scale. (b) The egg of* Cucumaria *(a sea cucumber) cutting off a polar body, in section. Compare* Figure 6.2 *for scale. (c) An immature egg of* Pomatoceros, *just 'fertilised'. Note the extended chromosomes in the 'germinal vesicle' and the fertilisation membrane. (d) Eggs of* Nereis. *The upper one has not yet shown fertilisation changes, the others have. The membrane has lifted, and yolk particles have fused into 'spheres'. (e) Pathological polyspermy in* Asterias *(starfish) egg. About 60 sperms are in this egg, and five sperm asters can be seen in this section. (f) An early oocyte, fertilised and just-penetrated eggs of* Pomatoceros. *The oocyte has a sperm head on it, the fertilised egg has the sperm 'debris' under the membrane (at about 3 o'clock) and the lowest egg has a sperm head swelling in its cytoplasm (at 12 o'clock)*

puzzling, and probably hides an important clue to the significance of sperm motility.

7.3 The Message

What does the sperm carry into the egg? The most important contribution is doubtless the paternal chromosome set, but there are other contributions which are by no means trivial. Nevertheless many species produce eggs which develop parthenogenetically, so nothing carried by the sperm is essential. Some of these species have lost the process of fertilisation altogether (for example, some gastrotrichs) and all individuals in these species are female (**thelytoky**). Others, like many rotifers, and the common stick insect (*Dixippus*) still produce occasional non-functional males. The eggs of these forms seem to have lost the need for sperms. More informative in the present context, however, are the male haploid species, many *Hymenoptera* and mites, in which fertilised eggs make females and unfertilised eggs make males (**arrhenotoky**). The male of these forms is just as anatomically perfect as the female, and is functional both vegetatively and sexually despite the lack of sperm-borne message in his construction. The female hybrid called *Mollienisia formosa* (*see* p. 40) is also informative. Her eggs require activation by sperms[15] but they have triploid nuclei; the sperm nucleus, which is from the sperm of a related species, does not contribute chromosomes but the eggs do require to be triggered into developing by sperm penetration. Thus we may consider a series, which begins with eggs that do not require fertilisation by a sperm but *have* nevertheless undergone meiosis. Then come those eggs, like *Daphnia* eggs, which if fertilised, take a very different path: they make a resistant shell and have a protected development period over the winter. (In this case it may be that two kinds of eggs are involved, and only one kind *needs* fertilising). In the honey bee, it seems clear than any egg can take either path, making a worker (or queen) if it is fertilised but making a drone if it is not. Such evidence as we have suggests that the double chromosome set makes female; that is to say, the path that is taken is not simply governed by activation by the sperm or some sperm components. The actual sperm chromosome set, or at least a doubling of one chromosome, is required for femaleness. This is unlike the situation in *Mollienisia formosa*, where the sperms are used solely for activation.

Frog eggs, although they can be activated by a needle, will rarely if ever give haploid frogs. Occasional apparent successes have turned out, on analysis, to have become diploid or chimaeric haplo–diploid (a mixture of the two cell populations). So, in this case, both activation and chromosomes are necessary for full development.

For some eggs, the sperms perform other functions also. Some frog eggs, for instance, have only their animal and vegetal poles determined in the ovary, and the antero-posterior plane of development is determined by the accident of sperm entry. The path of the sperm nucleus across the cytoplasm may also be an important feature in later development, as in the ascidian *Styela*. Those eggs with physiological polyspermy (*see below*) may employ the superfluous sperms in preliminary yolk digestion.

In some species at least, including some amphibia and mammals, the mitochondria which come in with the sperm are promptly destroyed; they do not seem to contribute, as pollen grain mitochondria may do to the ovular cytoplasm in angiosperms.

The sperm centriole may, however, produce a lineage; *not*, of course, one of the pair in every cell of the embryo, but both centrioles in half of the embryo's cells.

7.4 Polyspermy

It is usual for only one sperm to penetrate the cytoplasm of the oocyte, others being excluded by a more or less rapid fertilisation reaction. Eggs with a lot of yolk, though, especially those of cephalopods, selachians, some teleosts, reptiles and birds, always let many sperms in. Perhaps the rate of spreading of the fertilisation reaction is not fast enough to exclude sperms attacking the edge of the wide blastodisc. However this occurs, hundreds or thousands of sperms normally enter these eggs, but only one male pronucleus fuses with the egg pronucleus. The other sperms may, as in fishes, birds and reptiles contribute nuclei which then organise egg cytoplasm into 'cells' of the **periblast**, which commence the breakdown of the yolk at the margins of the blastodisc. This regular polyspermy is called **physiological polyspermy**, to distinguish it from the **pathological polyspermy** which may occur in mammals and usually results in defective development.

If a very dense sperm concentration is mixed with eggs, it might be expected that more than one sperm will enter before the block to polyspermy can spread over the egg surface. By comparing rates of such polyspermy at different sperm concentrations in the sea urchin, Rothschild and Swann[16] calculated the time that the eggs took to erect the barrier, and concluded that one to two seconds was required. In fact the fertilisation membrane can only be seen to lift at one to two minutes after insemination, so they proposed an invisible 'fast block' which was followed by the visible one commencing at 30 seconds. They made the assumption, however, that any sperm hitting the egg could penetrate. If this is not so, if only 1 in 20 sperms can penetrate, then we do not need to propose an 'invisible' fast block. Tyler[10], on quite other evidence, suggested that only 1 in 1000 could fertilise. Indeed, it is only rarely that the first sperm to arrive enters an *Echinus* egg[11], and our experience suggests that the same is true of *Pomatoceros triqueter*; it appears that only about 1 in 30 sperms can penetrate in *Pomatoceros*, even if they are the first to reach the egg. If only a proportion of sperms *can* penetrate, the visible lifting of the fertilisation membrane could well be the only prevention of polyspermy in these forms, and this would account for Rothschild and Swann's data.

Rabbit eggs normally allow many sperms through the zona, but only one into the vitellus. This mechanism, like the more usual fertilisation reactions, can also be swamped by excess sperms, as Adams has shown[17].

Sperms inseminated in large numbers directly into the fallopian tubes appear to destroy some eggs completely, perhaps by the excess proteases of their acrosomes.

There is a view of the female genital tract of mammals which sees it as a labyrinth to *restrict* the number of sperms at the fertilisation site[1]. The cervix of vaginal inseminators such as man and rabbit, and the utero-tubal junctions of uterine inseminators such as the pig, are seen as sperm reservoirs from which a small number of sperms is released over a long period[18]. This is supposed to give fresh sperms over an extended period after mating, but to avoid the dangers of polyspermy from too many sperms reaching the fertilisation site.

7.5 Capacitation

This is a phenomenon associated with internal fertilisation, especially of mammals. In 1951 Austin and Chang independently showed that mammalian sperms require to sojourn for some hours in the female tract before they can fertilise, and both these workers assumed that this time was needed for a kind of physiological development. Since then, there has been a tremendous amount of work in this area, but we are still left wondering exactly what capacitation is[19].

Most mammals investigated have been shown to need such capacitation of their sperms, whatever such capacitation may do. Possible exceptions, like man, guinea pig and wallaby are still in doubt because, although uncapacitated sperms have been shown to penetrate and activate eggs, unequivocally normal development has not been shown to occur. Capacitation of some sperms, especially those of mice, has been achieved in simple defined media[20]; these sperms have been tested by penetration and activation, and by their contribution to a normal embryo. In all other experiments, blood proteins, follicular fluid, or enzymes have been shown to be necessary for mammalian sperms to acquire the ability to fertilise. β-amylase has been proposed as a key enzyme in this system, but the evidence is by no means conclusive[21].

Investigation of the fine structure of sperms by electron microscopy has revealed an apparent association between fusion of the outer acrosome membrane with the plasma membrane and capacitation. But while it is obviously necessary for a sperm at the fertilisation site to have its enzymic teeth bared for action, so to speak, this change also occurs in many sperms much lower down in the female tract. Such 'false capacitation' could be considered to be part of the antibody-mediated destruction of sperms, which occurs in all mammalian females, and would then probably be a sperm selection. Capacitation may also coincide with the removal of a sperm coat acquired in the male tract, or from seminal fluid[21], and there is some rather puzzling evidence that sperms can be 'decapacitated' by re-immersion in seminal fluid[22].

The investigation of capacitation is further complicated by the usual fall in numbers of sperms while it is proceeding. For this reason it is very difficult to distinguish selection, survival of a small special population of sperms, from a change of *all* sperms most of which die leaving, not special ones, but a lucky few. If there is selection during this fall in numbers, then incapacitation of most sperms could be going on at the same time as capacitation of the minority, and this might explain our lack of understanding of the complicated situation[23]. There is, too, evidence that various sites in the female genital tract, or other organs and cavities, can serve for sperm capacitation to differing extents. The oviduct, for example, does not appear to be effective compared with the uterine lumen.

7.6 Survival of Sperms and Eggs

Once the sperms have been ejaculated, or the eggs ovulated, they have but a short fertile life.

Sea urchin eggs can only be fertilised for some eight hours, at most, after ovulation; and sea urchin sperms remain viable for a maximum of 24 hours after emission. Even this has been reported in only two cases and in each, only a small proportion of embryos developed normally.

Most of our information about gamete **'senescence'**, as it has been called, comes from the vertebrates, and for convenience we will deal with it in 'phylogenetic order'. Eggs stripped from salmon or trout, and perhaps from other teleosts, last for some hours if kept cool in the coelomic fluid in which they were expressed[24]. If they are expressed into water, however, they lose the capacity for fertilisation within two or three minutes. Similarly, the complete natural milt (semen) of the male fish may be stored for up to a day without dilution, but one minute after being added to fresh water the sperms are only about half as effective, and are totally ineffective after about 25 minutes. For this reason, the eggs and sperms are usually mixed 'dry' during artificial insemination, and water is added later. The same is done for marine flatfish, although in this case it may not be quite so necessary. Seawater is very hypertonic to marine fish tissues, but it may not harm the eggs or sperms as much as the extreme hypotonicity of fresh water harms those of the fish which breed in it.

We have little data for the selachians; fertile eggs have been laid by dogfishes isolated in our tanks for three months, so sperm survival for this time certainly occurs in female dogfishes.

The amphibians, too, afford only scattered data. Female urodeles will sometimes pick up spermatophores deposited some hours earlier, but it should be pointed out that other, fresh, spermatophores may already have been picked up. Certainly, some female newts and salamanders (e.g. axolotl) may not begin to lay their eggs for 24 hours after receiving spermatophores. Some terrestrial forms (plethodontids), indeed, are said to delay laying for several months after insemination. *Xenopus* eggs can be fertilised some hours after laying if kept in saline solutions, but they lose viability much more rapidly in fresh water; the same is true of the sperms. But even in Holtfreter's solution, only a small proportion of the eggs of *Rana temporaria* will fertilise after the egg jelly has begun to swell, an event that takes place about 10 minutes after laying. Sperms taken from the testis last well in Holtfreter's solution, but lose the ability to fertilise very rapidly, in about 10 minutes, when suspended either in fresh water or in Holtfreter's mixed with egg jelly extract. This may be because the acrosome reaction is elicited by jelly solution.

Reptiles have internal fertilisation and there has been little experimental investigation of this process. But many female snakes have laid fertile eggs years after isolation, so it is certain that the sperms can survive well in the female tract. We know that the eggs have not rested, fertile, in the female, as ovulation always seems to be followed rapidly by deposition of the shell membranes, and laying within days.

In birds, there is evidence of occasional sperm survival in females for up to six weeks. Again, the eggs are promptly coated in albumen and shell so there is no chance for later fertilisation, or for the persistence of fertile eggs in the females.

The senescence of gametes has been investigated mostly, as might be expected, in mammals. There is a classic instance, in the temperate bats, of very long-term storage of sperms in the female tract[25]. They mate in autumn and store the sperms till ovulation in the spring. Other mammals with such delays between copulation and the production of young have a delay between fertilisation and implantation, usually as 'delayed implantation', but these bats simply store the sperms.

The sperms of most other mammals have a very short fertile life in the female tract but since, even in spontaneous ovulators, females are only receptive to the male around the time of ovulation, there is no requirement for prolonged survival of the sperms. (This applies even more to non-spontaneous, induced, ovulators). Mouse sperms can still fertilise eight hours after insemination, but fertility seems to be lost by 12 to 15 hours[26]. In mammals with oestrus some time before ovulation, for example the sow, the sperms normally survive for up to 48 hours in the tract. Some fertility can persist for some 60 hours but it declines rapidly after this. Fertile sperms may last up to six days in the mare[26]. In induced ovulators, the ovulation normally follows some hours (12 to 14 in the rabbit) after copulation. Insemination up to 30 hours before a mating with a vasectomised male (to cause ovulation) gives small litters, but some of the eggs are found to be penetrated even by sperms which have sojourned 90 hours in the tracts.

In the senescence of mammalian eggs, there seem to be successive thresholds. Some fertility seems to be lost after a few hours, and polyspermy becomes more likely. Fertilisation after the natural loss of the corona cells (before the sperms arrive) causes a higher incidence of abnormal embryos. Fertilisation after some 36 hours, even in those species most tolerant of egg ageing, allows few eggs even to begin normal cleavage.

This senescence of gametes in humans has been suggested as a contributory factor to the higher incidence of abnormal embryos and babies born to older women. It has been suggested that less frequent sexual intercourse by these older couples gives a greater likelihood of senescent sperms if intercourse occurs well before ovulation, or of senescent eggs if it occurs after ovulation. The 'rhythm method' of attempted birth control has also been proposed as contributing to the incidence of abnormal embryos for the same reason. But most gametologists do not take these suggestions seriously because no correlation has yet been found between incidence of intercourse, or 'rhythm method', and abnormality.

7.7 Parthenogenesis

This, the *absence* of fertilisation, should be considered briefly here. Its chief interests are in the mechanisms of egg activation, and in the genetic consequences, which depend mostly upon ways in which diploidy is restored. The genetic consequences of temporary parthenogenesis, as in aphids and *Daphnia*, or of male haploid parthenogenesis as in the *Hymenoptera*, are not as serious as in those organisms which *only* practice parthenogenesis, and here all depends upon whether meiosis with crossing-over occurs or not. In those cases where it does not, for example various stick insects and the cockroach *Pycnoscelis*, there is essentially asexual reproduction using the egg as an asexual spore.

Those forms, like the lizards *Lacerta saxicola armenica* and *Cnemidophorus*, which fuse a polar body with the egg cell, have some recombinational possibility, so mutations can be shuffled, and gain or lose in dominance against different backgrounds[27]. These lizards probably fail to exclude the *second* polar body from the egg; in some races of turkey there is occasional production of males by this method. Male turkeys are homogametic ZZ, females are WZ, and the other possibilities, (no sex chromosomes or WW) are not viable[27]. Some organisms, in experimental conditions, may actually produce a haploid egg whose first mitotic division products fuse to restore diploidy; the resultant individual is then perfectly homozygous.

Experimental production of juvenile haploids by parthenogenesis has been achieved in several amphibians by the use of irradiated sperms, which can activate eggs but whose chromosome complement is so damaged that it cannot function. Mammalian eggs, too, have been activated in various ways, for example by chilling them in the oviduct, but haploid mammalian embryos apparently have never come to term; indeed they rarely develop beyond early cleavage[27]. Occasional reports in the medical literature are usually excluded by blood tests or, as might be expected, by the child being male; perhaps tests of a theological, rather than scientific, rigour might be appropriate in these latter cases!

REFERENCES

1. Austin, 1965
2. Colwin and Colwin, 1967
3. Rothschild, 1956
4. Monroy, 1973
5. Spirin, 1966
6. Bedford and Calvin, 1974
7. Austin, 1965 (pp.84–94)
8. Gibbons, 1975
9. Lillie, 1919
10. Tyler, 1962
11. Epel, 1975
12. Runnstrom, 1952
13. Metz, 1967
14. Mann, 1964
15. Hubbs, 1955
16. Rothschild and Swann, 1949
17. Adams, 1969
18. Braden and Austin, 1954
19. Austin and Bavister, 1975
20. Miyamoto and Chang, 1972
21. Johnson and Hunter, 1972
22. Williams *et al.*, 1969
23. Cohen, 1975a
24. Ginsberg, 1972
25. Racey, 1975
26. Austin, 1975
27. Beatty, 1967

7.8 Questions

1. Contrast fertilisation in a marine annelid or mollusc with that in a mammal.

2. When females are inseminated by several males, competition for fertilisation might occur at several stages; what are they?

3. What functions of the sperm must be taken over by the egg or female in parthenogenetic animals?

4. When does syngamy actually occur?

5. In what sense does fertilisation represent 'the start of the life of a new individual'?

8

MATERNAL EFFECTS IN EARLY EMBRYOLOGY

8.1 Polarity

The way in which the axes of the embryo form within the egg is determined by the position of the egg in the ovary, because of the necessary asymmetries of yolk, micropyles and so on. There are a few eggs in which the position of sperm entry determines the anterior-posterior orientation of the future embryo, but the dorso-ventral axis is nearly always determined by animal and vegetal poles of the egg, and hence by original yolk asymmetry in most cases. So the egg has acquired an architecture in the ovary which determines the whole symmetry of early development. The extent to which this architecture is shown by the presence of visible components does not reflect the complexity of such architecture. Some eggs, e.g. those of tunicates, have obvious differences in cortical plasm, deep cytoplasm, and the nucleoplasm from the germinal vesicle (*see Figure 8.1*). This is one of those eggs in which the position of sperm entry seems to determine the anterior-posterior axis, and indeed the visible differences may not be so important as invisible ones which can be demonstrated only by experiment. If this egg is centrifuged lightly (100 to 200 g for 15 minutes) the contents form layers, with the heavy yolk at the vegetal pole and the other plasms forming discs above

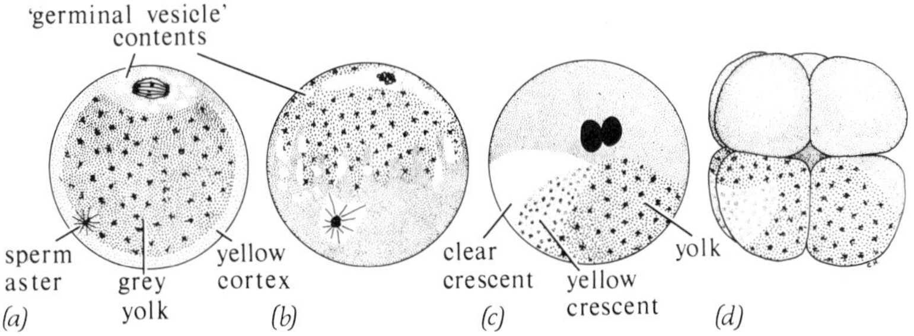

Figure 8.1 EARLY RE-ORGANISATION OF THE EGG OF STYELA *(from Nelsen, 1953, after Conklin). Many tunicates, including* Styela, *have coloured areas in their eggs which enable embryologists to follow the ooplasms (egg cytoplasm regions) before cleavage. (a) At ovulation; the sperm enters at M1 (see* Figure 6.5*). (b) Nuclear contents ('sap') and cortical plasm stream vegetally (Telophase I). (c) By the end of T2, when the male and female nuclei lie apposed, this streaming has resolved into a series of segments ('crescents') in the vegetal third. (d) During cleavage these areas maintain their integrity and form specific organs (nerve cord, notochord) of the larva*

it; but after standing for a further 30 minutes, the particles re-distribute themselves and moderately normal development may still occur. This shows that there is a cryptic organisation, perhaps in the egg cortex as this is least disturbed by the centrifugation, to which the visible particle distribution returns after the disturbance. The same phenomenon occurs after centrifugation of *Pomatoceros* or *Tubifex* eggs, but here development is substantially abnormal thereafter.

After centrifugation of amphibian eggs, however, no yolk returns to the animal half of the egg. Instead, it remains concentrated in the vegetal half of the egg to the exclusion of the cytoplasm; but the egg still gives a normal embryo. The development of this embryo, however, now resembles that of teleost fishes in that only the animal cytoplasm cleaves, the vegetal yolk remaining as an uncleaved mass which is engulfed by the gastrulating cellular mass 'blastoderm' (*Figure 8.2*). This demonstrates clearly that the yolk gradient is not a necessary prerequisite for development, and that some normal phenomena, like the slower-cleaving larger-celled vegetal hemispheres, are not integral to development but only a side issue, produced by the unimportant yolk gradient.

It happens, though, that in these amphibian eggs we can often identify a very important part of the egg's primary architecture, the **grey crescent**. Sometimes, as in many *Rana* species, it only appears after fertilisation, but in *Xenopus* and many newts it marks the posterior of the future embryo even in the unfertilised egg. These eggs always have a pigmented animal

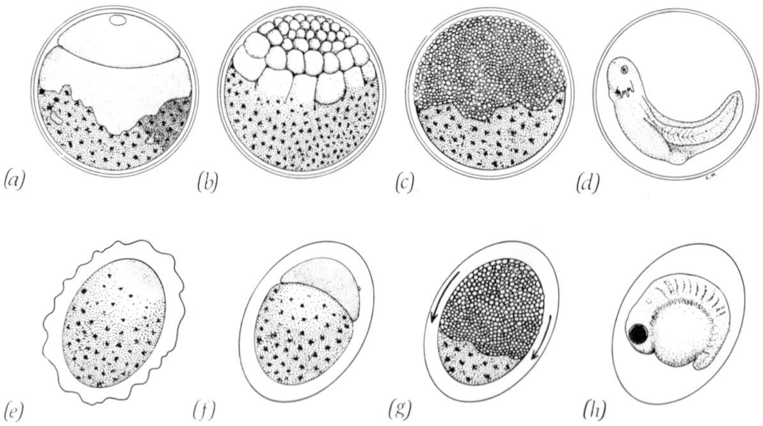

Figure 8.2 YOLK IN AMPHIBIA AND TELEOSTS (a) Egg of Xenopus *after gentle centrifugation, in a density gradient to avoid compression (pigment is not shown except the grey crescent at right). The clear cytoplasm and nucleus come to lie at the animal pole, and granular yolky cytoplasm fills the vegetal half of the egg. The first 'cleavage' sometimes cuts off an animal cytoplasm from the vegetal mass, as shown (compare f). (b) In this case the cytoplasm cleaved and the yolky mass did not, as in teleosts. (c) The cleaved mass of cells engulf the yolk (compare g). (d) A fairly normal embryo results, but with a very curved body axis. (e) The egg of a teleost (*Aequidens*) at laying: yolk and cytoplasm are not separate. (f) They separate. (g) The cytoplasm cleaves and the cell mass engulfs the yolk. (h) The embryo is formed (the yolk sac is drawn about half relative size for clarity). Compare* Plates 9 *and* 11

hemisphere (except for a spot at the animal pole where the first polar body has been cut off) and the yolky cytoplasm is visible as a white (or rarely yellowish) patch around the vegetal pole. This yolk, in *Xenopus*, may extend irregularly as high as the equator, but in most frogs it only comes to about the 'Tropic of Capricorn' and in newts and the axolotl only as far as the 'Antarctic Circle'. At this boundary there can often, particularly in some *Triturus* species, be seen a crescent which is neither the true brown or black of the animal hemisphere, nor white. In this region there is yolk over, i.e. outside, some of the pigment, and so this crescent-shaped area looks grey or buff-colour (cf. *Figure 8.2* and *Plate 11*).

It is in this region, after cleavage, that the foldings which build the embryo (**gastrulation**) begin, and we now have experimental evidence that this region of the egg cortex is very special. Curtis[1], in a series of very elegant experiments with *Xenopus* eggs, succeeded in transplanting this region of cortex to another uncleaved egg (*Figure 9.3*, p. 168). When this host egg developed, the graft made it produce a second embryonic axis from the local host tissue. Curtis also showed that damage to the grey crescent cortex, which overlies the cytoplasm to be included in the germ cells, is transmitted in a surprisingly faithful way to the cortex of eggs made in the ovary of the animal whose egg cortex he injured. The damage found in the eggs of their progeny, if any, was much less specific; presumably this was because severe, and so easily recognisable, damage also sterilised so no second generation eggs appeared from some experimental toads. At the time, Curtis argued that *any* transmissions over two generations showed the existence of a replicating, but cortical, hereditary system maternally transmitted.

This was not a new idea, for such systems had been known for decades in the ciliate protozoans[2], but it was criticised very strongly at a time when all heredity was considered to be DNA-borne. Damage to the complex networks of the infraciliature of the ciliates *Stentor* or *Paramecium*, especially in the oral region, was often reproduced through many cell divisions, and Raven[3] had for many years considered the egg cortex, like the ciliate cortex, to be a system contributing as much to the organisation of the egg as the nucleus. Nevertheless, the inability of workers at that time, or since, to suggest a plausible mechanism for the replication of cortical structures has quite unjustly retarded the integration of Curtis' results, as well as the work on ciliate morphogenesis, into our knowledge of early embryology. There are, of course, many cases like this one, where we must accept experimental evidence while remaining ignorant of the mechanism which underlies it.

There is a mutant in *Drosophila*, **grandchildless**, whose action is very interesting in this connection, for it throws light on whether such apparently extra-chromosomal inheritance as the grey crescent of frogs, the mouth infraciliature of ciliates, and the morphogenetic cortical patterns of eggs are all ultimately controlled by the DNA genome. **Grandchildless** female *Drosophila* have ovaries in which the polar (germ) plasm of the eggs is not segregated[4]; this plasm in normal eggs contains the special polar granules which determine that nuclei which enter their cytoplasmic vicinity become germ cell nuclei[4]. The eggs of **grandchildless** females make perfectly ordinary-looking offspring, but these have no germ cells and are of course sterile. They have rudimentary gonads but look like the normal males and females, because in insects the secondary and tertiary sex organs are chromosomally determined in each cell, not via hormones. In this case, as in cleavage in *Limnaea peregra* (*below*) the dependence upon the chromosomal genome is clear. But the other cases we have mentioned *may* not, in ciliates certainly do not, depend on the chromosomes of a previous generation except in the (probably) trivial sense that their continuity may depend upon gene-prescribed enzymes, though their patterns may not (*see below*).

The animal-vegetal polarity, then, is certainly determined by the ovary in which the egg developed; even though it may be seen as yolk or mitochondrial gradients, these may be less important than invisible gradients. Anterior-posterior polarity of the future embryo is usually determined by cortical factors in the egg, occasionally visible and transplantable like the grey crescent of amphibian eggs. Many fish and bird eggs have thickened cytoplasm at the posterior edge of the blastodisc, which derives from their ovarian environment and probably depends on the chromosomal genome of their mothers.

8.2 Effects on Cleavage

All metazoan eggs must cleave to produce the cellular organism. For nearly all organisms this cleavage is obviously a true *division* of the original egg, with no increase in substance. The exceptions, gastropod molluscs and mammals, both have unusually small eggs whose food material is not donated within the cytoplasm but from outside during the whole of development; their growth during cleavage is exceptional and is not a clue to the general case.

Cleavage clearly must not disrupt the architecture of the fertilised egg, upon which the early and basic development of the embryo depends. What started as cytoplasm underlying the grey crescent should become cells, part of whose surface is this grey crescent cortex; that cytoplasm which received the nucleoplasm, when the nuclear membrane of the germinal vesicle disappeared, becomes a group of very basophilic cells that have a special fate. The thickened blastodisc edge of fishes becomes several cell layers deep.

The organisation of the original cytoplasm thus comes to be the organisation of a cellular mass. Cleavage should be regarded as a process parallel to the development of the embryo, and not as a process integral to this development. Nevertheless there are three examples we can use to show how the process of cleavage illuminates the maternal contribution to the embryo's heredity. These are determination state, the control of cleavage rate, and control of direction of spiral cleavage.

Firstly, let us consider the **'determination'** state relative to cleavage. Many organisms, like tunicates and nematode worms, have eggs each part of which is determined in its fate before cleavage starts. Each cleavage product or **blastomere** can only produce that part of the organism which

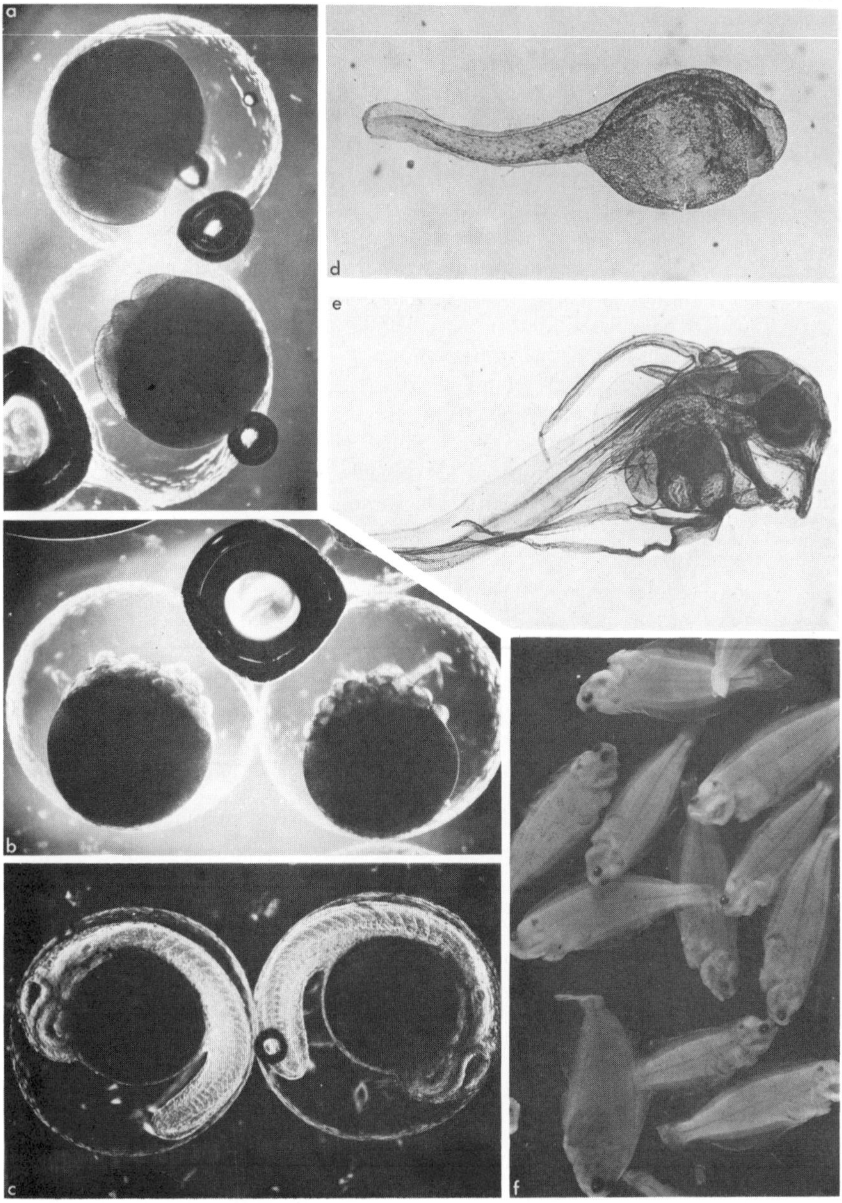

Plate 9 FISH DEVELOPMENT (a) Two Betta *eggs. The upper has separated cytoplasm from yolk and the cytoplasm of the lower has cleaved twice, giving four blastomeres. Bubbles are from the nest. (b) At four hours cleavage has progressed to about 64-cell. The centre dark-rimmed object is a bubble. (c) At 18 hours the little fish are well-formed, with eyes, ears, heart and a dramatic blood circulation. (d) A newly-hatched (28 hour post-laying) Paradise fish alevin, with rudimentary eyes and ears. (e) A young angler fish (*Lophius*) larva, which feeds in the plankton. The long first dorsal fin ray will become the 'lure' of the adult fish. (f) Small wild-caught plaice larvae. Two have gut abnormalities, and stages of metamorphosis can be seen from left or right sides (compare* Figure 14.1, *p. 236)*

it would have produced in the entire egg, or less. There is no 'making up' for loss. These eggs are said to have **'determinate'** cleavage. Other organisms, like sea urchins and frogs, show a very different situation; if the first two blastomeres are separated, they nearly always make twin complete embryos. Even the eight blastomeres after the third division can sometimes be persuaded to produce up to four fairly normal embryos; and implantation of grey crescent or comparable material may also change the fate of parts of the host egg (*Figure 9.3*). These eggs are said to have **'indeterminate'** cleavage, i.e. their parts have not yet been determined at early cleavage stages, unlike the nematode and tunicate which have been determined by the time of the first cleavage.

The actual spatial organisation of the cleavage planes differs between animal groups and relates to this state of determination, which itself must depend upon the structure given to the egg in the ovary. Most indeterminate animals show **'radial cleavage'** which is not very regular in geometry after the initial cleavages. The first cleavage is animal–vegetal and along unique lines of longitude, e.g. those with the point of sperm entry or grey crescent; it divides the embryo into right and left halves. The next division is usually at right-angles to this, also through animal and vegetal poles. The third cleavage is at right-angles to the previous two, i.e. parallel to the equatorial plane, and again cuts the cytoplasm into two approximately equal halves; but the presence of yolk in the cytoplasm of the vegetal part may lift this division above the equator, as in the frog. It may be as high as the 'Arctic Circle', in some *Dipnoi* and *Gymnophiona* with very yolky eggs (*Figure 8.3*). This results in eight cells, the upper (animal) four usually smaller and directly above their vegetal counterparts whose cytoplasmic volume is augmented by yolk. Divisions proceed but soon lose regularity, so that by the sixth cleavage they are not usually synchronous divisions; the vegetal yolky cells lag behind, and a spherical blastula of irregularly shaped and positioned cells is formed. The grey crescent of amphibian eggs can now be seen to spread through a whole group of cells.

In contrast, determinate embryos usually show very 'organised' cleavage, as they must if they are not to lose, during cleavage, the complex structure built up during oogenesis and modified at fertilisation. Nematodes show very peculiar cleavage with cells moving over each other, e.g. at the four-cell from a T-shape to a rhombus, in a very precise way (*Figure 8.4*). The larva is built up by a series of very precise divisions, at odd angles to each other, followed by cell movements often across several cell diameters; each organ even finishes up with an exact number of cells, and any cell removed or killed during development leaves a precise gap. This condition of extreme determination is called **'eutely'**. More usually, as in annelids and molluscs most of which have determinate eggs, a tactic such as 'spiral cleavage' is adopted (*Figure 8.5*). Here the mitotic spindles for all cleavages are 'tilted' with respect to the major axes of the egg: the first cleavage plane is oblique and results in two cells opposed as in two hands clasped; the second results in four cells not in a plane but forming a tetrahedron; and the third cleavage results in eight cells of which the upper (animal) smaller quartet, of **'micromeres'**, alternates with the larger

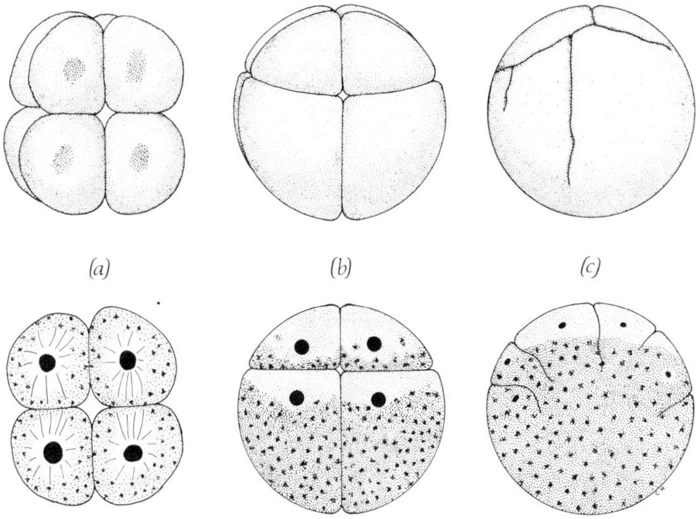

Figure 8.3 THE THIRD CLEAVAGE AND YOLK *(a) The eight-cell stage of a homolecithal egg like the sea urchin, whole and in vertical section. The animal and vegetal cells are of equal size. (b) Eight-cell stage of the frog or newt egg. The cytoplasm has probably been equally divided, but the presence of vegetal yolk makes the vegetal cells much bigger (compare* Plate 11*). (c) The egg of a lungfish undergoing 'third' cleavage. The mass of yolk presents great difficulty to the cleavage planes, and only the animal cytoplasm is actually divided (compare with* Plate 11*)*

vegetal **macromeres** (*Figure 8.5*). Usually each cell of this 'first quartet of micromeres' lies clockwise to its macromere sister when viewed from the animal pole. At the fourth cleavage the obliquity of the cleavage spindles is reversed, so that the macromeres 'give off' the second quartet of micromeres in an anticlockwise direction. The first quartet divide at the same time, the more animal daughters lying anticlockwise to the more vegetal. Synchronous divisions continue, alternating in obliquity, so that the progeny of any cell relates to that of its neighbours in interlocking fashion.

The blastula that results from spiral cleavage has each cell exactly positioned, like bricks in a wall. Every cell, too, has its specific fate, and there is a complex labelling system used by embryologists which recognises this fact. After radial cleavage, in contrast, the borders of clones are fairly smooth; the progeny of each of the four-cell blastomeres relates to the others like the quarters of an orange. There would be no point in labelling the blastomeres of the frog's egg, because they differ in pattern from egg to egg and their fates are mutable. But the pattern of the 128-cell spirally-cleaved blastula is consistent across a major part of two phyla, and the larva produced, the trochophore, usually uses the same cells to make its important organs.

What does this tell us about the maternal role? Up to the formation of the trochophore, or the gastrula of the frog or dipleurula of sea urchin (p. 160), only the maternal organisation of the egg cytoplasm has

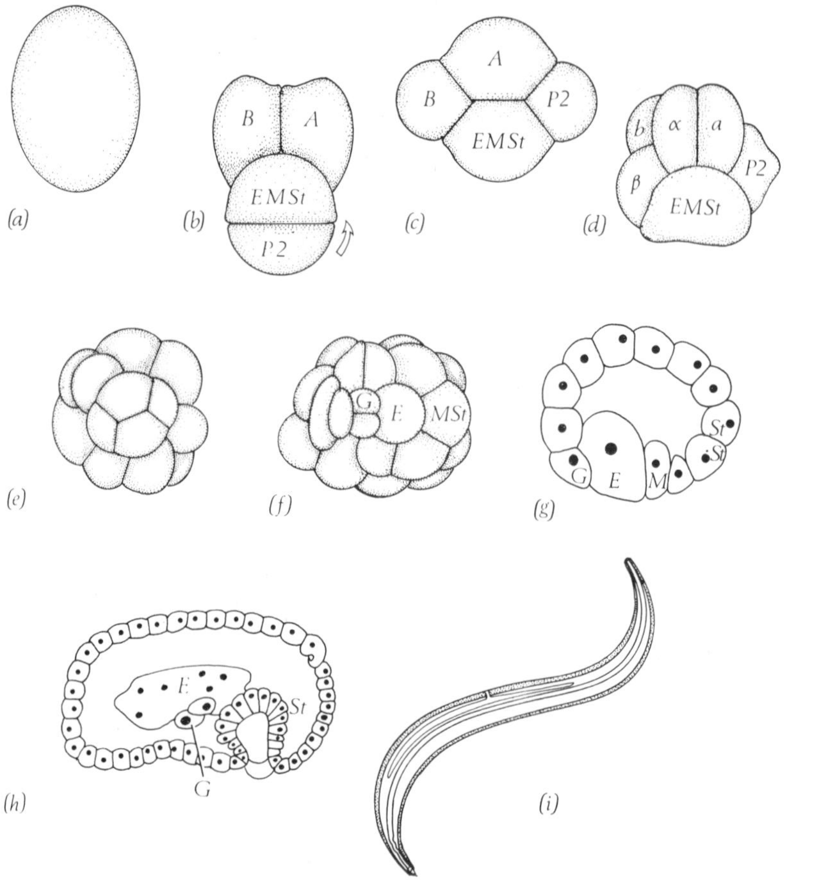

Figure 8.4 DEVELOPMENT OF NEMATODE WORMS (compare Plate 10*). (a) The egg is usually oval. (b) The four-cell stage is initially T-shaped, from two angled mitoses (*Plate 10*). (c) The P2 cell moves round to make a rhomboid. (d) A → a and α, B → b and β, giving six cells. (e) The 16 to 18-cell stage. (f) The germ cells have been segregated from P2 and E (endoderm) has separated from MSt (mesoderm and stomodeum). (From vegetal.) (g) A vertical section of (f). (h) The basic body form (phyletic stage). Stomodeum (St), germ cells (G), and the endodermal gut mass (E) have all appeared independently (compare* Plate 10*). (i) The young worm. If cells are deleted earlier in development, there is no making up for loss or even wound healing; all the rest of the developing organism is unchanged*

determined the patterns of the cleavage which distributes the embryonic architecture into the separate cells. So the larva, as we will see in other forms as well, is an unfolding of the **egg** cytoplasmic organisation; the zygote nuclei are essentially passengers up to this stage. Zygote DNA (usually recognised by its paternal contribution) has not yet made significant specific m-RNA (*see the next chapter*) except in oddities like the mammal egg.

The second kind of information which cleavage gives us about the maternal determination of embryonic events comes from its rate. Closely

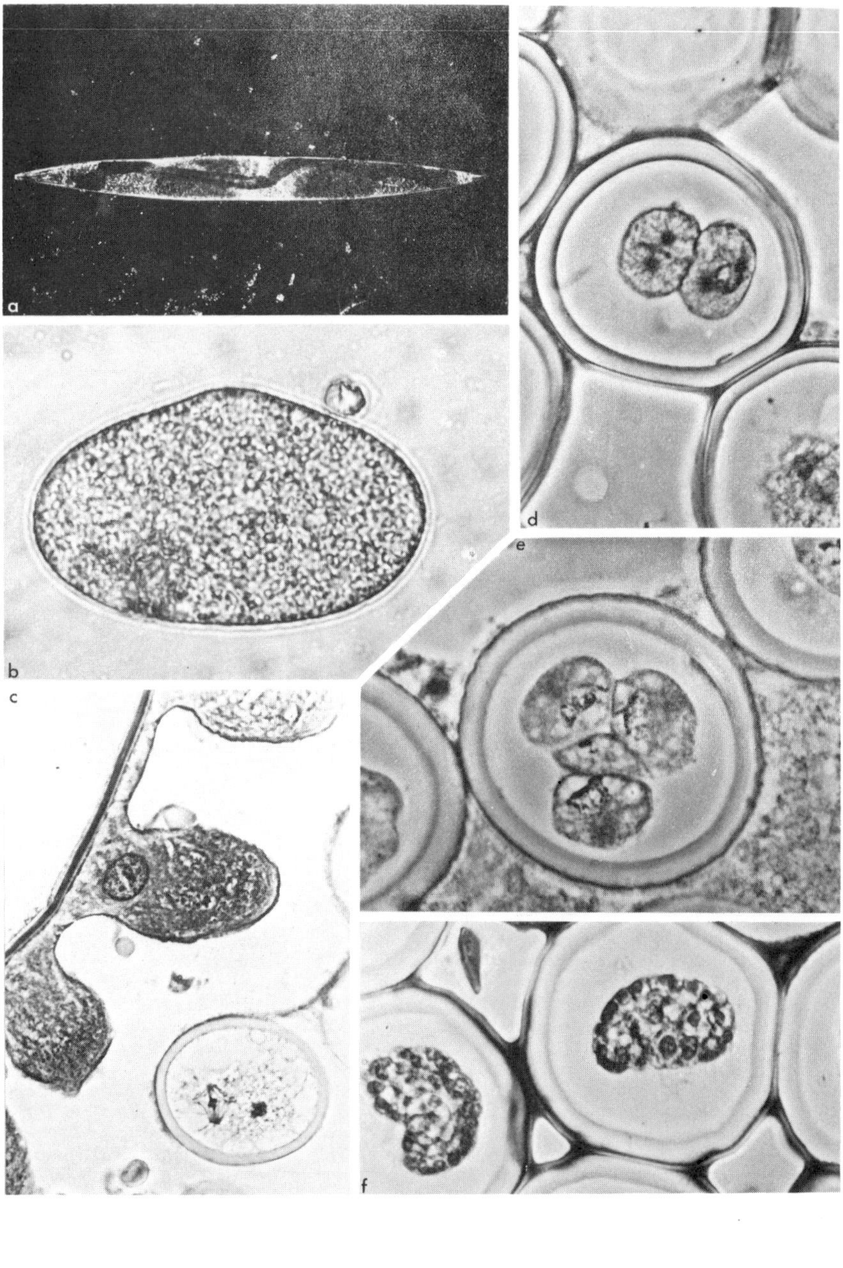

Plate 10 NEMATODE DEVELOPMENT (a) A female Rhabditis, *with eggs (compare* Figure 4.2*). (b) A living egg of* Ascaris, *with a sperm on its membrane. The meiotic spindle can be seen bottom left. (c) Part of a section of* Ascaris *oviduct. An egg in section shows a sperm nucleus and a second meiotic spindle. (d) The division from two- to four-cell in* Ascaris *(compare* Figure 8.4*). (e) The T-shaped four-cell of* Ascaris. *(f) A 'gastrula' of* Ascaris. *Note the large nuclei of the future germ cells (compare* Figure 8.4*)*

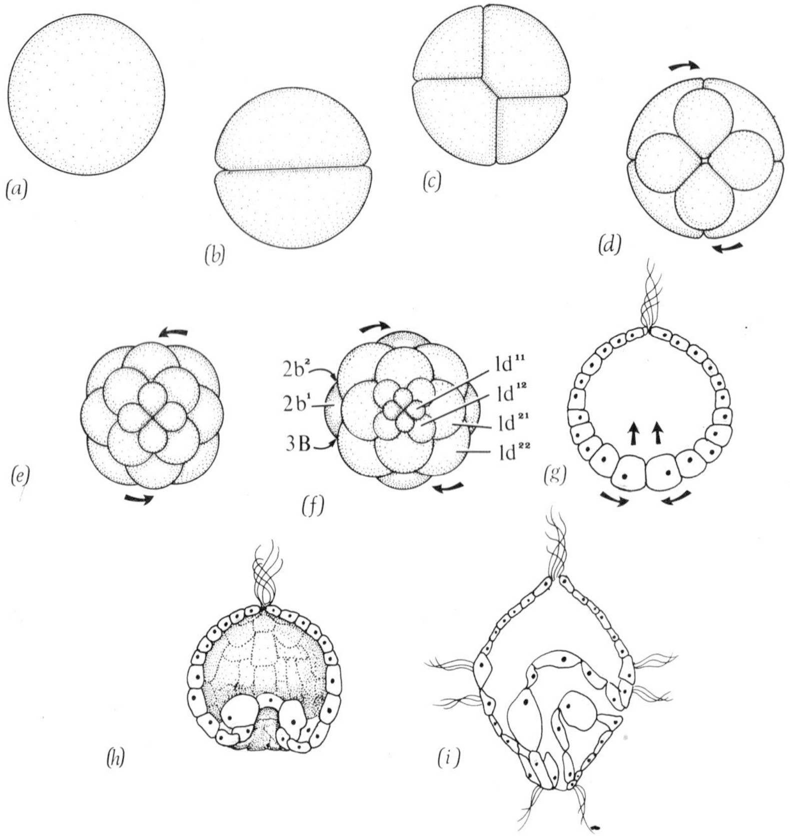

Figure 8.5 DEVELOPMENT OF ANNELID WORMS (a) Uncleaved; some organisation is usually visible, by uneven distribution of pigment or yolk. (b) Two blastomeres. (c) Four blastomeres; two meet at the vegetal, two at the animal pole forming a tetrahedron. These are called A, B, C, D. (d) The tetrahedron flattens, and division spindles appear obliquely, cutting off the 'first quartet of micromeres' above (animal) to the macromeres (1A–D) and clockwise to them. First quartet of micromeres are called 1a–d, or in general 1q. (e) The next division is oblique, anti-clockwise. 1q cells become $1q^1$ and $1q^2$, 1Q become 2q (second quartet of micromeres) and 2Q (macromeres). (f) The 5th division, to produce 32 cells, is again clockwise. All cells divide, and some cells of the B and D quadrant are labelled: first quartet progeny in D, more vegetal cells in B. 3B, $2b^2$ and 3b are in the vegetal hemisphere and invisible from the animal pole. (g) At the 64-cell stage a groove appears on the vegetal face, 4D → 4B with 4A left and 4C right. (h) The groove deepens and becomes an enclosed gut, and 4d drops into the blastocoel, later to form all the adult mesoderm, when it divides into two, one each side of the rear of the gut. (i) The trochophore larva, with sensory apical ciliary tuft and two ciliated girdles

related species often differ in cleavage rates, and in the effects of temperature on these rates. It is often possible to fertilise eggs of one species with sperms of another, closely related, species. In these cases it is always found that development up to early gastrula is that characteristic of the egg species. Other experiments, using irradiated sperms or eggs so that only one genome contributes to the zygote, are also informative: hybrid **andromerogones** (embryos formed from sperm nucleus and an enucleate egg) usually develop right up to the blastula stage according to the programme of the **egg** species, despite the fact that the egg nucleus is absent or totally inactivated[5]. So we can be sure that, at least in the amphibian species Moore[5] worked with, and presumably also in other species, the egg cytoplasm controls the rate of cleavage and presumably the whole pattern of early development.

The third example we shall use, to show the way in which studies of cleavage illuminate extra-chromosomal inheritance via the egg cytoplasm, concerns the water snail *Lymnaea peregra*. Most races of this bisexual gastropod have shells which curl clockwise, viewed from the spire; this is called dextral. A few locations have been discovered where occasional sinistral (anti-clockwise) specimens can be found. The inheritance of this character is very interesting[6]; the gene is a simple Mendelian system with dextral *D* dominant to sinistral *d*. But the gene is not expressed in the individuals carrying it; instead it affects all the progeny of the individual, whatever *their* genetics. So a snail's coil sense is dependent on its mother's genes, *not* on its own zygote nucleus. Consider one snail, whose mother was genetically *Dd* but with sinistral coil because *her* mother was *dd*; the individual we consider could be *dd* but would have a dextral coil *whatever* its own genetics, because its mother was *Dd* (*Figure 8.6*). This gene acts by affecting the organisation of eggs in the ovary, determining whether their third cleavage spindles will be oblique in a sinistral or dextral sense and so whether a mirror-image embryo will be produced. The effect probably derives from an asymmetry of follicle cells around the oocyte (*Figure 8.6*) in mother's ovary[3]. Here we have a clear case of embryonic symmetry (in this case asymmetry) being determined by a maternal gene, but many other maternal gene effects are known. The case of 'grandchildless' in *Drosophila* is a good example. Flies carrying this gene fail to put 'polar plasm' into their eggs[4]. This allows the eggs to develop, but no nuclei are protected by this polar plasm so that they become germ cells; so the progeny of flies carrying the gene are sterile because they have no germ cells.

There is a symmetry-related problem in the vertebrates which is nearly certainly an effect of ovarian oocyte organisation on the future embryos. In all vertebrate embryos, as the gut lengthens that part which will be stomach is tethered anteriorly in the transverse septum behind the heart, and tethered at its posterior end by the bile duct, itself connected to the liver in the transverse septum; so as it lengthens it must form a curve. This curve is, with very rare exceptions, into the left side of the body cavity and determines the other visceral asymmetries, for example the heart being towards the left side in man, and the liver lying mostly in the right side. It probably also determines which gonad will develop the left ovary in female birds. In **dextrocardiac** fowl (birds with reversal of the normal asymmetry) the right gonad becomes an ovary. There is some evidence that in some dextrocardial humans (true *situs inversus*), left-handedness is much more common than right-handedness, which suggests that this is not a 'basic' left- or rightness, but really a 'same-sidedness' or 'opposite-sidedness' which uses the primary asymmetry adopted by the stomach bend as its referrent. This is not to suggest that the visceral

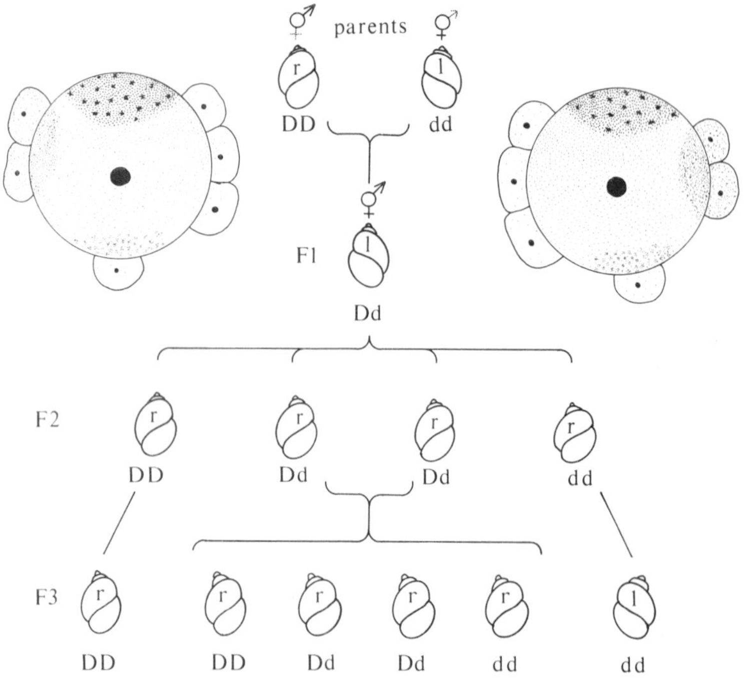

Figure 8.6 THE INHERITANCE OF SHELL COIL SENSE IN LIMNAEA PEREGRA *Two oocytes are shown, with some of their surrounding follicle cells and with an indication of their ooplasms; both are viewed from the animal pole, and they are mirror images. One cleaves clockwise at the 4 → 8-cell (*Figure 8.5*), the other cleaves anti-clockwise. A possible lineage, with genotypes, is shown. Note that the parents, though bisexual, are shown as if male and female for simplicity, whereas the homozygotes of the F2 are supposed to self-fertilise (or mate with a snail of the same genotype). D is dominant to d, and each snail has a shell reflecting its* mother's *genotype, not its own (r is dextral, l is sinistral for shells)*

asymmetry is primary itself but that, like the side on which chick embryos lie, it is a function of original egg structure. So it could be a part of the embryo's heredity which is independent of nuclear factors, or it might derive from maternal genes like sinistrality in *Lymnaea* or sterility in fruit flies whose mother is **grandchildless**. Even if all these cases can be referred back to mother's genes, we must not then suppose that this means that the mother's DNA *determines* the early development of the embryo by itself.

8.3 Instruction and Prescription

Most attention has been paid to the DNA 'instruction for development', so that it has been likened to a magnetic computer tape with a programme for the organism's development. It must not be forgotten that the chromosomes, or even the whole nucleus, cannot build an embryo by itself; the

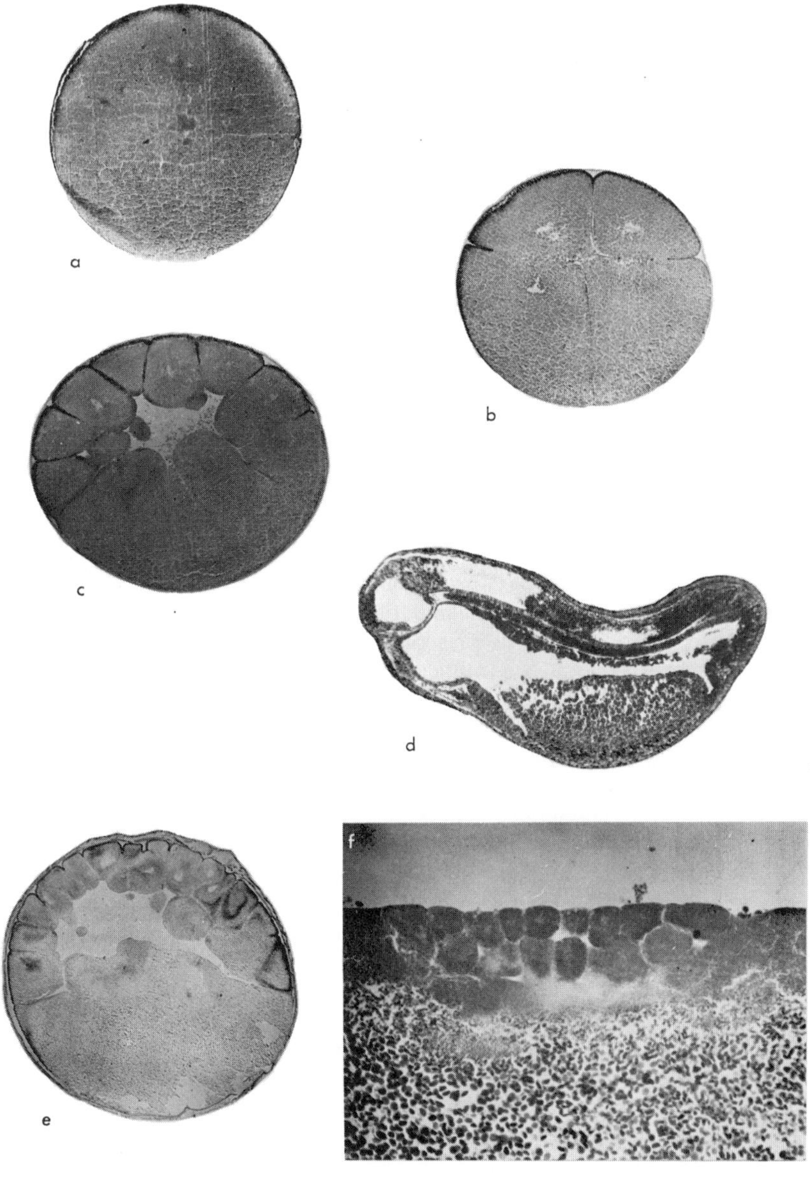

Plate 11 FROGS AND DOGFISH (a) Vertical section of a frog egg at fertilisation, showing the two nuclei (pale) meeting above the yolk, and the grey crescent (pigment below the surface) at the left (compare with Figure 4.1c*). This print, and (b), (c), (d), (e), have been trimmed out to avoid confusion by agar or liver section used in preparation. (b) Vertical section of an eight-cell frog egg (compare* Figure 8.3b*). (c) Vertical section of a frog blastula. (d) Vertical (sagittal) section of a frog neurula (compare* Figure 9.3b*). (e) Blastula of a frog egg which has been centrifuged, before first cleavage, to separate the yolk from cytoplasm (compare* Figure 8.2*). (f) Section of the early blastoderm of a dogfish egg. The surface of the cytoplasm has cleaved, the yolk has not. The pale area under the centre is the 'sub-germinal cavity' an artefact of the solvents used in preparing the section*

machinery to read the tape is essential. This is the cytoplasmic machinery of the egg and, as in the computer example, it is much more complicated in structure than the tape itself. To a large extent, what we have been considering in this chapter is the switching on of the machine (egg activation); the loading of the tape so that its parts pass the reading head appropriately (cleavage); and the putting of motors into gear for playing the tape forward or reverse (the determination of polarity – this is not, of course, to suggest that sinistral *Lymnaea peregra* play their genes backwards!). This analogy is close, and useful in that it directs attention to the tape recorder as well as the tape; the playing deck, amplifier, speakers and supply voltage as well as the gramophone record; the cortex, informosomes, mitochondria, Golgi apparatus and yolk of the egg as well as its nuclear chromosome assortment. Even if it was a previous gramophone record which gave instructions on how to build this amplifier (mother's genetic programme acting through her ovary), that too had to be 'played' through an amplifier, and both, even if interdependent, control development.

In a series of penetrating articles in the early 1960s, Gustavson and Wolpert[7] analysed the early development of sea urchins in a new way; instead of listing all the complex movements and foldings which resulted in the complex armed larva, they asked how *few* processes would be necessary to produce this larva. There is a distinction here of the utmost importance to the whole field of reproduction. It is the distinction between **'prescription'** and **'description'**. Much more information is needed to describe than to prescribe. Let us consider a woollen sweater and the way it is 'built'; a full description of this must not only consider the three enormously complex knots which are the body and sleeves, but must describe the twists of each hair in each wool thread, the adhesion of the cortical scales and links of the wool, the detail of the linkage at the shoulders, and so on. A conservative estimate would put the number of words required to describe this at about the number in two complete sets of the *Encyclopaedia Britannica*. But a full *prescription* for such a sweater is contained in a knitting pattern, usually four pages, which specifies wool grades and needles and continues 'knit one, purl one' and so on.

The egg obviously does *not* contain a 'description' of the adult, a **'blueprint'**, though it is often said to do so. Its structure is a prescription, and if this is filled (made up) in the usual way a normal embryo and juvenile will result. But if, for example, thalidomide or German measles virus (rubella) are present at crucial stages in human development the end-product is tragically different. The human egg does not *become* the baby; it organises the food materials supplied by the mother into an embryo which in turn organises more food into a fetus and so on into senility.

Gustavson and Wolpert[7] showed that very few, perhaps only four, processes need be prescribed in the cleaving sea urchin egg to account for the formation of the complex pluteus larva (*Figure 8.7*). Part of the process will be described here as an illustration. By cleavage a hollow ball of about 1000 cells, the **blastula**, is formed. Cells at the vegetal pole, primary mesenchyme cells, bulge upwards into the cavity, the **blastocoel**, and soon move out of the layer on to its inner surface, where they form

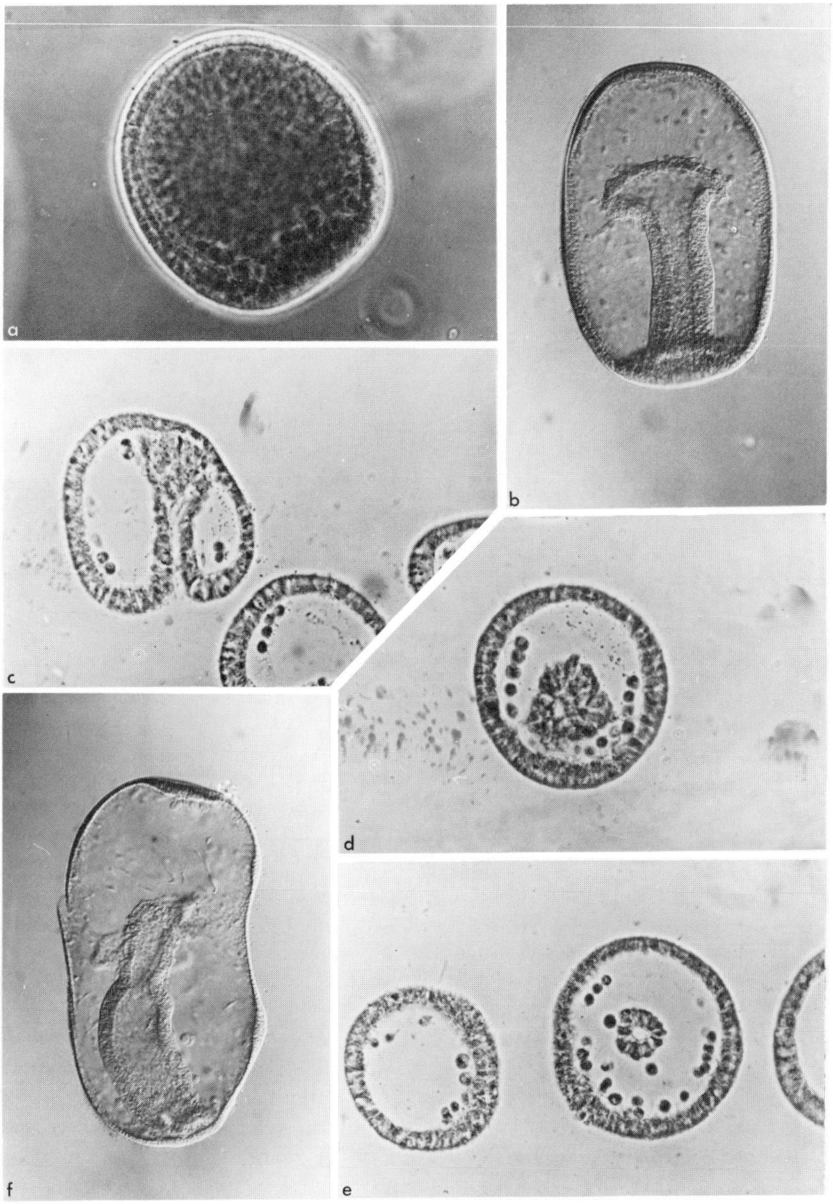

Plate 12 SEA URCHIN GASTRULATION (Compare with Figure 8.7*). (a) A late blastula; note the cells moving into the blastocoele from the vegetal end, bottom right. (b) The enteron (future gut) has been drawn in. The mesenchyme cells can be seen, but their pattern is not clear. (c) Vertical section of a gastrula like (b). (d) Transverse section near the top of the enteron. (e) Transverse section near the base of the enteron. (f) A later gastrula, called a dipleurula; the mouth has broken through (just above centre) and lines of mesenchyme forming the spicules (arm skeleton) can be seen. Compare* Figure 10.9f *and* Plate 15c, *which is a later stage of this larva*

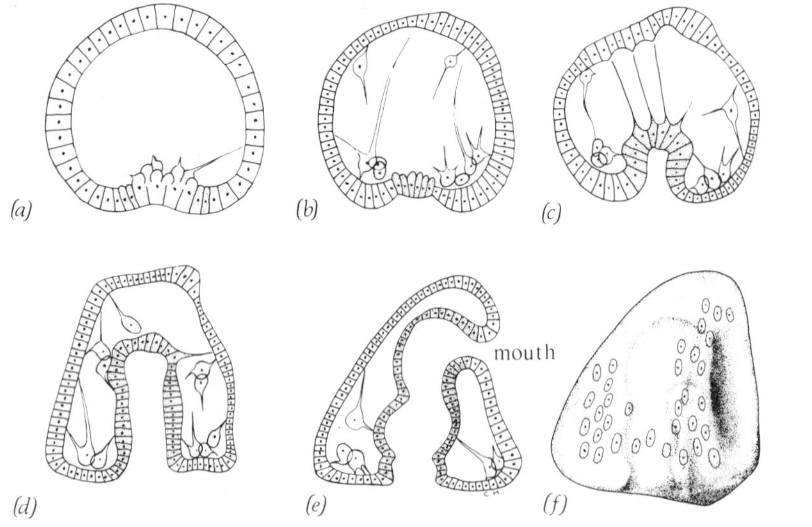

Figure 8.7 THE PRESCRIPTION OF THE ECHINODERM LARVA (see Plate 12*). (a) Cells at the vegetal pole of the blastula, made of original vegetal plasm, change their 'stickiness' and bubble up into the blastocoel. (b) They are the primary mesenchyme cells, which have long pseudopodia with which they move actively. (c) They settle in a pattern on the inner face (see text and f). Now the vegetal pole bulges in again, but this time retaining cell contact; its long pseudopodia attach near the animal pole, at a patch perhaps related to original sperm contact. (d) The pseudopodia are firmly adherent, and when they contract the vegetal hemisphere is tugged in as the mid-gut. (e) The mouth breaks through. (f) An imaginary transparency of a simplified echinoderm larva like (e); the pattern of primary mesenchyme cells on the inner face of the outer layer determines the skeletal rods. These determine the shape of the planktonic larva (see* Plate 16*), and derive mostly from egg organisation, not from zygote DNA instruction*

a complex pattern which determines the shape of the larva and the number and position of its arms (*Figure 10.8g*, p. 189 and *Plate 12*). Perhaps these cells could be programmed to follow specific paths to make this pattern; but Gustavson and Wolpert showed that this is not so. They filmed development, and showed that these cells throw out long pseudopods which attach randomly to the inner aspect of the wall of the blastula. New connections are made and old ones broken continually as the cells pull themselves about like little octopi. But gradually they accumulate in a ring about the equator, reaching towards the animal pole at four points. Careful analysis of the films showed that pseudopodial connections made to these regions were less likely to come unstuck than those made to other parts of the surface, so this differential 'stickiness' accounted for the pattern. There must have been a similar distribution of substances in the uncleaved egg, perhaps from the oocyte, or derived from new organisation of the egg cytoplasm at fertilisation. No genetic programme of the primary mesenchyme cells was involved in this apparent genesis of the arm pattern. It was not in fact a genesis of the pattern, but a visualisation of a cryptic pattern already present by making it a more complex one, adding the primary mesenchyme cells on the inner aspect of cells which were already different because the cytoplasm they were cut out of was different.

DeHaan[8] has shown that the path of the migrating cells which will form the heart of the chick embryo is determined by properties of the cell layer they migrate over, and that these may derive from similar simple properties of the original blastodisc cytoplasm.

8.4 The Phyletic Stage

Many of the basic structures of larvae, such as the ciliary girdles of trochophores or the arms of the basket larva of the sea urchin, and many basic structures of vertebrate embryos such as the notochord and heart, can all be referred to cytoplasmic not nuclear instruction in the egg. For this reason, it is profitable to separate development into two overlapping phases whose control is different.

The first phase, that which is cytoplasmically ordered by the egg architecture derived from the maternal ovary, culminates in either an embryo with the basic characters of the phylum concerned or sometimes in a characteristic larva. During the second phase, of control by the zygote nuclear genes, this embryo or larva differentiates further into the specific juvenile of its species. Because the intermediate stage is more or less characteristic for each phylum, I have called it the **'phyletic stage'**[9]. Eggs of different but related organisms converge upon this organisation in their early development, and then diverge again as organogeny occurs. In vertebrates

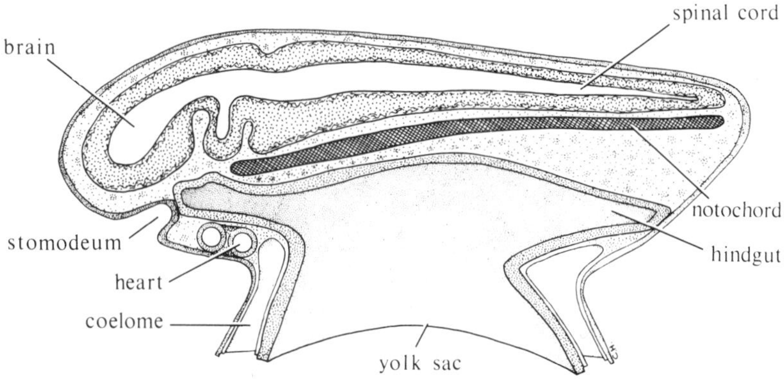

*Figure 8.8 SCHEMATIC PHYLETIC STAGE OF THE VERTEBRATES (chordates?). All vertebrates pass through a state formally comparable to this, but they arrive at it from different directions. Teleosts, for example, form all their organs in a pile of cells without a blastopore, while selachians, lungfish and amphibia roll their future notochords over the blastopore dorsal lip; reptiles, mammals and birds have a 'primitive streak' into which tissue dives to emerge anteriorly as the embryonic layers (*Figure 9.3*). Teleosts have yolk in the peritoneal space, while others have it in the gut lumen (selachians and birds) or the gut cells (some amphibia, like the frog* Rana*).*

After this stage the organisms diverge again toward their adults. Phyletic stage structure is largely dependent on egg structure, but cells of the different regions have started to express their own genomes too

the phyletic stage is the **neurula** (*Figure 8.8*); for annelids, nemertines and molluscs it is three kinds of **trochophore** (*Figure 10.9*, p. 192); for echinoderms it is the **dipleurula** (*Plate 12*), and so on.

This view of development illumines such generalisations as von Baer's Law, and is not subject to the same criticisms. This law stated that early stages of related organisms resemble one another more than later stages, the organisms diverging toward their adult state as they developed. The critics soon pointed out that eggs, such as those of rabbit and chick, differed more than did the early embryos, and went on to claim that creatures were equally diverse at all stages; but this is manifestly not so either. By considering and comparing developments in two parts, separated by a common stage, these opposing views can be reconciled. Eggs are very specialised and different, but all those of one phylum or class converge in their early development towards a common phyletic stage, represented in the vertebrates by the neurula, and in the gastropod molluscs by the veliger. Then, as their nuclear genes come into operation, they again diverge toward their own specialisations.

REFERENCES

1. Curtis, 1965
2. Tartar, 1960
3. Raven, 1966
4. Mahowald, 1972
5. Moore, 1955
6. Boycott *et al.*, 1930
7. Gustavson and Wolpert, 1961
8. De Haan, 1964
9. Cohen, 1963

8.5 Questions

1. In what ways may deficiency in maternal effects prevent or modify development?

2. Why do determinate eggs frequently have very complex cleavage patterns?

3. Distinguish between instruction and prescription in the control of development; to what extent does the DNA serve as a 'blueprint'?

4. What is the phyletic stage? How does it simplify the organism's adaptations at different phases of development?

9

THE ZYGOTE HEREDITY

9.1 Activation of Zygote Genes

It has been emphasised in the last chapter that the oocyte has most of the structures and machinery required to produce a heterogeneous cellular organism of some complexity. Although in a few forms, like mammals and some insects, the zygote genes are turned on during early cleavage, this is unusual[1]. The trophoblast of mammalian embryos is precocious so we might presume that the zygotic genes used for its elaboration are comparable with those used for the homologous embryonic membranes of reptiles, which develop much later in embryonic life.

Usually, but with the important exceptions of mammals and some insects, development may be considered as occurring in two distinct phases. The first phase, under the control of the oocyte organisation using m-RNA and ribosomes elaborated by the oocyte ('informosomes'), builds an embryo whose cells have different cytoplasms and different environments and soon acquire different neighbours. Then, in late cleavage, often during a dramatic period of tissue movement and folding called **'gastrulation'**, the nuclei begin to synthesise new m-RNA and the second phase of embryology begins, controlled by zygotic genes. There are some differences associated with determination state of the embryos of different organisms (p. 148) but these are not basic.

The work of Neyfakh[2] on the loach is a classic analysis of these two phases. *Figure 9.1* shows the comparison of m-RNA being synthesised during development; the m-RNA was taken from the adult organs, the early cleaving egg, or the gastrula. This technique has given us a most powerful tool for analysing the control of protein synthesis during development and its technical principles will be considered at this point. It has recently proved possible, largely through the work of Paul[3], to re-attach m-RNA to DNA, and to separate the hybrid strands on cellulose or hydroxyapatite; it has even been possible in rare cases, by the autoradiographic use of labelled m-RNA, to localise the sites of certain genes on chromosomes by sticking the homologous RNA back on to them.

There are many problems with the chemical isolation of long DNA and RNA molecules and with sticking them together again; in practice, a useful technique of estimation of the degree of re-association of m-RNA and the *active* DNA in nuclei depends upon the kinetics of chromatin from the nuclei, not extracted DNA (*Figure 9.2*). This is because short lengths of m-RNA and DNA will, by chance, have nucleotide sequences in common; even long lengths will have short sequences which will fit.

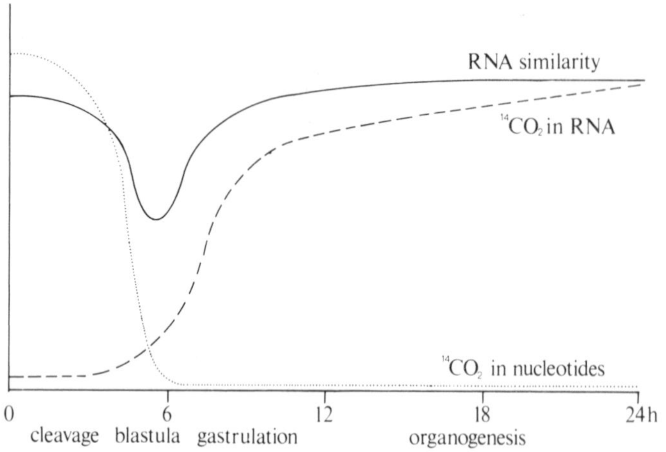

Figure 9.1 RNA IN EARLY FISH DEVELOPMENT (after Neyfakh, 1971). The loach is a favourite with Russian embryologists because of its rapid development and prolific egg-laying. The dotted line shows that little labelled CO_2 carbon finds its way into RNA until six hours after fertilisation; until then it all remains as nucleotides (dotted line); the 'y' axis is increasing radioactivity of nucleotides or RNA extracted from batches of eggs. The solid line shows a comparison between the kind of RNA produced during development and that produced by late (pre-hatching) embryos. Up to six hours the RNA in the eggs (maternal, because we know from the $^{14}CO_2$ experiment that little is made) is quite like late embryo RNA; that is to say, it competes successfully with it for the specific sites on DNA. At about six hours, however, new RNAs appear, unlike the test sample; then by 12 hours much the same spectrum is again produced. It is during the process of gastrulation that the RNA is different, just before and during the phyletic stage (see text)

There are many short repeated sequences, too, on all chromosomes. So there will be considerable association even between unrelated DNA and RNA. How, then, can the extent of 'real' homology be established?

This is usually done by an experimental design called 'result reversal' which we will first illustrate by an unrelated example, before returning to m-RNAs and DNA. Although this is nothing to do with DNA/RNA *per se* a good example is the modern pregnancy test.

The urine of pregnant women contains a polypeptide hormone, human chorionic gonadotrophin (HCG) elaborated in the placenta, as well as a great variety of other substances. In addition, there is a good possibility that the urine samples will be contaminated with organisms during collection or storage even if originally sterile. Although antibodies can be raised by injection of rabbits with human chorionic gonadotrophin, and are very specific giving a 'precipitin' reaction (a precipitate of the HCG-antibody complex) with very dilute solutions, the rabbit serum extract may also give precipitates with other constituents of the urine sample. For example, if *streptococcal* antigens are present, there may well be antibodies to these in the rabbit serum and these will give a precipitate even if no HCG is present. How can we distinguish this 'false positive' from a true but weak precipitation? In practice, if there is doubt about the direct reaction, the experiment is result-reversed. A *small* quantity of rabbit antibody is added to the urine sample, and the mixture is then filtered or centrifuged to *remove* any precipitate. The presence or absence of precipitate at this stage is unimportant. Then the quantity of gonadotrophin which reacts with this antibody is added; if there is *no* precipitate, i.e. all the antibody has already been precipitated by gonadotrophin in the sample, then the woman *is* pregnant. If there *is* a precipitate, the antibody has not reacted and the woman is *not* pregnant.

The same principle can be used with m-RNA/DNA association. It can be shown experimentally that only a small part of the DNA is actively making m-RNA in a particular cell type (*Figure 9.2a*). For example, m-RNA can be extracted from a tissue and the 'chromatin' (DNA associated with histones and acidic proteins of various kinds) can be extracted from another similar tissue sample. The chromatin can then be incubated with radioactive RNA nucleotides in the presence of RNA-polymerase, and the radioactive m-RNA formed can be isolated. A complete DNA preparation from the same organism, but from which the inhibitory histones and acidic proteins have been removed, will have sites to which either the natural or the *in vitro*-synthesised RNA can attach. So the experimenter can compare the kinetics of the increase in radioactivity of the 'hybrid' molecules (RNA/DNA) that are formed, *after the DNA has been saturated with another RNA*. If this other RNA is *very* effective in preventing attachment of the radioactive m-RNA, the m-RNAs are to a large extent homologous; they have most nucleotide sequences in common (*Figure 9.2c*). m-RNAs extracted from other organs, or made *in vitro* from the DNA rather than chromatin, can be compared for homology in this way by their effectiveness at preventing attachment of homologous m-RNA. *Figure 9.2* shows one of Paul's comparisons of kidney, liver and spleen m-RNAs. It is clear that a range of different sequences of nucleotides are present in each sample, and that the three samples differ. This is the common finding, and illustrates the basic mechanism of differentiation of the various cells of the embryo: the DNA, identical in all cells, has its transcription restricted in different ways in different tissues so the spectrum of m-RNAs produced is somewhat different.

The cells are different because of the segregation of cytoplasmic regions during oogenesis, fertilisation and early cleavage and their reorganisation during gastrulation. This difference is expressed during and after gastrulation by the different degrees of restriction of the common DNA, so that some of the genes expressed in liver cells, for example, will be different from those in skin or nerve cells.

There are, of course, many genes which remain active in nearly all cells. Most of these are in the nucleus but a few are mitochondrial DNA; they make common enzymes required for basic cell processes, which have been called the 'housekeeping' enzymes. As cells specialise, those genes which will make the proteins characteristic of that kind of cell may be *amplified* (i.e. many copies may be made so that maximum production is achieved). Usually proteins are required for *activation* of genes, which sometimes also require **hormone** to attach before they become active. The cells which make silk, some kinds of keratin, myosin and actin in muscle cells, haemoglobin in erythrocytes, all seem fairly autonomous, with the **structural genes** active, sometimes only for a short time. But cells making milk protein, thyroid cells making thyroglobulin, epithelial cells making mucus or keratin, and the array of syntheses in the mammalian uterus all seem to be controlled by substances such as hormones and vitamin A, which act via receptor proteins **regulating** the structural genes directly or indirectly.

We may summarise early animal embryology, then, as being a progressive **'regionalisation'**[4] whose commencement is in the heterogeneity of the oocyte structure, somewhat modified by entry of sperms. Later the nuclei of different regions begin to produce specific m-RNAs and make specific enzymes; these change cell metabolism in such a way that further

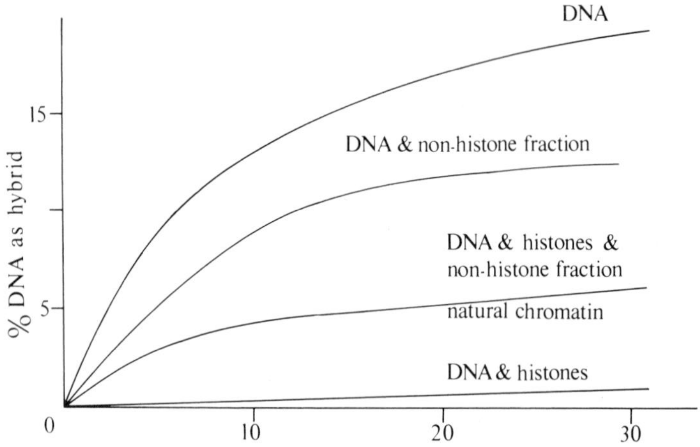

(a)

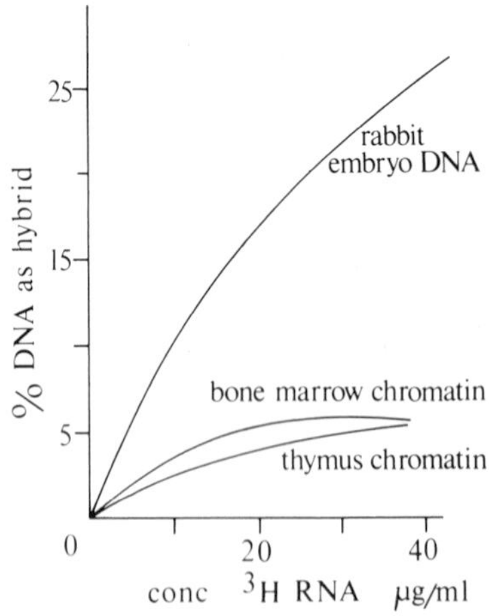

(b)

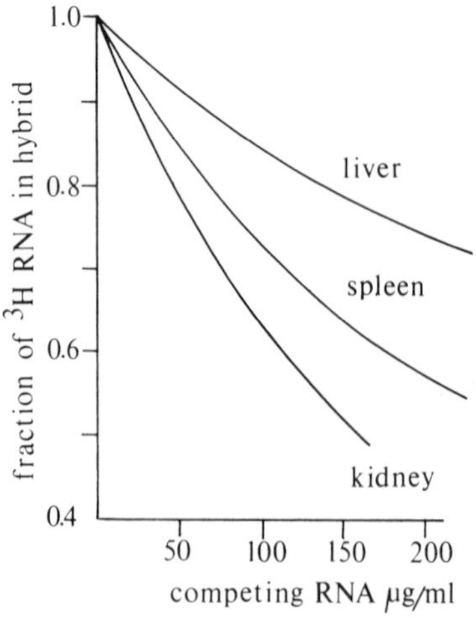

(c)

changes can occur either in the same cell or in neighbouring regions, as a result of chemical induction. The parts of each region border other regions or may be approached by folding cell layers of different kinds; they may then show **'individualisation'**[4] into sub-fields, with nerves and hormones having some integrative effects.

Determinate eggs rely largely on the original pre-cleavage pattern, set up by the interaction of oocyte structure with the events of sperm entry, and maintain this complex architecture through cleavage. Indeterminate eggs start simply, with only a few 'regions' remaining after cleavage, and use intrinsic nuclear machinery to react and interact at gastrulation and so achieve complexity. Two kinds of nuclear genes are turned on in the individual cells: those for general enzyme machinery, common to all cells of the organism, and **organ-specific gene arrays** characteristic of each developmental stage of each tissue.

9.2 Restriction of the Active Genome

There are many examples of this restriction of the m-RNA-transcribing activity of the DNA to a small, specific, array of genes. In many cases, as with activation of the egg, apparently simple triggers will cause such **differentiation** of cells. A system which has received great attention is that **induction** of outer layers of cells, in amphibian and chick embryos, which restricts their fate to neural tissues, the spinal cord or brain. This induction normally comes from the notochord, moving anteriorly

---

Figure 9.2 RESTRICTION OF DNA ACTIVITY IN DIVERGENT TISSUES (after Paul and Gilmour, 1968 and more recently). (a) RNA is transcribed from DNA (e.g. calf thymus DNA) using a DNA-dependent RNA-polymerase, and radioactive precursors. Then this assortment of radioactive RNAs is offered in increasing concentration to various DNA preparations of the same tissue. It re-associates with 'bare' DNA in a concentration-dependent fashion, as expected. The natural 'chromatin', however, is saturated at low concentration and never shows a high recombination; probably only few sites are available for RNA to attach. If the 'chromatin' is disassociated into histone, DNA, and 'the rest' (mostly acidic proteins, unlike histones which are very basic), the mixture behaves very like natural 'chromatin' (same line on the graph). But if histones alone are added to DNA, they virtually prevent any re-association with homologous RNA; they are thought to be a non-specific block. On the other hand, re-association of all but the histone results in re-association well above chromatin but below DNA levels; it is supposed that acidic proteins link to DNA, preventing direct attachment of histones and so allowing controlled transcription in some specific regions, which these proteins 'recognise'. (b) A similar experiment, with radioactive RNA transcribed from rabbit DNA in various conditions. Rabbit embryo DNA-transcribed RNA associates readily and increasingly with rabbit DNA; however, RNA made from rabbit DNA in bone marrow chromatin, or in thymus chromatin, sticks to only part of rabbit DNA (about 7%) and increasing the concentration of the reacting RNA does not help, above some 20 μg/ml (experiments of R.S. Gilmour). (c) A result-reversal experiment (see text). RNA samples from mouse kidney, spleen and liver interfere differently with the re-association of radioactive RNA with its 'own' chromatin. In this experiment (of H. Thomou) radioactive RNA was transcribed from mouse kidney chromatin. Increasing concentration of natural liver RNA does not prevent re-association very well, but natural kidney RNA competes very successfully; natural spleen RNA is intermediate. Liver competes best with liver, and spleen with spleen, however. It is supposed that different RNA populations are being produced on the chromatins of the three tissues

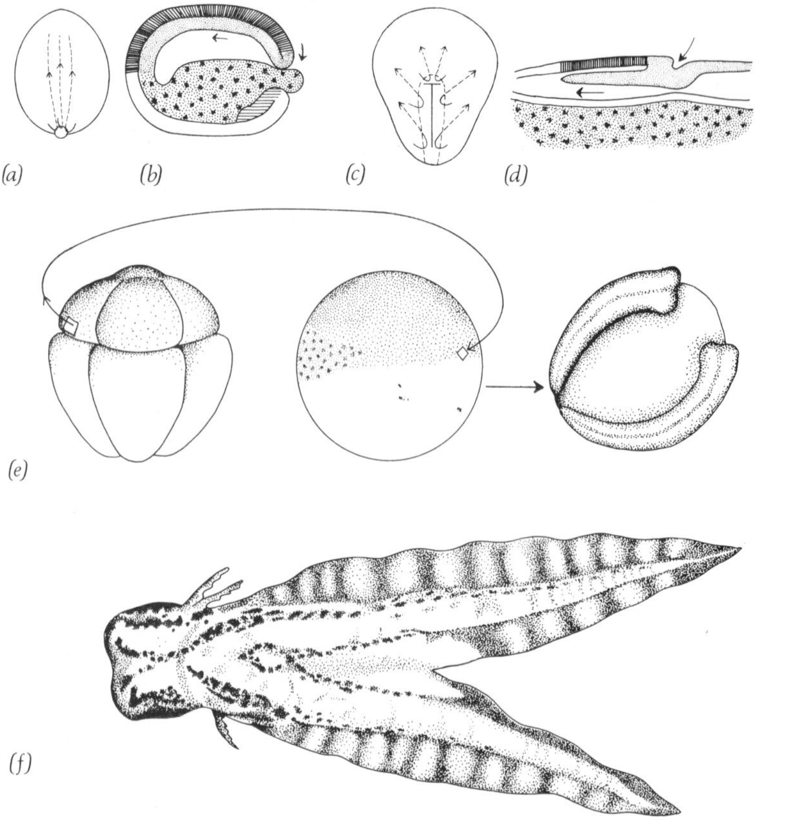

Figure 9.3 GASTRULATION IN THE VERTEBRATES (a) The blastula of amphibians becomes folded in a way which closely resembles that of echinoderms. Compare Figure 8.7, Plates 11 *and* 12. *This view, from dorsal, shows a late stage, with outside tissue moving posteriorly, dropping in over the edge of the blastopore, and moving forward inside the embryo as differentiating internal organs. (b) Vertical longitudinal section of a late gastrula like (a). Notochordal tissue (fine dots) moves forward in the dorsal mid-line (sometimes as the centre of the roof of the gut), and influences the outside layer above it (radial lines) to become neural plate. (c) The 'primitive streak' of a reptile, bird or mammal. This is in many functions comparable with (a): outside tissues are moving back, rolling over an (apposed) edge, and then moving forward inside the future embryo. (d) Vertical longitudinal section of this primitive streak. Again, notochord (stippled) is moving forward, and inducing the outer layer above it to go neural. The upper arrow shows the position of the 'notochordal pit' in reptiles and primates, comparable with the dorsal lip of the blastopore of amphibia. (e) Curtis' experiment with grey crescent cortex. Transplanted to a second (uncleaved) embryo, grey crescent cortex caused the production of a second gastrulation centre and a second embryonic axis. (f) A newt larva from a newt blastula into which has been implanted a second dorsal lip of blastopore, from another older embryo. Again, two embryonic axes have formed; the vast majority of the secondary embryo is made of host embryo tissue which has changed its developmental fate*

from the dorsal lip of the blastopore or the primitive streak, and underlying the dorsal region of the embryo (*Figure 9.3*). Ever since Spemann and Mangold[5] showed that transplantation of the cellular dorsal blastopore lip induced a new embryonic axis in a host embryo[4], attempts have been made to duplicate the effect in a more defined system. The aim was discovery of the natural chemical **'evocator'**, the substance which evoked differentiation of neural tissue. An enormous range of boiled, frozen, extracted, lyophilised and otherwise denatured plant and animal tissues and fluids proved able to evoke **neuralisation** and heroic attempts were made to find order in this chaos, often by contrasting evocation of brain and spinal cord by different factors or extracts.

Analysis of other essentially morphological systems had as little success. Even the discovery by Beerman[6] that the chromosomes of dipteran flies showed a characteristic array of **'puffs'**, specific for organ and developmental stage, added little to the idea of induction. It was shown that insect hormones could elicit specific arrays of 'puffs', but the mechanism of this evocation remained mysterious. Again and again, complex patterns of differentiation were elicited by simple and apparently non-specific agents, so that attention came to be focused on the capability of the cell rather than the stimulus. In speaking of induction, one spoke of the **'competence'** of the cell, or the reduction in **'potency'**, and the model underlying these thoughts was of a branching railway network. The parts of the egg started out from the station, and the division products took various lines, binary decisions being made at forks by sets of 'points'. These points were set to left or right forks by the passage of other cells on neighbouring lines, or by calendar time, or by messengers (hormones and nerves). The metaphysics of differentiation was complicated by such mathematical ideas as those of Turing[7], about the origins of patterns from homogeneous systems with special rules, and by the Waddington school who wished to produce an inclusive theory of the genetics of development[8]. From this attempt many interesting models arose, and some explanations have come from biochemistry and especially from the suggestions of Jacob and Monod[9] about enzyme induction in bacteria.

Their investigation of the **lac** gene system of *Escherichia coli* had led them to propose the **'operon'** theory[9]. Bacteria not normally possessing a particular enzyme, in this case β-galactosidase, would produce it when the medium was altered so that it was required, i.e. when lactose was present but not other sugars. The theory suggested that the **structural gene**, i.e. that DNA sequence complementary to the m-RNA from which the enzyme was translated, was linked to a control system which normally prevented transcription of m-RNA from it. The repression was achieved by another gene product, the **repressor**, which in turn could only be inhibited in the presence of the enzyme substrate. This repressor gene *i* has now been identified, and its protein, the **lac** repressor, acts on a DNA sequence close to the **lac** structural gene, called the **lac operator**, and so inhibits transcription of m-RNA from the **lac** region. When a β-galactoside, like lactose, binds to the repressor, this no longer binds on to the **lac** operator and the **lac** region m-RNA is transcribed. Many such interactive gene units, or operons, have been found in bacteria and a few in animal cells in culture, and this mechanism has been proposed to account for expression of only a small number of genes in differentiated cells. The others, it is assumed, are in the 'repressed' state.

Despite the wide acceptance of the general nature of the Jacob–Monod hypothesis, this simple form of the operon theory cannot account for the complex differentiation of the somatic cells of animals, because it does not explain three major properties of this differentiation.

9.2.1 DIFFERENTIATION IS PERMANENT

When tissues are taken from chick embryos and cultured separately, they lose those differentiative characters which are recognised by the histologist. They lose their characteristic shapes as they spread on the glass, they lose actin and myosin 'stripes' if they were striped muscle cells, and the cells come to resemble generalised 'fibroblast', 'amoebocyte', or 'epitheliocyte' tissues found in early embryology. They are then said to have 'de-differentiated'. But experiment shows this not to be the case physiologically even if it is the case morphologically. When returned to the organism, all the evidence denies that they can differentiate into other kinds of cells[10]. Their differentiation is maintained through many cell generations.

The operon hypothesis of differentiation can explain the induction of a new state, but not its maintenance through many cell generations in the absence of the original stimulus. In other words, it can explain acute change, but not chronic change.

9.2.2 NECESSITY FOR CELL DIVISION

There are 'acute' changes, called 'modulations', in which differentiated cells (e.g. epidermal cells) produced from a basal dividing population take different paths according to prevailing conditions, for example, hair cortex cells or keratin of skin. Hormones, too, can act to modify the differentiation. But these kinds of modification require a cell division[11], and usually one daughter cell does not differentiate but remains to divide again. Such a requirement for cell division by the responsive cell is the opposite of the Jacob–Monod prediction, for the repressor/operon system acts via the normal DNA/RNA-protein routes of the metabolic cell, and DNA replication is not part of the theory.

9.2.3 TERMINAL DIFFERENTIATION

Some kinds of differentiation of the somatic cells of metazoa are obviously more radical than an acute chemical control of gene transcription whose basis is essentially regulatory; other kinds exhibit long-lived m-RNA, which cannot be part of the framework of a simple regulatory control system. An example of a radical differentiation is the red blood cell of mammals, which extrudes its nucleus as it differentiates. The red blood cells of other vertebrates do not discard their nuclei, but they have nuclei which neither divide nor transcribe m-RNA. These red blood cells use very long-lived m-RNA to guide their ribosomes in the manufacture of haemoglobin. Many other cells, too, have an apparently inactive nucleus and depend upon long-lived messengers[12] whose very existence precludes simple Jacob–Monod control.

Harris has been a consistent critic of the Jacob–Monod view of differentiation, and his experiments have demonstrated that there must be some **translational controls**, i.e. of protein synthesis, as well as the **transcriptional** ones, i.e. of m-RNA synthesis[10]. His primary technique has been the use of virus envelopes to fuse cell membranes, and so to make **hybrid cells**. These need not be of the same species, and the combination Hela cell (a human cervical carcinoma cell) with chicken erythrocytes has demonstrated many effects of cytoplasm upon nuclei. In this case the Hela cytoplasm causes the chicken erythrocyte nucleus to become active, making chicken haemoglobin m-RNA and even dividing. This is positive control, for the erythrocyte cytoplasm does not prevent the Hela nuclear activity as would be predicted by repressor theories, and such positive action is also found with other pairs of cell types.

Positive control of nucleus by cytoplasm is also found in another kind of experiment which illumines differences between the germ line and somatic cells in restriction of the genome, and shows that the restrictive agent is again the cytoplasm. Gurdon[14] has implanted somatic nuclei from the differentiated cells of various organs and stages into the enucleated egg cytoplasm of *Xenopus*. He has found that these may contribute to the formation of normal embryos which can develop into fertile *Xenopus* adults. In this case the nuclei are truly **de-differentiated** by the cytoplasm of the egg, unlike whole cells implanted into tissue cultures. Presumably the daughter nuclei react in the normal way to the various cytoplasmic environments into which they are segregated during cleavage. Again we have cytoplasm affecting the nucleus, but Gurdon has produced an even more dramatic example. If adult *Xenopus* nuclei are implanted into oocytes, instead of activated eggs, these nuclei swell and become 'germinal vesicles' like the resident nucleus. Even nuclei from adult neurones, which would ordinarily never divide again, can be persuaded to do so by the oocyte cytoplasm, and they will then undergo division with the egg nucleus.

9.3 The Somatic Cell Chromatin

This control of nuclear activity by the cytoplasm emphasises the major difference from *E. coli*, and is the greatest block to argument from gene control in prokaryotes to differentiation in eukaryotes. This interaction between cytoplasm and nucleus maintains the differentiated state, perhaps by many gene repressions, but more likely by control of the chemical structure of the chromatin itself, particularly when the nucleus divides and the chromosomes contact the cytoplasm.

Because the histones are the proteins most closely associated with the DNA in chromosomes, it was thought for many years that specific gene repression was due to the specific blanking off of unused gene sequences by special histone molecules, preventing access to RNA polymerases (which copy DNA onto RNA) and so preventing transcription of these genes. Recent work has shown that histones are apparently not selective themselves, although they may be the actual molecules which prevent access of RNA polymerase (*Figure 9.2a*). There is a very diverse and unstable group of acidic proteins which probably function selectively to block specific genes by histone attachment to them. The variation in side-chains of these histones, as well as the tightness of coiling of DNA varying with histone density, is supposed to account for **banding** with biological stains and for the **'heterochromatin'** so characteristic of most differentiated cells. Only the **euchromatin** seems to be involved in synthesis.

The enormous quantity of DNA present in animals, compared with prokaryotes (*Table 15.1*, p. 258), has provoked many theories of chromosome structure. The actual number of genes may be fairly small, if the numbers which can manifest themselves by mutation are a significant proportion of the population. There are certainly long lengths of **reiterative** (short) **sequences** of DNA too, whose function *may* simply be to separate the genes. But it is becoming increasingly probable that the chromosome is not simply a string of genes, or even a string of operons; at least part of the function of the repetitive sequences is concerned with the function of the centromere[15], and probably they are involved in genetic crossing-over too (*see* p. 118).

Modern theories see the 'gene' as a complex set of DNA sequences, with repetition not only of the short sequences but also multiple controlling and structural sequences, so that control is not binary, off–on, but is graded.

Controls may not only be spatial in nature. Goodwin[16] has pointed out that if control of gene activity depends on the activity of enzymes specified by that gene, as in the Jacob–Monod system, then inevitable delays in the feedback loop make the system a non-linear oscillator. The kinetics of production of several important enzymes leads to sawtooth waveforms of various shapes which represent the necessary oscillations of chemical concentrations in such a system. If all the m-RNAs concerned are short-lived, then these may interact to produce longer-term cell rhythms; Goodwin explains circadian (about 24 hour) rhythms in this way. Goodwin and M.H. Cohen[17] have proposed that such 'cell clocks' may also determine synthetic capabilities, i.e. the differentiation, of the cells concerned because only certain enzymes can fit the cell rhythms. Such cell rhythms may act to organise tissues into spatial patterns of differentiation, by coupling between the cells' rhythms via their diffusable products. Rhythm-determined differentiation of this kind has been found in regenerating *Hydra*. Such temporal organisation of cell associations could explain many difficult problems, notably some of the organisation of the vertebrate nervous system.

9.4 Somatic Cell Mitosis

Germ cells are effectively immortal. The continued production of germ cells in each generation is evidence of this, although there is the occasional interpolated meiosis and fertilisation. It used to be believed that this was the case for somatic cells too, and there was evidence for this from the chick heart culture at Strangeways Laboratory in Cambridge, maintained since the early years of this century. But there is now new evidence, from the work of Hayflick and his school, that there is a limit to the number of mitoses of which mammalian somatic cells, at least, are capable.

Hayflick[18] attacked the problem from both ends. He cast reasonable doubt on the longevity of the original Strangeways chick heart, by pointing to the intermittent addition of 'embryo extract' which could easily have contained living cell contaminants. He also cultured human lung cells from donors of all ages, from embryos to the very old (85+). He found that embryo lung cell cultures could undergo about 75 to 85 mitoses, then most cultures died; those few which survived often had a great dip in viability followed by a resurgence. Cytological examination of these 'continuing' cultures always revealed chromosomal abnormality. In order to distinguish 'cell time' from 'calendar time' Hayflick deep-froze some cells at 170°C and, after thawing two years later, these made up their mitoses before degenerating – the time spent frozen had not counted. Apparently adequate controls for the multiplication of infection, or depletion of essential nutrients by dilution, were employed: for example, late media were used for early cultures, and the latter did not suffer. Further, it was found that, while embryo lung cells could divide some 80 times, cell lineages from older donors died after fewer divisions, until those of old people (60 to 70) only divided 20 to 30 times.

These experiments, while they may not illumine senility as Hayflick believes, nevertheless do point to apparent restrictions of continued division by these somatic cells. Those cultures which changed and continued are, in his view, tumourous: they have over-ridden their internal controls. There are many 'domesticated' (i.e. continuous) cell lines available commercially which have apparently normal chromosome complements, but of course the anomaly need not show as a gross defect, as aneuploidy. (**Euploidy** is the name given to normal haploid, diploid or multiple chromosome sets, **aneuploidy** to anomalous numbers.)

So we must suppose that part of the somatic programming in many, if not all, somatic cells of mammals includes built-in mortality. This should not surprise us, because we know that many somatic cells differentiate terminally, i.e. have no further mitotic possibility, e.g. erythrocytes and neurones. Different tissues, of course, require to divide at different rates; epidermal and gut cells divide about every eight hours in humans, for instance, but nerve and muscle nuclei never divide. In some other organisms, notably the aschelminths and especially the nematodes, this is carried to extremes as the phenomenon of **eutely**. Here each organ except the gonad (and occasionally gut and epidermis) has a specific cell number; the somatic cells of nematodes cannot divide after the fourth larval stage is reached. They only enlarge, and even minor wounds cannot be healed, just as mammalian neurones or muscle cells cannot divide.

On the other hand, some organisms, such as coelenterates and ectoprocts, seem able to continue vegetative reproduction by budding, and produce large numbers of polyps and other kinds of 'person'. Whether this is really an indefinite process, or whether senility of the whole clone occurs, as in some ciliates, is not known. But it is suggestive that most, if not all, animals separate their immortal germ cells from their mortal soma (p. 10). We should keep an open mind, and consider the possibility that built-in obsolescence of the soma is not a necessary consequence of this separation of germ cells and somatic cells, for there seems to be a complete spectrum from coelenterates to nematodes with mammals about half-way; coelenterates may have 'immortal' somatic cells, and nematodes have somatic cells which cannot even divide.

9.5 Gastrulation and the 'Germ Layers'

The processes by which development moves from a cleavage phase into a phase of genome restriction in somatic cells are usually dramatic. It is at this time that many embryos use directed cell migrations, tissue foldings and the invagination of parts of the surface, to change basically spherical cell masses into tubular, cylindrical or toroidal forms with, usually, three layers. The processes are collectively known as **gastrulation**, and details will be found in nearly all embryology textbooks. In older books gastrulation, especially of vertebrates, is seen as the separation or genesis of the 'three primary layers', **ectoderm** on the outside, **endoderm** lining the gut, and **mesoderm** forming the muscular, connective and other tissues between them; the neural tissue, sometimes called **'neurectoderm'**, was usually lumped, rather unjustly perhaps, with the ectodermal derivatives and not given separate status. In the new view of the two phases of development presented here, these great divisions are seen as deriving from the architecture of the oocyte. The cytoplasm of these groups of somatic cells was originally different and therefore their nuclei started differentiating in three rather different directions from the beginning, if they were in a determinate egg. But if they were in an indeterminate egg like a frog's then the processes of gastrulation pushed these tissues in different directions.

Whereas the older generation of embryologists worried about 'anomalies' like the development of muscle from the 'neurectodermal' iris tissue when muscle *ought* to be restricted to mesoderm, we do not worry about these. Restriction to similar gene arrays could easily be achieved by very different routes, and it would not be surprising if some cells 'mimic' the differentiation patterns of others. In the vertebrates many of the **neural crest** cells (also 'neurectodermal') disperse, and these cells contribute to head mesenchyme, including dermal papillae of teeth and hair, and cartilage of gill bars, as well as nerve and epithelioid cell types. This may be a 'takeover' of some of the functions of the oocyte organisation, by the later differentiation of a special population of somatic cells which modify, and add to, this pattern and give the development still more, controlled, sophistication.

General questions about the increase of complexity in developing embryos have recently found a new class of answer in the topological theorems of Thom[19] ('catastrophe theory'). This is particularly true of the very puzzling quantitative-to-qualitative transformations of gastrulation. Zeeman has proposed new views of amphibian gastrulation[20], based on Thom's idea, which may turn out to be illuminating in many other embryological contexts too, for they are testable.

9.6 The Extra-genic Heredity

In the last chapter we considered maternal effects in early embryology, mostly restricting description to those effects, prior to phyletic stage, which lay down the basic framework of embryonic organisation so that zygote genes may be turned on differentially. There are many other maternal effects on the young animal which form an important part of its inheritance, notably yolk and other food reserves, protection by shell membranes and nest, release at a time when food will be available. All these use the mother's complex physiology to free the developing organism from developmental constraints. In the next chapter we will discuss larval adaptation for larval problems; a maternal inheritance of yolk, protection and good timing *avoids* such adaptation in the very young embryo, which is therefore free to take developmental routes not possible to forms whose inheritance is less ample. Young adders (*Vipera berus*) in England are born in late autumn, with a mass of yolk and a fully-functional nervous system which allows them to find territory and hibernate during the first winter without feeding; some do feed, but they need not because they have been well-fed all summer inside their mothers. Young mantises (*M.religiosa*) from a well-fed mother can live actively for eight weeks before feeding, moulting twice and possibly moving hundreds of metres. Some teleost eggs, like Siamese fighting fish, hatch after only 30 hours at 27 °C, the young fish having non-functional eyes and ears, barely functional vascular system, but an enormous yolk sac and care by father to span the next two days (*Plate 9*). Chapter 12 lists many more examples.

All such tactics of carry-over of the profit of one generation as investment in the next operate independently of the zygote genome, though it is of course true that each offspring thus favoured carries half of each parent's genes.

Animals with many generations each season provide another example of variation in inheritance without reference to the genome (although it must be constantly emphasised that the utilisation of extra-genic benefits or handicaps depends considerably upon the zygote genome). Organisms produced early in the season may have less yolk but a richer food supply (e.g. the planktonic copepod *Calanus*) while those produced at the end of the season may be well-provided with yolk or other reserves, to last over difficult times like winter or the dry season (*Daphnia*, *Mantis*). Parasites with several hosts give eggs very different starts according to the timing of their deposition. More subtly, large broods may impose more stringent conditions upon each member, so that the resulting young all fail in competition with young from smaller broods which have each had more yolk or parental attention. There is therefore an optimum brood size[21], balanced between flooding the gene pool and providing sturdy competitive offspring. Social animals provide other variables; larvae of bees may be fed high-protein diets and turned into queens, or termites may attain different castes according to the recent history of the colony. These lives are all determined, in a sense, by maternal or parental timing rather than by internal factors, and this is clearly true for all organisms. The bitterling, laying her eggs in a living mussel mantle cavity, gives them only as much protection as the male stickleback, but much more dramatically.

However, the social primates do add another factor, **status**[22]. The offspring of low-status mothers are themselves low status, even though most offspring are fathered by the high-status males, who get most chances at all females at each oestrus. Some of these initially low-status animals do achieve highly in later life, and it may be that this is the effect of a 'high-status' genetic endowment. But, like the coiling of *Limnaea* shells (p. 155) mother's status determines that of her offspring in its early months.

REFERENCES

1. Gurdon, 1974
2. Neyfakh, 1971
3. Paul *et al.*, 1968 and 1972
4. Waddington, 1966
5. Spermann and Mangold, 1924
6. Beerman, 1956
7. Turing, 1952
8. Waddington, 1957
9. Jacob and Monod, 1961
10. Harris, 1970 (p.157 *et seq.*)
11. Bonner, 1974 (p.168 *et seq.*)
12. Whittaker, 1974 (p.168)
13. e.g. Harris, 1970
14. Gurdon, 1974
15. Whitehouse, 1973
16. Goodwin, 1963
17. Goodwin and Cohen, 1969
18. Hayflick, 1965
19. Thom, 1972
20. Zeeman, 1976
21. Lack, 1954
22. Brown, 1975

9.7 Questions

1. What is 'cell differentiation'?

2. How are the 'operon' theories inadequate to explain eukaryote differentiation?

3. Cells cannot be made to de-differentiate, but their nuclei can. Why is this?

4. What is gastrulation, and how does it differ in determinate and indeterminate eggs?

5. Does the expression of zygote heredity differ according to season, status or other privilege conferred independently of its genotype?

10

LARVAL FORMS

10.1 Larvae as Survivors and as Evolutionary Successes

We saw in Chapters 1 and 2 that the adaptations required for vegetative survival of organisms are usually very different from those that a few of them will require in order to breed. Let us now consider the growing juvenile and those adaptations which may allow it to survive to become a breeder.

All animals change their proportions as they grow (p. 222). There are very few exceptions to the general rule that young and adults of the same species lead very different lives. Even some of these apparent exceptions, like earthworms and some mammals, are brought into line by consideration of the latter part of the 'embryonic' period as a juvenile adaptation: baby guinea pigs, earthworms and ostrich chicks can fend for themselves only because the helpless phase is passed in mother or egg. It is commonly found that the juvenile and adult adaptations diverge, and it is not rare to find four or more changes of life style during development. As these diverge from each other in the course of evolution, so the transformations, or **metamorphoses**, become more dramatic. Often, as in the axolotl, one of the earlier stages comes to develop gonads and supplants the previous 'adult phase' as the breeding stage. There is good evidence[1] that this has occurred many times during the evolution of animals, and that old larval structures now form adult organs. Then new larval stages have been acquired, emphasising that the whole relationship between juvenile and breeding stages is in a state of continuous flux (*Figures 10.1* and *15.1*, p. 250).

Figure 10.1 AN IMAGINARY SERIES TO ILLUSTRATE THE IMPORTANCE OF LARVAL FORMS IN THE HISTORY OF MAN Hollow arrows link larvae to their adults; solid arrows link (hypothetical) larvae to the successive larval forms. Modern forms have often been used to represent comparable organisms in our evolutionary history. Lower left: ancient echinoderm and its larva, whose structure is the basis for the larval tunicate, whose (modern!) adult is irrelevant to our history. Larval urochordates serve as basis for larval cephalochordate, agnathan, lobe-finned fish, amphibian (what did the larva of Eryops *resemble? Was it perhaps viviparous?). Even the adult early mammal-like reptiles were not a structural pattern for later forms; their young were a more versatile design. Adult specialisations of early mammals were ephemeral, but their juveniles may have had useful adaptations toward a pro-simian. Again, the adult lemuroid forms were not modified towards the anthropoids; only the juvenile structure could be modified toward the large apes and man. Again, adult apes and early man (*Ramapithecus?*) were specialised for particular habits; the hominoid characters of curiosity, long lower limbs, flat face, generalised hands, are all characteristic of* juvenile *modern apes and presumably of our juvenile ancestors too*

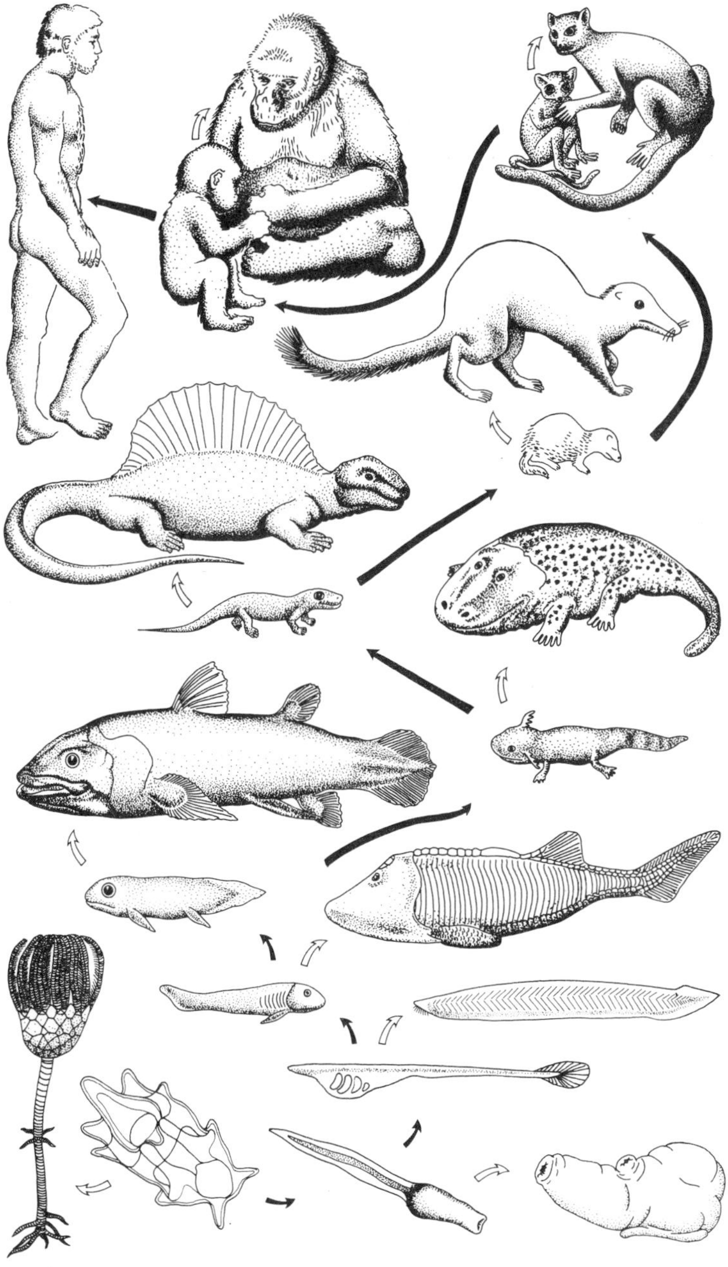

There are a few organisms, however, whose breeding has become 'tied' to a specific adult form and whose metamorphoses cannot therefore be omitted by precocious sexual maturity. This seems to be the case in the modern echinoderms, even though the chordates probably arose as one ancestral echinoderm group who escaped this constraint; the holometabolous insects, with very few exceptions (*Heteropeza*, p. 183) must all metamorphose to the adult in order to breed.

The arthropods all have a chitinous exoskeleton, and this has been cited many times as the reason for their immense success in terms of number of species (insects) and individuals (crustaceans). This has the grave disadvantage, however, that moulting is required for growth to occur, and the animal has no effective skeleton for short periods. In consideration of their life histories, however, such stepwise growth can be seen to have been useful, for the **'instars'** (periods between moults) form natural stages whose adaptations may diverge. This divergence is seen very clearly in the succession of planktonic instars of marine crustacea, discussed more fully below. Here we will consider the insects[2], whose early moults are modified by the presence of juvenile hormone, and whose terminal moult lacks this hormone, allowing the gonads and **imaginal** (adult) characters to appear together.

10.2 The Insects

Some insects hatch as little **'nymphs'** with six legs, head, thorax and abdomen, and most of the other characters of the **imago** except for wings and sex organs. This kind of development is called hemimetabolous or paurometabolous. Successive moults reveal larger wing buds and proportions approaching those of the imago, until at the final moult gonads appear with copulatory, ovipositing, and other secondary sex characters, and the wings are pumped up to spread their membranes and are allowed to dry out. Familiar examples of such paurometabolous insects are the cockroaches and locusts.

Other insects, notably those with aquatic juvenile forms like mayfly and dragonfly, also have a succession of nymphs but these do not resemble the adult. In the case of aquatic larvae they are called **naiads** and have at least some modifications for taking oxygen from water; these may be gills or other special adaptations. There are frequently much grosser special adaptations, like the protrusible 'mask' of dragonfly larvae and the extreme flattening of some mayfly larvae living in torrents, like *Baetis*. In these insects the final moult may be a one-step metamorphosis as in dragonflies, or may have two stages as in the mayflies, with a **sub-imago** succeeded by the true winged imago.

The *Diptera* show a wide variety of aquatic larval stages (*Figure 10.2*). The culicids (mosquitoes) hatch as long-bodied larvae with more superficial resemblance to annelids than to their adults. Their anal tufts may serve for respiratory exchange as they hang head-down in water from the surface film and filter-feed. Some dipteran larvae are midwater predators (e.g. *Corethra*), while others (*Chironomus*) are 'bloodworms' curled up in muddy

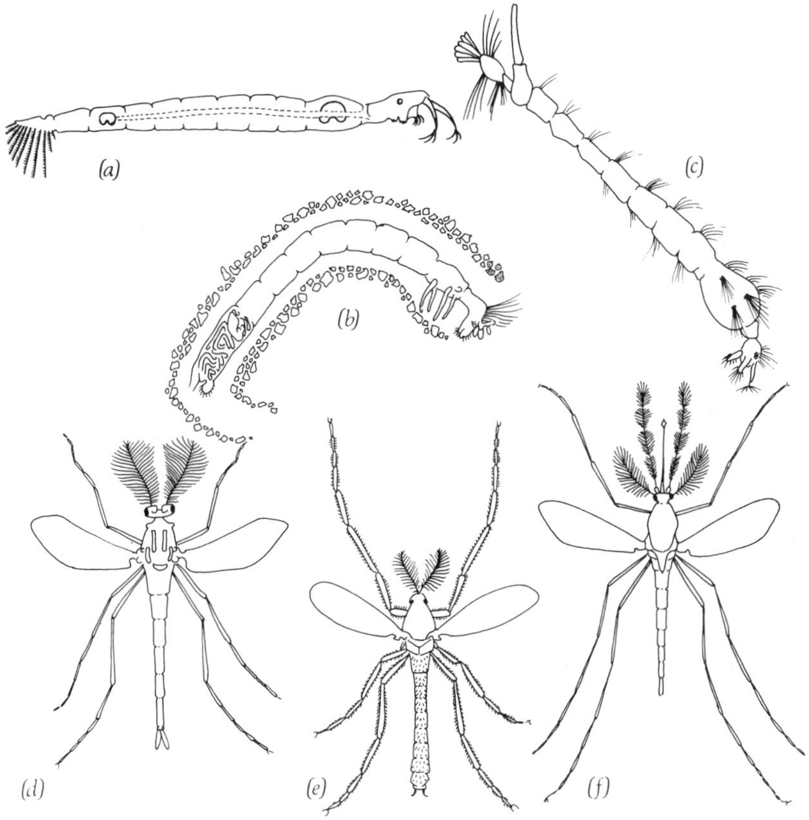

Figure 10.2 LARVAE AND ADULTS OF SOME MOSQUITOES (a) The predatory larva of Corethra *(*Chaeoborus*), the 'ice worm'; the two air-filled floats keep the animal neutrally buoyant and it hangs, transparent, in mid-water. (d) is its adult. (b) The filter-feeding larva of* Chironomus, *the 'blood worm'; it makes a mud-and-mucus tube, and has haemoglobin which may assist in its (posterior) respiration. (e) is its adult. (c) The larva of* Culex, *the 'wiggler'; they hang from the surface film, filter feeding and breathing atmospheric air via posterior spiracles on a 'periscope'. (f) is the (male, non-biting) adult*

tubes on rotten wood on the bottom of ponds. After a succession of moults the penultimate instar is revealed as a **'pupa'**. This is frequently comma-shaped, with respiratory horns on the head-thorax. It leads a fairly active life before finally metamorphosing into the winged adult.

Other *Diptera* (flies), and the *Coleoptera* (beetles), *Hymenoptera* (bees, ants, wasps) and *Lepidoptera* (butterflies and moths), have larvae which do not resemble the adults in the least (except in basic things like presence of chitin, of course). Maggots, mealworms and caterpillars are all extremely specialised larvae which moult and grow but do not approach the adult structure as they do so (*Figure 10.3*). Instead a passive **pupa** or **chrysalis** is formed at the penultimate moult, within which extensive breakdown of larval tissues occurs. Adult organs are formed from **'imaginal discs'** and the whole organism is remodelled to a more characteristic insect pattern.

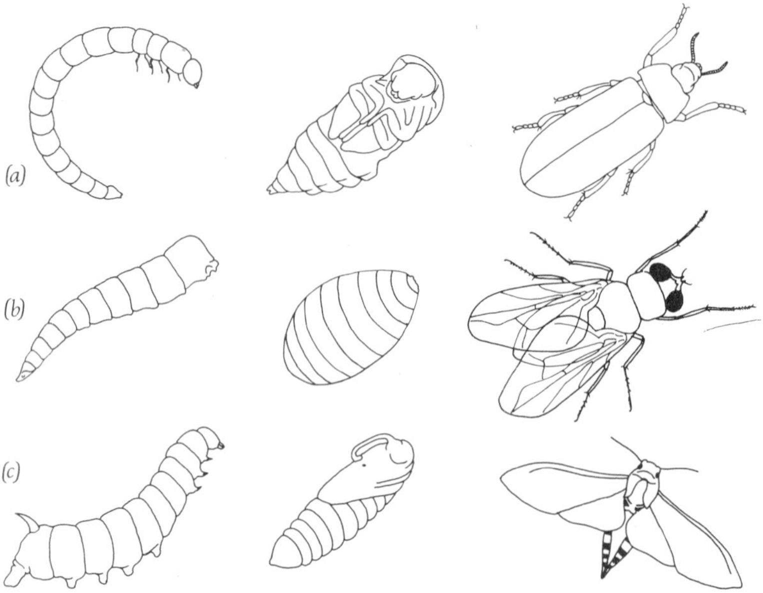

Figure 10.3 STAGES IN THE LIFE HISTORY OF INSECTS (a) The mealworm Tenebrio. *The larva has a long brown body, tiny legs and the ability to starve for up to a year. They lived, probably, in birds' nests and squirrels' dreys before man produced granaries and their populations exploded. The pupa is soft and whitish, and the beetle usually weighs less than the last instar larva. (b) The fly* Musca. *Maggots, pupae and adult houseflies are less familiar now than the larger* Calliphora, Phormia *and* Lucilia *bred for anglers. (c) The convolvulus hawk moth,* Herse. *The larva (caterpillar) has a set of posterior pro-legs as well as the anterior true legs. The pupa has a visible proboscis sheath. This adult moth does feed, on nectar*

Figure 10.4 THE LIFE CYCLES OF HETEROPEZA, *A PAEDOGENETIC FLY (after Engelmann, 1970). (a) The larvae ('female mother-larvae' FM) live in fungus-rotted tree-stumps, where their ovaries produce, parthogenetically, more larvae which hatch directly through the body wall of the mother-larva. When the food supply declines, other kinds of larvae appear. Some larvae become female flies by an orthodox life cycle, and produce eggs on new stumps whether or not fertilised. Other larvae (mfm) produce more larvae, which in turn become male or female flies; some of them (mm) produce larvae which can only become males, and this may be nutrient-determined. (b) The chromosome cycle is even more peculiar. The usual eggs have (about!) 77 chromosomes, and the male-producing eggs grow much larger before ovulation. The nuclei of female-producing eggs divide, and one nucleus degenerates. The other divides, and one of its products moves into the pole plasm, where its chromosome complement is protected and remains at 77. Other nuclear progeny, however, lose chromosomes during successive mitoses till most somatic nuclei contain only 10 (or sometimes 11). In the male-producing egg, the first two nuclear divisions result in four nuclei, three with 38 (or 39) chromosomes, the other with only 10. It is this last which enters the pole plasm as the germ cell nucleus and, as if regretting its losses, fuses with all the polar body nuclei; the resulting composite, 3/4 to 7/8 female derived, has about 58 chromosomes (or less). The somatic cells of the male, however, have only five chromosomes (sometimes six), being descended from early cleavage nuclei with 38 or 39. The lost chromosomes are probably* not *B-chromosomes, and are received by each fly from its mother. The statement that cecidomyids show male haploidy fails to give an adequate impression of this incredible chromosome cycle*

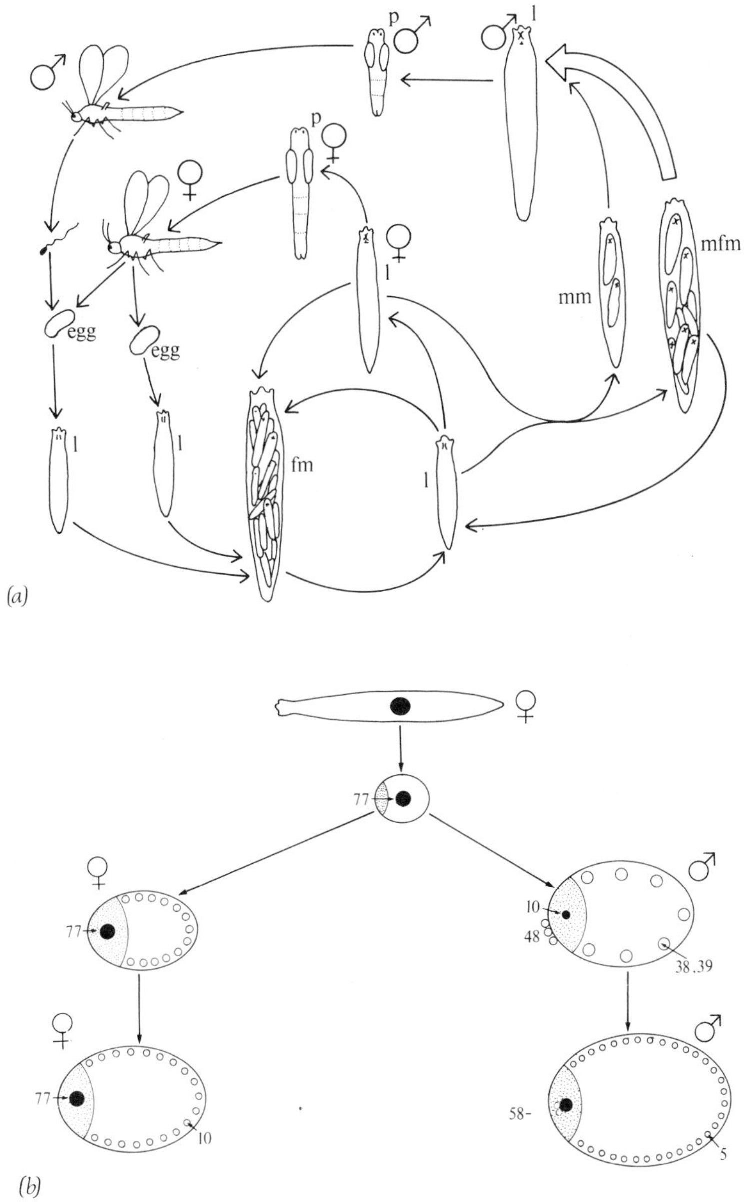

Presumably divergence of the larval stages could have proceeded further but for the necessity of the formation of the adult form for breeding. On the other hand, it is difficult to imagine a greater morphological difference between animals within a single class than that between a maggot and the fly it will become. A few *Diptera* have even struggled successfully against the constraint of producing the imago for breeding; *Figure 10.4* shows the life cycles of *Heteropeza*[3].

10.3 Planktonic Larvae

Impressive as the variety of insect larvae appears, the marine crustaceans can produce examples whose succession of oddities can match the most extreme insects. Rather than producing a list of oddities, I will give three examples in detail, which demonstrate amply the possibilities of divergence toward vastly different ways of life at successive instars. I will consider the life histories of barnacles, *Sacculina*, and the mantis shrimp (*Figures 10.5*, *10.6* and *10.7*).

Barnacles, like many other crustaceans, hatch as **nauplii** (*Figure 10.5* and *Plate 13*). The nauplius larva has only the antennae functional as limbs, plus mandibles, maxillae, and a single eye. It usually does not feed in the first instar, but after moulting is a **metanauplius** with thoracic limbs for catching food. Several instars may be passed as metanauplii of increasing complexity and limb number, then a cypris larva is produced (*Plate 13*). This resembles *Cypris*, an ostracod, and has a pair of lateral shells around the whole body. These can be closed by adductor muscles, but usually gape to allow the anterior thoracic, locomotory limbs to protrude. The more posterior limbs set up filter feeding currents in those crustaceans whose cypris larva feeds. The barnacle cypris finds a settling place

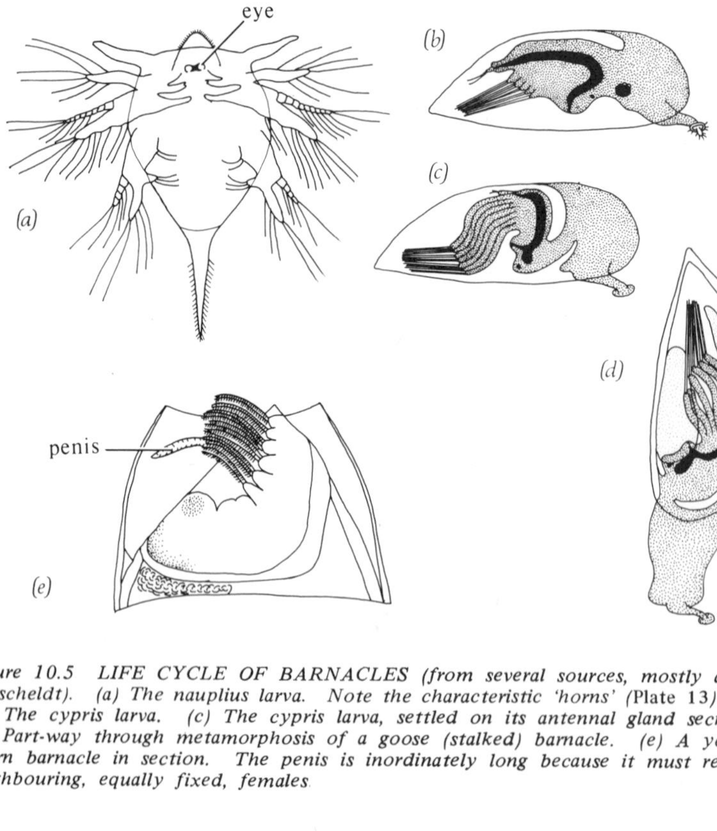

*Figure 10.5 LIFE CYCLE OF BARNACLES (from several sources, mostly after Korscheldt). (a) The nauplius larva. Note the characteristic 'horns' (*Plate 13*) (b) The cypris larva. (c) The cypris larva, settled on its antennal gland secretion. (d) Part-way through metamorphosis of a goose (stalked) barnacle. (e) A young acorn barnacle in section. The penis is inordinately long because it must reach the neighbouring, equally fixed, females*

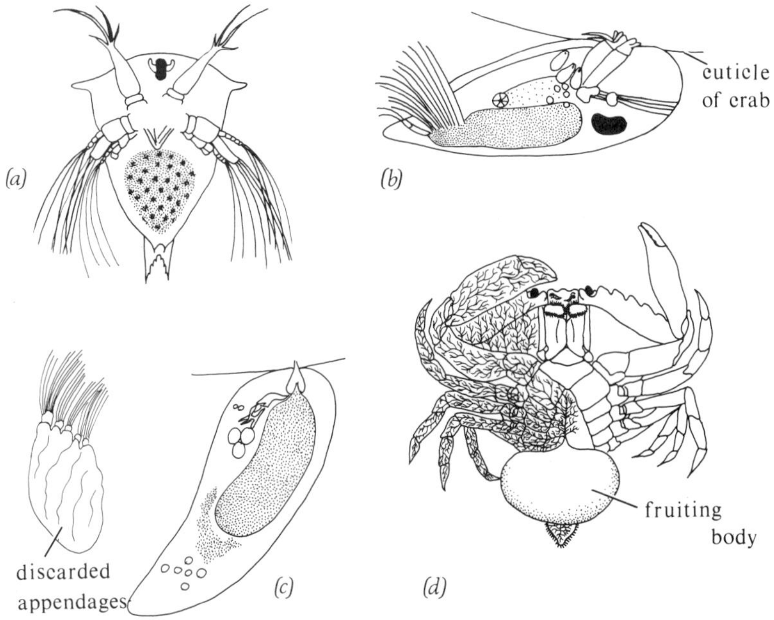

Figure 10.6 LIFE CYCLE OF SACCULINA (modified from Hickman, 1973, after various sources). (a) The nauplius larva. Note the 'horns', showing relationship to barnacles, and the yolk. (b) The cypris larva, attached to the cuticle of a shore crab by its antennal secretion. (c) The cypris moults, becoming a kentrogon which injects a cell mass, via *the antennal connection, into the haemocoele of the crab. (d) The cell mass attaches to the gut, and spreads filaments throughout the crab. After several moults, a fruiting body (*Plate 2a*) appears under the abdomen of the crab, which releases nauplii*

(p. 194) and produces an adhesive disc from a cement gland by the first antenna. After some modifications an acorn barnacle secretes its calcareous plates and a goose barnacle its long 'neck', and they remain 'standing on their heads, kicking food into their mouths with their feet' and growing, via several moults, from this juvenile into adult barnacles.

The parasitic rhizocephaline crustaceans include *Sacculina*[4], commonly found as a large smooth body under the abdomen of the shore crab *Carcinus* (*Figure 10.6* and *Plate 12a*). This body is the reproductive part of the adult, whose vegetative phase extends as rootlets, resembling fungal hyphae, through the body of the crab. It produces vast numbers of eggs in recesses in its surface, and these hatch into typical nauplius larvae, which become metanauplii, then cypris larvae of barnacle type called 'kentrogons'. These have a special aggregate of cells near the base of the antenna. These kentrogon larvae settle on their heads, very like barnacle juveniles, but their settling place is the cuticle of shore crabs. However, only the odd antenna group of cells has any future. This penetrates into the crab and forms the rootlet system of the adult *Sacculina*. The rest of the larva probably falls off the crab and dies after its mission has been accomplished. The rootlet system penetrates all the tissues, castrates the

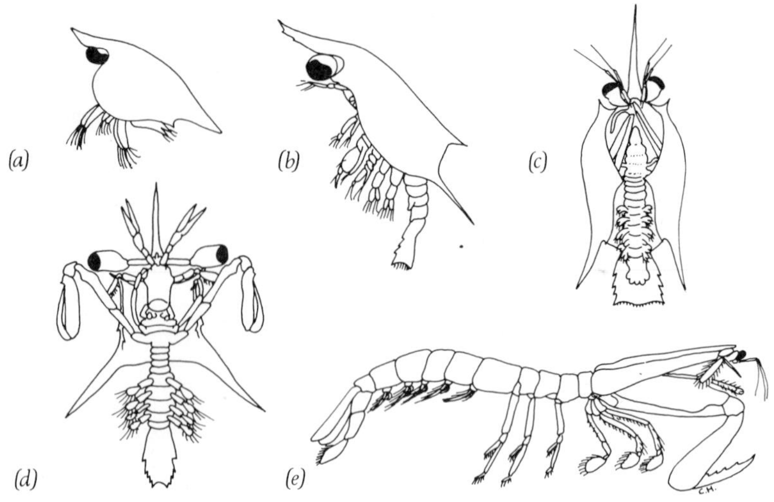

Figure 10.7 THE LIFE CYCLE OF THE MANTIS SHRIMP SQUILLA MANTIS
Like many large marine crustaceans a series of different larvae succeed each other. (a) The metanauplius hatches, but does not feed. (b) The pre-zoea larva is a planktonic filter feeder. (c) So is the protozoea, which has abdominal limbs. (d) The antizoea is predatory, among the plankton. It has very drawn-out eye-stalks, spines, limbs and rostrum presumably to aid in flotation by slowing the sinking rate. (e) The adult, a mud-dwelling predator and detritus feeder

crab and later produces its own fruiting body. This is the female; kentrogons which find this stick to it, metamorphose into tiny males, and inseminate it. *Sacculina* demonstrates retention of the larval stages of its barnacle-like ancestors, but substitutes a totally new juvenile instead of the young settling barnacle. This in turn produces a wholly new kind of adult crustacean. Such a change of final form is apparently not possible to insects, which depend on the final moult, in the absence of juvenile hormone, to stimulate the gonadal maturation and the adult organs which go with it. Among the crustaceans on the other hand, there are many cases of 'dropping' the terminal stages of a life cycle, the new adult form resembling the old larva and producing a radiating group of organisms of the new pattern (cf. *Figure 10.1*). Although the ostracods and the copepods are ancient groups, there is a good case for considering them to be representative of the **larvae** of their ancestors, the old adult form, perhaps a trilobite, having been lost. The alternative, that new and different sexual adult forms have been added to the later stages of ancient ostracods to produce barnacles, is much less likely[1].

Figure 10.7 shows the life history of the mantis shrimp (*Squilla*) which begins with a kind of nauplius. The free-living part of the life cycle of the advanced decapod crustaceans, crabs and lobsters, does *not* begin with a nauplius. This stage, with only a few anterior appendages, is spent in the yolky egg, held on the swimmerets of the mother. A **zoea** hatches; this has thoracic appendages for locomotion and frequently has long cephalic and posterior thoracic spines for flotation (*Figure 10.8*). There

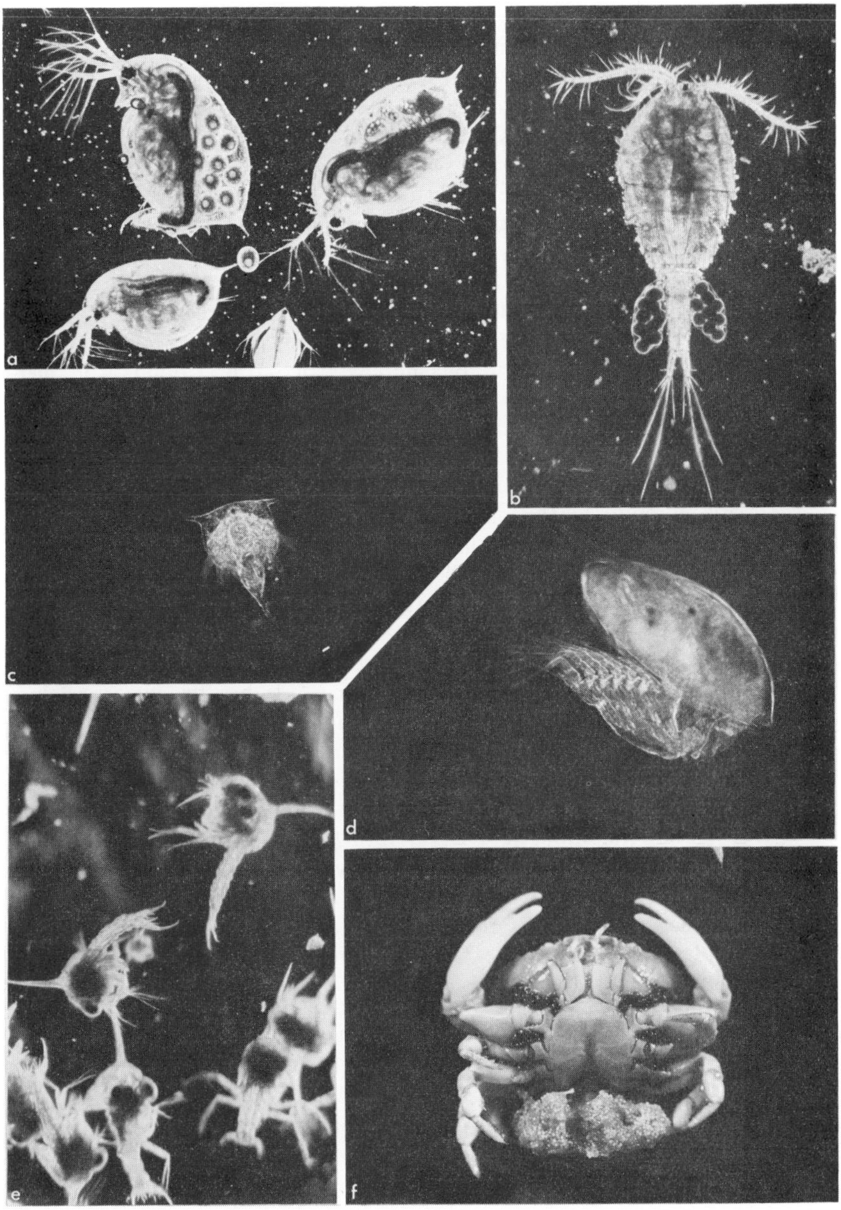

Plate 13 REPRODUCTION OF CRUSTACEANS (a) Some Daphnia, *taken from a pond in September. Female, upper left, has resistant winter eggs. Female, upper right, has five yolky baby* Daphnia *in the pouch (compare* Figure 16.9*). (b) Female* Cyclops *with a pair of egg-sacs. (c) Nauplius larva of a barnacle (compare* Figure 10.5*). (d) Cypris larva of a barnacle, at the same magnification as (c) (compare* Figure 10.5*). (e) Zoea larvae of* Carcinus, *the shore crab (compare* Figure 10.8*). (f) A female* Xantho *'in berry' i.e. carrying eggs on the abdominal appendages*

may be two or three zoea and **metazoea** instars, perhaps followed by a shrimp-like **mysis** larva, whose life style is very like actual mysids which live in the plankton. Crabs become **megalopa** larvae, the little crawfishes become **palinurus** larvae, whose enormously attenuated appendages, including eyestalks, retard sinking. From this fantastic creature (*Figure 10.8*) moults a tiny but ordinary crawfish about 2 cm long, which lives the rest of its life on the sea-bed.

Many other kinds of marine animals have planktonic larvae[5]. Diagrams of some of the commoner forms are shown in *Figure 10.8*, and some photographs in *Plates 14* and *15*.

The plankton has a number of advantages over other habitats in the sea, especially for those organisms whose reproductive strategy involves the production of enormous numbers of larvae. Food is in plentiful supply, basically from the primary productivity of the phytoplankton but often actually ingested in the form of unsuccessful larvae of other species. When the adult organism requires a minimum size for effective life (e.g. many special carnivores like *Squilla*, and some filter feeders like oysters) larval growth in the plankton removes competition from the adults until the young are large enough to compete successfully (*see* Chapter 13). So the young may attain this size without losing out to parental competition, and the adults do not compete with enormous numbers of juveniles for food or territory.

The zooplankton is not composed entirely of temporary residents, the larvae of pelagic and benthic creatures. It has its own special fauna, and these usually also have planktonic larvae. Examples are the copepods, whose nauplii always abound, and *Sagitta*, the arrow worm, whose juveniles are found seasonally. But many organisms which feed in the plankton as adults do *not* have planktonic larvae, and they emphasise the larval/adult separation function of a larval form. Most coelenterate medusae have a polyp stage (or a **scyphistomal** equivalent) on the sea-bed, from which the adults are produced asexually; some polychaetes have temporarily planktonic terminal sexual forms produced asexually from benthic juveniles, but these in turn may have come from planktonic larvae (e.g. the palolo worm *Eunice*).

10.4 The Structure of Larval Forms

Larval forms become more modified and specialised, i.e. diverge more from the ancestral larval form, as generations pass and the better-adapted larvae remain as the survivors and some become breeders. Any stage of the life history can diverge and produce a larva. There is indeed no firm line which can be drawn between oocyte specialisations, which affect the 'behaviour' of the fertilised egg, and early zygote specialisations. This is because the latter, if before the phyletic stage, depend upon this same oocyte structure. For example, the oil droplet in the planktonic eggs of many pelagic fish like the herring is actually put into the oocyte while it is in the ovary, and this is easily seen as a maternally-derived character; but so is the ciliated girdle and apical tuft of the trochophore larva, for

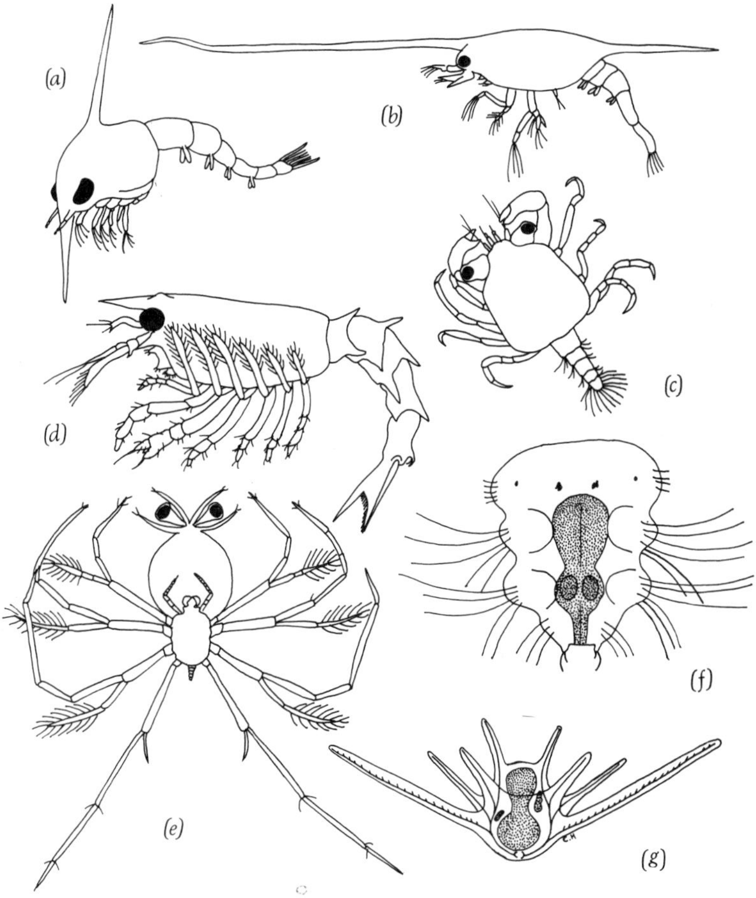

Figure 10.8 A VARIETY OF PLANKTONIC LARVAE *(a) The zoea of a shore crab,* Carcinus. *(b) The zoea of a swimming crab,* Portunus. *(c) Megalopa larva of a crab. (d) Mysis larva of the lobster. (e) Palinurus larva of the crawfish. (f) Late trochophore larva of a marine polychaete. (g) The ophiopluteus or basket larva of a brittle star (ophiuroid)*

it is produced even if the nucleus and cytoplasm never cleave[6], and the nucleus in this case seems totally inactive.

It would seem simplest, then, to restrict the term 'larva' to forms which appear later than the phyletic stage, so that we might reasonably suppose zygote genetics to control larval development. This restriction would relegate oil droplets and other pre-phyletic adaptations to the same category as yolk, egg-shells and so on. Unfortunately for this simplicity, the most famous larvae, those which exhibit common features across a phylum, *are* effectively the phyletic stage (*Figure 10.9*); the various species approach this from different eggs and diverge again towards their several adults.

So there must be, for our purposes, two kinds of larvae (*Table 10.1*). Those which appear earlier in the life history, like the trochophore,

Table 10.1 Larvae

Phylum	*Class or order*	*Larvae found*	*Kinds of larva (*phyletic)*	*Referred to*
Protozoa	*Suctoria*	Always	'Ciliate'* swarmer	*Figure 10.10*
Mesozoa		Always	'Infusoriforms' nematogens	
Sponges	*Calcarea*	Always	Amphiblastula*	
Coelenterates	*Hydrozoa*	Nearly always	Planula*?	p. 221, *Figure 13.2*
			polyp	p. 188, 221
	Scyphozoa	Usually	Scyphistoma	p. 188
			ephyra	
	Anthozoa	Usually	Planula	
Flatworms	*Monogenea*	Nearly always	Onchomiracidium*	*Not* Gyrodactylus
	Digenea	Always	Miracidium	*Figure 4.10*, p. 192, *Plate 14*
			Redia	*Figure 4.10*, p. 89
			Cercaria	*Figure 4.10*, p. 89
			Metacercaria	*Figure 4.10*, p. 192, *Plate 16*
	Cestoda	Always	Onchosphere	*Figure 4.11*
			Hexacanth	*Figure 4.11*
			Cysticercus	*Figure 4.11*
			Procercoid	*Figure 4.11*
			Pieroceroid	*Figure 4.11*
			Hydatid	*Figure 4.11*, p. 90
Nemertine		Usually	Pilidium*	*Figure 10.9*, p. 90, *Plate 14*
Aschelminthes	*Nematoda*	Usually	3rd 'instar'*	p. 174
	Acanthocephala	Usually	Acanthor*	
	Gordiacea	Always	'False acanthor'*	
Ectoprocta		Usually	Cyphonautes	
Phoronida		Usually	Actinotroch*	
Molluscs	*Polyphacophora*	Usually	Trochophore*	
	Gastropods	Nearly always	Trochophore*	p.151
		Possibly in egg capsule	Veliger*	*Figure 10.9*, *Plates 14, 16*
	Lamellibranchia	Always	Trochophore*	
			Veliger*	*Figure 10.9*
			Glochidium	p.192, *Plate 16*
Annelids	*Polychaeta*	Usually	Trochophore*	*Figures 10.8, 10.10*, pp.41, 151, 188, *Plate 14*

Phylum	Class or order	Larvae found	Kinds of larva (*phyletic)	Referred to
Arthropods	*Merostomata*	Always	Trilobite larva*	
	Crustacea	Usually	Nauplius	*Figures 10.5, 10.6,* p. 184
			Metanauplius	*Plate 14f*
			Zoea	*Figures 10.7, 10.8,* p. 186
			Megalopa	*Figure 10.8,* p. 188
			Palinurus	*Figure 10.8,* p. 188
	Insecta	Nearly always	'Nymph'	p. 180
			'Grub'	*Plate 16*
			Aquatic larvae	*Plate 16,* p. 180
			Caterpillars	*Figure 10.3*
			(Pupae?)	p. 181
Echinoderms	*Asteroidea*	Nearly always	Dipleurula*	pp. 151, 162
		Always	Bipinnaria	p. 192
			Brachiolaria	*Plate 15*
	Echinoidea	Nearly always	Dipleurula*	*Plate 12*
			Echinopluteus	*Plate 15,* p. 158
	Holothuroidea	Always	Dipleurula*	
			Auricularia	*Plate 15*
	Crinoidea		Various complex larvae	
	Ophiuroidea	Always	Dipleurula*	
			Ophiopluteus	*Figure 10.8*
Hemichorda		Nearly always	Tornaria*	*Plate 14*
Chordates	*Urochordata*	Nearly always	'Tadpole' larva or appendicularia	*Figure 10.10, Plate 15,* p. 192
	Teleostei	Often		p. 192
	Amphibia	Usually	Tadpoles	p. 195
			'Newt- and salamander-poles'	
	Birds	Many	Altricial young (nest)	
	Mammals	Many	Babies 'nestlings'	p. 196
			Infants children	p.196

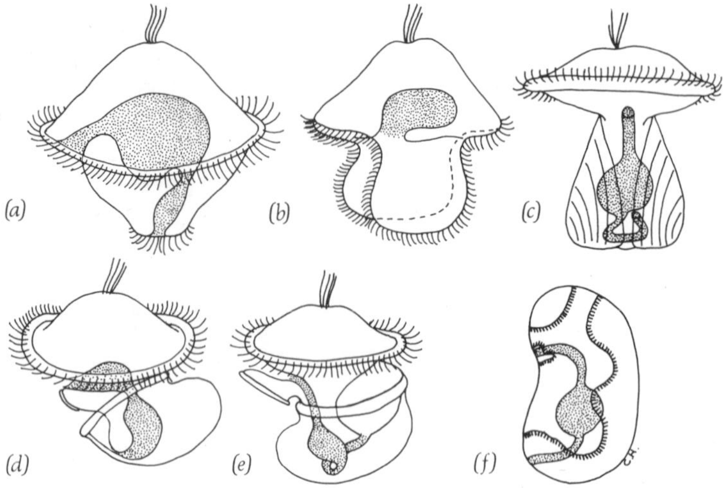

Figure 10.9 SOME 'PHYLETIC' LARVAE (a) Early trochophore larva of a marine worm like Pomatoceros. *(b) Pilidium larva of a nemertine worm. There is no anus, and the juveniles are formed in an 'amniotic cavity'. (c) The 'veliger' larva of a bivalve (lamellibranch) mollusc like the oyster, viewed from the anterior end to show the paired shells on either side of the gut. The prototrochal girdle of cilia remains to support the organism in the plankton. (d) The veliger larva of a marine or freshwater gastropod, for example* Littorina *or* Limnaea, *before torsion. The gut is straight, and the prototroch cannot fit into the shell. Some veligers have a pair of velar lobes with cilia instead of retaining the trochophore's girdle. (e) A gastropod veliger, after torsion. The soft ciliated portion can now be pulled into the shell, and the operculum can close it. (f) The 'dipleurula' stage of echinoderms (*Plate 12*). There is a ciliary band around the mouth, one pre-oral and one dorsal. The chordates could have arisen from such a larval type*

dipleurula, miracidium and tadpole larva, represent the phyletic stage of their phylum, and are a developmental *constraint* on their organisms; divergence has occurred before and after this stage. The later larvae, like the glochidium of *Anodonta*, bipinnaria of starfish, metacercaria of digeneans and ammocoete of lamprey represent later divergences, adaptations to peculiar modes of life (*Plate 15*).

The two kinds can be distinguished easily if the life history of several related forms is known. The first kind, a **phyletic larva**, is morphologically very similar in related forms, although this may be confused by differences in lecithy (yolk distribution and amount) of the eggs; topological equivalence not morphological identity should be sought, as in the trochophore of polychaetes and the similar stage of the earthworms. The second kind of larva will usually differ as much as, or more than, the adults, like the various later echinoderm larvae, tadpoles of frogs, or the larvae of mayflies.

No detailed description of the great variety of larvae can be presented here. Zoology textbooks[7], or works on the plankton[5] should be consulted, and the organisms themselves; if possible the living organisms or a cine-film should be used, rather than a preserved specimen. Few planktonic larvae, especially phyletic larvae, preserve well.

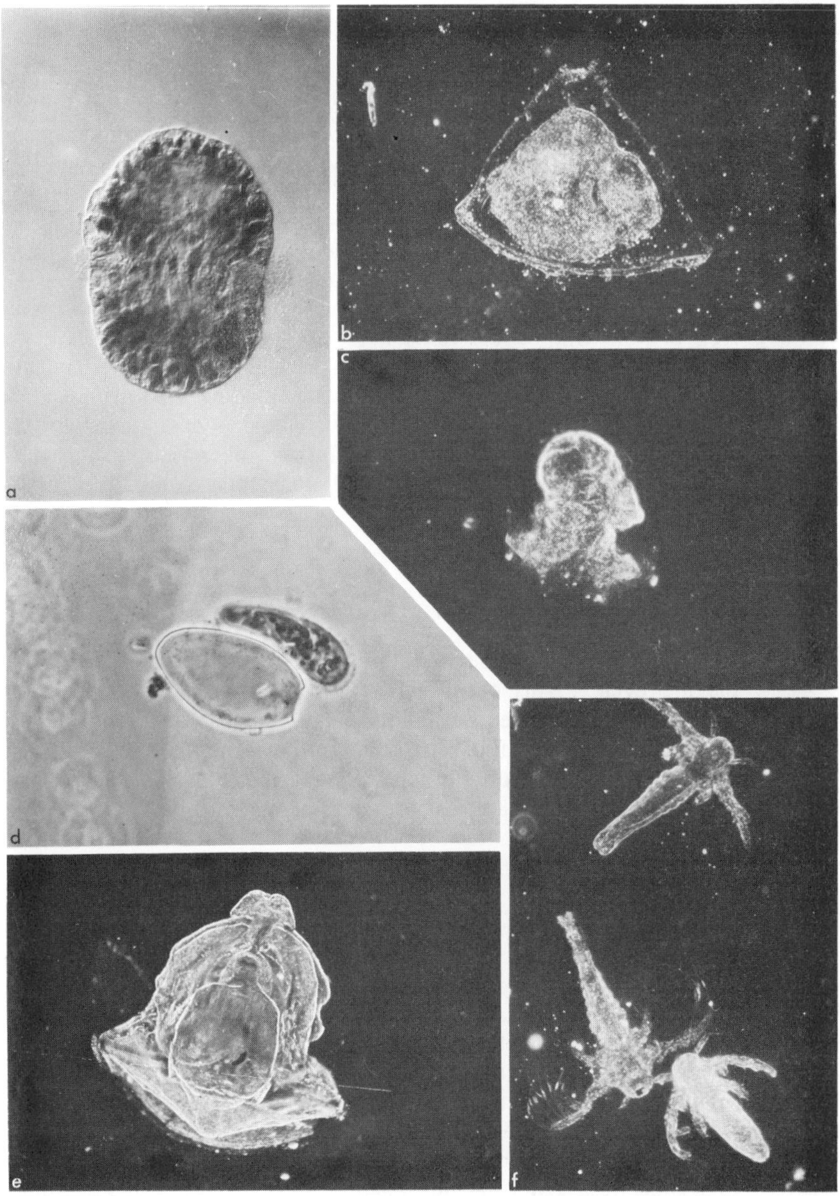

Plate 14 PHYLETIC LARVAE (a) A trochophore of Endalia. *Note the central girdle of cilia, the prototroch. (b) The pilidium larva of a nemertine. (c) The early veliger of a marine gastropod. The yolk-filled visceral hump is at the top and the two velar lobes below (compare* Figure 10.9d*). (d) Miracidium of the liverfluke* Fasciola, *with the egg case from which it has just hatched. Note the cilia, and eyes (at right) (compare* Figure 4.10*). (e) Tornaria larva of* Balanoglossus, *a hemichordate. (f) A nauplius (bottom right) and two metanauplii of* Artemia, *the brine shrimp*

10.5 Metamorphosis

'Metamorphosis' said Bernard Miles[8] of the Greek Gods 'is the art of changin' yourself into somethin' else in order to get around the women!' and it is this among more earthly organisms too. Although there may be several dramatic changes of morphology during the life span, as in most arthropods, only the final change to the breeding form, the **imago**, should be referred to as **metamorphosis**. Again, there are innumerable examples and we will give but four. These are a protozoan *Dendrocommetes*, a tapeworm *Schistocephalus*, a polychaete worm *Pomatoceros*, and a sea squirt, *Styela* (*Figures 4.11*, p. 90 and *10.10*). Neither an insect nor a frog tadpole will be described here, as their metamorphoses are described in detail elsewhere[9], and description here must be limited in space. Therefore I have chosen four less well-known examples which illustrate facets of metamorphosis; the reader may draw on his own knowledge of the better known forms, while bearing in mind the problems of changing from a highly adapted vegetative, to a highly adapted sexual, form in order to breed.

10.6 Site Selection

It is frequently the case that the larval form is the distributive phase of the life cycle. This occurs where the adult is sessile and the larva planktonic, and also where the adult is restricted to a territory and the young are excluded from adult territories as they mature. Site selection by the young, then, both 'tops up' the present range of the species and is a constant attempt at colonisation of marginal, or new, environments. Only one example will be described in detail here, because much of Chapters 13 and 14 are concerned with this general area. The settling of the planktonic acorn barnacle larvae of two genera, *Balanus* and *Cthalamus*, shows the settling technique used as a major weapon in the competition for sites on tidal rocky shores. Each cypris swims actively near the surface except for short spells, of some minutes duration, near or on the substrate. Preferred settling sites are near, but not on, members of their own species, but they will also settle on the actual valves of adults of the rival species[10]. The consequence of this settling tactic can be seen on any rocky beach (*Figure 10.11*). Competition for sites produces rocks completely covered with one or other species; where they meet there is no merging, but a 'no-man's-land' with a few tiny specimens only. The adults differ in that *Cthalamus* resists drying, but *Balanus* seems to be a better competer with the other filter feeders of the littoral zone and can tolerate predation by dog whelks etc. (*Figure 10.11*).

10.7 Children

Most larvae do show a dramatic metamorphosis to the adult stage, because divergent evolution has occurred. But in some cases there is no sudden

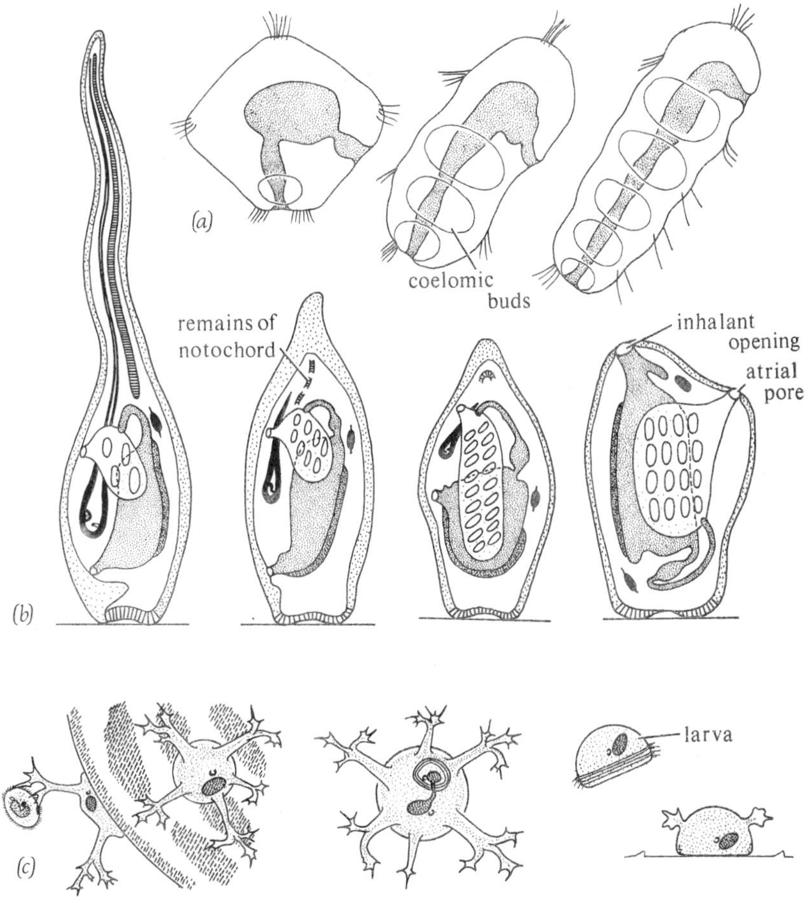

Figure 10.10 METAMORPHOSIS OF THREE LARVAE (a) The trochophore of Pomatoceros *adds to its animal–vegetal length by division, in the skin, of the progeny of cell 2d; in the gut, by division of the fifth quartet of micromeres; and between gut and skin by the production of a series of syncytial 'bubbles', the coelomic 'buds'. These expand till they meet to form the septa and the dorsal and ventral mesenteries supporting the gut. (b) The transformation of the tadpole larva of an ascidian, like* Styela, *into the sessile adult. Notochord and nerve cord degenerate and the atrium expands around the pharynx, whose gill openings multiply (compare* Figure 4.9*). (c) The protozoan suctorian* Dendrocommetes, *found on the gills of* Gammarus *the freshwater shrimp. Some are found on the gill-edge, some on the face (*Plate 3*). An internal bud becomes ciliated, escapes and swims about like its ciliate relatives before settling down, casting off its cilia and growing hollow tentacles to catch its ciliate food*

metamorphosis because a lengthy transition period is possible. This may be stepwise, as in such arthropods as the hemimetabolous insects and the crabs whose metazoea produces a megalopa larva, which is morphologically a tiny crab with extended abdomen whose every moult takes it nearer the adult form. Or the transition may be gradual, as in most vertebrates except lampreys and anuran amphibia. The urodeles provide a prime example: external gills slowly lose function and atrophy, legs are used as well as tail for locomotion, and the adult size is slowly attained; nevertheless the juvenile is certainly a true larva. Again we cannot draw a firm line, demarcating all forms with 'true' larvae. In many ways baby humans should be considered larval: they have different proportions from the adult, a different diet, and indeed are highly specialised for a parasitic existence not only before but also after birth. As all successful parasites evolve to assist their hosts, so primate babies confer status upon their mothers and so gain privilege.

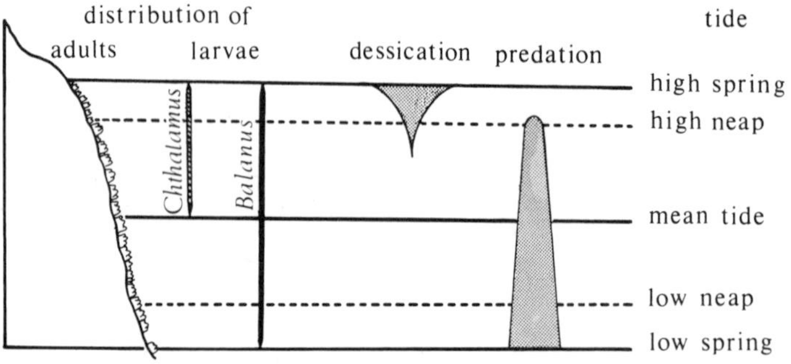

Figure 10.11 DISTRIBUTION OF THE ACORN BARNACLES BALANUS *AND* CTHALAMUS *(from Salthe, 1972, after Connell). Settled* Cthalamus *are found only above high neap tide, presumably because they tolerate physical stress (desiccation) but not biological competition with* Balanus *or predation by dog whelks. The larvae are found to overlap the adult distribution considerably, and this argues for continuing competition (see text). So does the population density of settled barnacles and the high ratio of juvenile (tiny) to adult (large) individuals, not shown in this diagram*

Among the lower primates the mother with her newborn infant attains a status only slightly below that of dominant males, and they are constant subjects of grooming and apparently of sheer curiosity by the entire group. There is evidence of this in human society, where babies are frequently passed around among women until they cry, when they are promptly returned to the mother. Much advertising and several industries are devoted to this parasitic subgroup of western society. With the passing from fashion of human lactation, baby-foods have filled the gap in great profusion; and with the acquisition of a new set of sanitary habits, the nappy has become the symbol of the western human larva.

The decline in neonatal mortality as a result of public health measures, and perhaps of medical practice, has given each child a new value. Emotional investment can begin at birth, whereas in less advanced communities the child may not even be named until he attains his first birthday. The probability that children will mature has already 'caused' a great decline in birth rate in western society, and it is hoped by many that the same will be true in South America and North Africa where too many babies are now maturing.

The special status of **children** in society is very interesting and reminiscent of a larval state in many ways. In more primitive societies they become involved in adult activities as they become capable; they may help mother gather crops or prepare food, from three or even two years

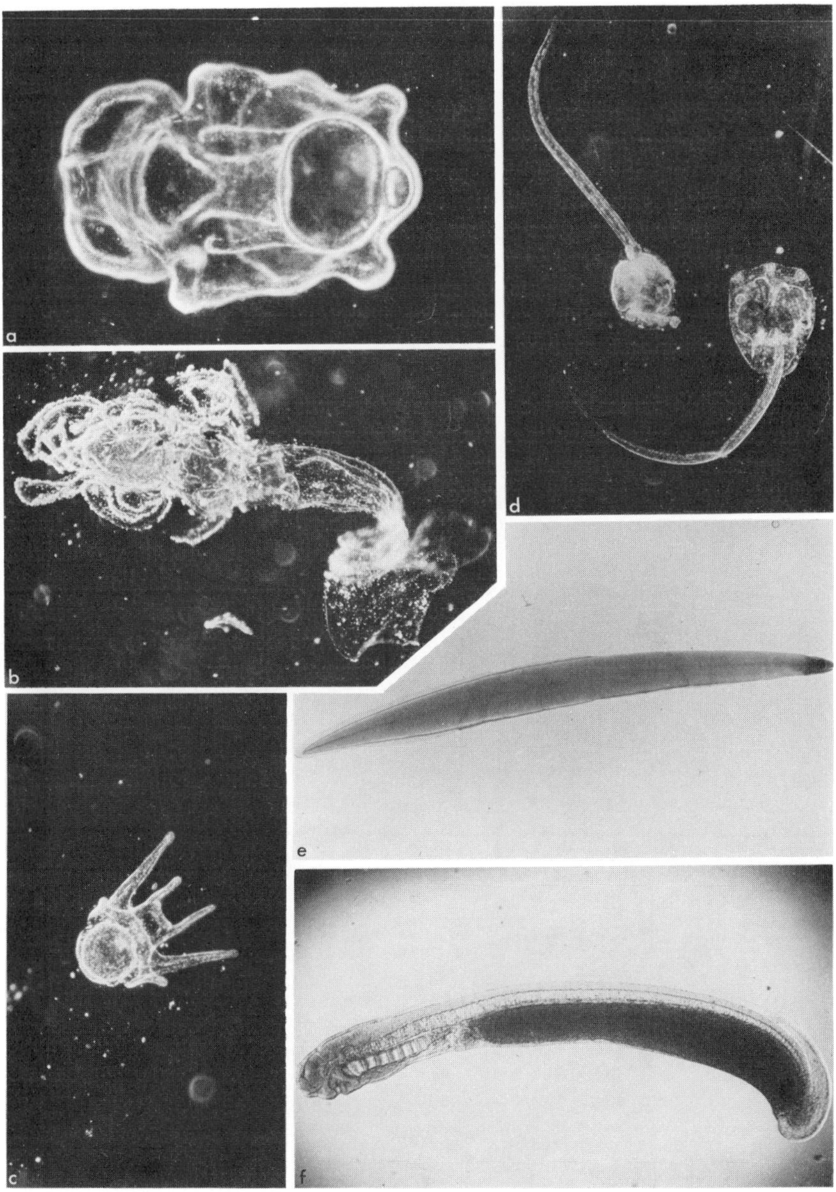

Plate 15 LATER LARVAE *(a) Auricularia larva of a sea cucumber. Note the tortuous ciliary bands (white). (b) Brachiolaria larva of a starfish. Note the extension (right) which carries even more cilia. (c) Echinopluteus ('basket') larva of a sea urchin (compare* Figure 10.8g*). (d) Two tadpole larvae of an ascidian. There are already pharyngeal buds of two new individuals in the older specimen (compare* Figures 4.9 *and* 10.10b*). (e) Leptocephalus larva of a conger eel (compare* Figure 3.1*). (f) Ammocoete larva of a freshwater lamprey. The gills can be seen (left) and the yolk-filled gut (dark)*

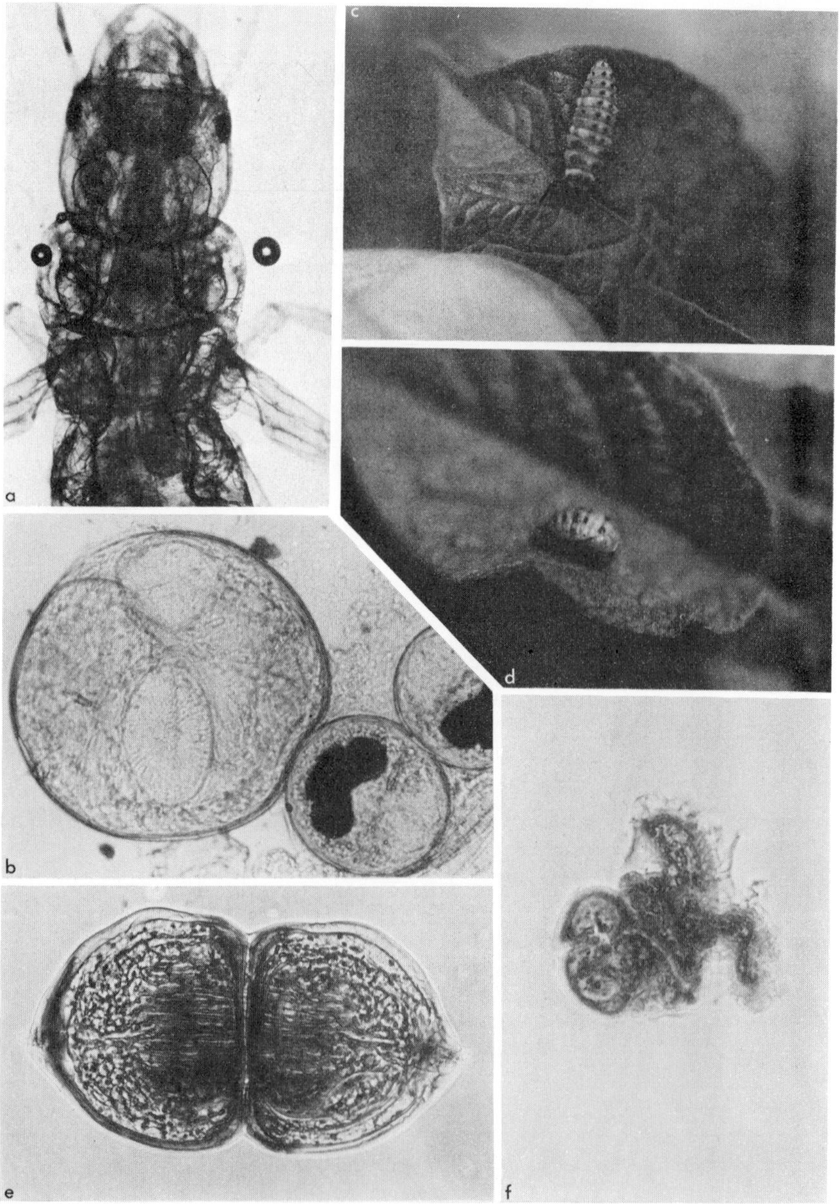

*Plate 16 PECULIAR LARVAE (a) Larva (naiad) of the mayfly, (*Sialis lutaria*). Note the three spheres: one behind the left eye, one central in 'segment' 2, and one to the right of midline in 'segment' 4. The two dark rings are bubbles. (b) These spheres in more detail. They are metacercariae of the flukes* Phyllidistomum folium *(left) and of a plagiorchid. (c) Larva of the seven-spot ladybird. (d) Pupa of the same species. (e) Glochidium larva of* Anodonta, *the fresh water mussel. This develops in mother's gills then swims up to a fish, to whose skin it attaches (note the hooks on the tips of the shells). Weeks later, it drops off to start adult life away from the parent. (f) Late veliger of a gastropod. Torsion has occurred. The visceral mass (and transparent shell) are to the left, ciliated velar lobes at right (compare* Plate 14c *and* Figure 10.9e*)*

old. Toys in such societies usually have direct relevance to adult roles either in 'work' or 'games'. There are traditional toys in our society too, which have presumably lasted from this stage of our culture, like dolls and 'Wendy-houses' for girls and rocking horses and toy weapons for boys; but some modern toys are 'educational' or entertaining without direct links with adult life.

Children have special clothes as they grow up, and pass through very circumscribed environments, in which nearly all activities are mandatory or proscribed, as in tribal life. Several such environments usually coexist in our complex society, so that a child must have roles for home, for school, and for his peers which are different in a number of important ways. His language usually reflects these differences, as well as more subtle behaviour. These parallel lives have their own cultures, too, which are to a great extent independent of adult occupations. The Opies[11] have collected many examples of the peer-group culture, in jingles and reiterative parables whose roots go back many centuries. Nursery rhymes are sometimes from such child culture, transmitted through mother or babyminder as well as between children, and edited in the process.

Each generation of adults fails to comprehend the motives, the rules, the constraints of their young. It is fashionable now to blame rapid technological change for this misunderstanding, but it is certain that Shakespeare knew of it, and Dickens. Even in times of political and social stability the adolescents, at least, were occupied with a continuous juvenile culture, running parallel with the adult culture but with its own rules and history.

REFERENCES

1. De Beer, 1958
2. Chapman, 1969
3. White, 1954
4. Hickman, 1973
5. Hardy, 1956
6. Lillie, 1902
7. Barrington, 1967
8. Miles, c.1950
9. Etkin and Gilbert, 1968
10. Knight-Jones and Moyse, 1961
11. Opie and Opie, 1959

10.8 Questions

1. Why do the adaptations of larvae and adults diverge?

2. What characterises a phyletic larva?

3. Why are phyletic stages resistant to evolutionary divergence?

4. What are the advantages of planktonic larvae for benthic adults?

5. Many parasites use larvae, sometimes multiple larval forms, as infective stages. What advantages has this over use of egg or adult stages?

6. Which organisms use larval multiplication as a reproductive tactic?

11

VIVIPARITY

11.1 Oviparity and Viviparity

This is a classical distinction made by all books which deal with reproduction, and it does bring to the reader's mind a useful contrast. Oviparous animals lay eggs, viviparous animals 'bring forth living young' i.e. development has started when they are released from the females. But further examination shows that the distinction is not really very clear. The egg of birds, the prime example of oviparity, always develops during its passage through the female tract, while it is acquiring tertiary membranes; the chicken's egg blastodisc has hundreds or thousands of zygotic nuclei when it is laid. *Table 6.1* (p. 120) shows Amoroso's series[1] of oviparity through ovoviviparity to viviparity in the elasmobranchs. Other groups doubtless could provide similar series, in which related species differ in the relative amounts of nutrient they supply via yolk or via accessory structures.

Some groups, like the insects, rarely leave all yolk out of the eggs, even in viviparous forms, probably because their development requires it for architectural reasons; those few which have very little yolk, like some parasitic *Hymenoptera*[2], show great modification of development usually leading to polyembryony. Even mammals, viviparous *par excellence*, retain the yolk sac of the telolecithal reptilian ancestor, for it is still necessary for the architecture of the blood vessels of the hepatic portal system.

A group which has not evolved viviparity is the birds, although the emperor penguin achieves some of the advantages by leaving the male incubating the egg, and additional advantages by leaving the female free to feed at this time unencumbered by egg or young. It seems obvious that birds would be encumbered by internally developing young, and this is the reason usually given for their oviparity. But some flying insects, like the tsetse fly *Glossina*, contrive to carry a relatively enormous larva inside, which pupates as soon as it is laid so that the whole period of growth is intrauterine. The bats seem to manage viviparity, and they fly very well even when pregnant. Birds, on the other hand, are restricted by their breeding activities, and the female hornbill plastered into the hollow tree is tied down with her progeny much more effectively than any female mammal. So another explanation should perhaps be sought for the universal oviparity of birds.

Glossina, the tsetse fly has already been mentioned; *all* the prepubertal growth depends on nutrition acquired via the mother. The maggot when born simply burrows and pupates, then weeks later the adult fly, full-size of course, emerges. This would seem to be viviparity carried to greater extremes even than some hystricomorph rodents like the guinea pig, or ungulates like the gnu, in which the young are almost completely capable at birth. But if this is extreme viviparity, then *Gyrodactylus* (*Plate 2* and p. 42) must be considered hyperviviparous. Adults are bisexual, and in the 'uterus' each normally carries a complete offspring, in whose uterus is usually a further generation[3]. Newly delivered animals with an empty uterus seem much more rare than simple arithmetic would suggest – most individuals are triple Chinese boxes. It is probable that the sperms from one copulation inseminate all three generations, and that self-fertilisation is not usual. Birth may occur seasonally or in response to rare stimuli, perhaps the transfer of the parent to another fish. This is a very useful reproductive strategy, for even if only one *Gyrodactylus* transfers to the skin of an uninfected fish, it will carry genetic contributions from three or more sources; these are its own genetics, that of the (probably different) fathers of its first and second generation offspring, and that of the sperms it carries; a genetically heterogeneous group is transferred to start the colonisation of the new host. (It has now been suggested, however, that the series of apparent generations may be 'internal siblings' produced by larval multiplication; is this viviparity?)

11.2 Ovulation Rhythms

It is obvious that the presence of developing young puts some constraints on ovulation, and we will consider some mechanisms by which this can be brought about when we examine mammalian systems in more detail (Chapter 16). Here several examples will suffice to show possible interactions of ovulation and pregnancy.

Some organisms like the shark *Lamna* continue to ovulate even during pregnancy, and the yolky eggs are used as food (**embryotroph** or trophic eggs, *see* p. 121) by the young growing in the oviduct. Livebearing fishes of the family *Cyprinodontidae* show a variety of patterns. Some, like *Gambusia* (mosquito fish) and *Platypoecilus* (platies and swordtails) continue to produce more eggs while embryos are developing; eggs mature but remain in the ovarian follicles where they are fertilised, so many stages are to be found in each ovary. This is called **superfetation**, where young of several ages are carried together, and the tiny *Heterandria formosa* (another 'mosquito fish') can have up to eight batches. *Lebistes* (guppies) and *Mollienisia* (mollies) only seem to have one brood of developing young at a time; the growing oocytes must delay maturity, or at least fertilisation, until birth of the present incumbents. This occurs about every 30 days, and one fertilisation may provide sperms for 12 such broods each of 10 or more young. So the delay in starting the next batch of eggs is not due to shortage of sperms, nor is it simply a question of delaying mating until after parturition; there must actually be control of egg maturity and so fertilisability, as sperms are always present.

Some viviparous animals use pregnancy, within control by exogenous factors like season, to suppress ovulation. Even in those cases like the mouse or woman where the female might cycle and ovulate regularly, under natural conditions she would probably conceive on the post-partum oestrus or immediately after the suppression of ovulation during lactation (p. 244). Most of those forms which control ovulation by pregnancy make it an annual event, so it may be difficult to recognise the nature of the control. Because development **(gestation)** usually is much shorter than a year, delays are incorporated. Often, as in seals and polar bears, implantation is the stage of the delay[4] so that one mating and one pregnancy occur in most females every year (*see Table 16.1*, p. 280). The same end is achieved in temperate bats which store sperms over the winter, having mated in autumn[5], and which ovulate in spring and give birth at the beginning of the temperate summer; lactation continues till the next mating time so only one cycle usually occurs per year. It may be that in this case hibernation delays the ovulation associated with this mating, making the reproductive cycle break in an odd place.

It might be expected that those viviparous animals which store sperms and which control ovulation by pregnancy would be permanently pregnant, but there are only few cases where this happens, as in the cyprinodontid fishes mentioned above. The hare should be mentioned in this context, and one odd breed of laboratory mouse[6]. In these animals sperms are stored during one pregnancy and a second, usually smaller, litter is born after a slightly abbreviated gestation period without further fertile mating. It may be that in the hare[7] mating is required to elicit the second ovulation, but this can be with a (tested) vasectomised buck.

Some macropid marsupials (kangaroos) have avoided most problems associated with ovulation into a genital tract already occupied by the developing fetus, and have developed an efficient production line. A female kangaroo may have a 'joey' in the pouch, an embryo ready to come out through the tiny birth canal, a blastocyst awaiting implantation in the uterus as soon as it is vacated, and a Graafian follicle ready to ovulate when all move on one place. Later there may be two joeys in the pouch of different ages and receiving different milk. This strategy of fast breeding, by a 'series' rather than 'parallel' arrangement, is made possible only by the development of the pouch. This was itself necessitated by the tiny birth canal, which resulted from one way of developing Wolffian ducts and ureters in the female (*Figure 4.7*, p. 85) to solve the *male* retrapod's problem that he must urinate as well as ejaculate. The price paid for the eutherian solution was to loop the vasa deferentia over the ureters in the male, and to avoid complication in the female who may therefore have a very large or distensible birth canal (*Figure 16.5*, p. 270). Female eutherians can have several young brought almost to self-sufficiency in the uterus, as in guinea pigs, a tactic that would be impossible to a marsupial. The pelvic opening in female bats is said to stretch during birth to a circumference more than 100 times the normal. This would not help a kangaroo, as the birth canal itself, and the uterus, are relatively small; the juveniles are passed on to the next nutrition point, the pouch, when they are grotesquely underdeveloped but for the forelimbs and their muscles and nervous connections.

11.3 Protection of the Young

This is probably the most important function of all kinds of viviparity. The young escape with the female, and benefit by her superior physiological and behavioural repertoire for nutrition, fluid and salt balance and especially locomotion or defence.

These young therefore are not subject to as many traumas as those of externally developing forms, and their development is free of two major constraints as a result. They need not form osmotic, defensive, or locomotory structures early in development but can undergo developmental sequences impossible to organisms which must regulate their own physiology from an early stage. It is common to find, as in mammals, that **extra-embryonic** structures are developed for the mother–embryo interface by which the nutrition, excretion and often growth of the young are controlled; they often appear, as in human eggs, before the embryo itself. The embryo may then develop parallel with these structures and have the minimum of developmental necessities for its own embryology.

The second major freedom gained by the retained embryo concerns the 'reliability' of the developmental environment. The enzymic and other processes of development are, for example, temperature sensitive. Their changes with temperature must be linked in those animals, like the common frogs, whose eggs may complete normal development over a great range of temperature (about 7 °C to 32 °C in *Rana temporaria*). This must limit the possibilities of genomic prescription in external developers to those more stable routes to which the organism can be committed over a great range of temperatures, oxygen tensions and so on. For internally developing forms, however, more 'sensitive' developmental routes may be embarked upon, for the environment is less changeable, more reliable. Even in poikilothermic females like teleosts, large temperature variations are not encountered by internal embryos because they are avoided by the mother. This is probably the main reason why ' 'higher' forms are usually viviparous; the classic exception, the birds, take care of this uniformity of developmental environment in another way. 'Higher' organisms are complex and need this reliability of developmental milieu, in order to develop their complexity reliably.

There are many quasiviviparous organisms which protect the young, and the developmental environment, in this way, but do not give continuous nutrient so are commonly called ovoviviparous and are not credited with sophistication in their reproduction. These include seahorses, some frogs where the male carries the eggs in a pouch or on his back legs, mouthbrooding fishes, and perhaps even those spiders which carry the cocoon of eggs about with them. Not only freedom from predators, but also from many other environmental variables, is achieved by this use of the adult behavioural repertoire to protect the young directly.

11.4 Nutrition, Excretion and Respiration

These are commonly achieved by exchange with the mother's body fluids across a cellular membrane, as in the **placenta** of mammals (*Figure*

11.1). In ovoviviparous forms, by definition, yolk or other food material is supplied initially, but oxygen, water, salts and excretory products must be exchanged, and perhaps important hormones may cross the placenta too. Only the route of food supply may differ in these ovoviviparous organisms, and to separate ovoviviparity and viviparity as is customarily done probably directs attention away from important similarities (*Table 6.1*, p. 120).

It is common for the network of blood vessels used for yolk absorption (in the ancestor if the present form has lost the yolk from its eggs) to be used also for gas and metabolite exchange, by their close approach to the fertilisation membrane or chorion from the inside; mother's tissue fluids or blood vessels are in close contact on the outside. Such a **yolk-sac placenta** may be the only kind or may precede a more complex structure as in some of the mammals (*Figure 11.1*).

The intimacy of contact between the maternal and embryonic circulation is very variable. Some sharks are effectively free in the oviduct (*Figure 4.6*, p. 84), and may use long gills applied temporarily to oviduct wall for gaseous exchange (*Plate 2*); some rays produce villi (**trophonemata**) which in *Pteroplatea* enter the spiracles of the young and contact the gut wall directly. Egg investments of assorted thicknesses commonly separate parental and embryonic tissues, but in a wide variety of organisms there is direct contact of embryonic and parental tissues. These include *Pipa*, a toad with the eggs buried in pits in the back skin of the male (*Figure 11.1*), and many insects like some aphids as well as the mammals. In

---

Figure 11.1 ADAPTATIONS FOR VIVIPARITY (a) The maggot of the tsetse fly Glossina *fits snugly against the anterior end of the uterus, and is fed by uterine milk secreted by modified genital glands (*Table 12.1*). All its growth is intrauterine, through four instars. When laid, it burrows and pupates without feeding. (b) Part of the dorsal skin of the toad* Pipa. *Eggs are pressed into the female's back, where vascular ridges and horny lids develop to oxygenate and protect them. The embryos have very vascular yolk sacs and tails, presumably for gas exchange (after Amoroso). (c) Young cyprinodont fishes, like the guppy, platy and molly develop in the ovary before ovulation. The ovarian vessels press close against the chorion and the embryos rely on these for gas exchange (and in* Gambusia *and* Heterandria *for nutrients). The embryos perform a very odd anatomical manoeuvre, passing their heads between the vessels entering the posterior end of the heart, the sinus venosus, completely through the extra-embryonic pericardium. The result is a very vascular 'neck strap' used for gas and nutrient exchange. (d) 'Zonary' placenta of a cat or dog; note that apposition to uterine wall and development of villi only occurs in the central girdle beneath which the allantoic vessels contact the external chorion (partly after Amoroso, 1952). (e) Fetus of cow, with patches of villi which contact 'cotyledons' developed as patches inside the uterus (partly after Amoroso, 1958). (f) Human fetus, with one large patch of intimate contact between fetal and maternal tissues, the 'discoidal' placenta. (g) The variation in intimacy of the placenta of mammals. The early contact of all species, and throughout pregnancy in a few like the pig, is between the uterine epithelium and the chorion layer of the embryo. The uterine epithelium disappears in most mammals, exposing the connective tissue (syndesmium) to the chorion; maternal blood vessel walls (endothelium) are pressed against it in bovines and some carnivores. Most carnivores and primates, including man, show rupture of these vessels, allowing maternal blood (haemo) contact with chorion. The rabbit and rat erode away part of the chorion, allowing embryonic vessel walls to be bathed by maternal blood. Equally intimate is the placenta of some shrews, where it is possible that maternal capillaries are pressed against fetal capillaries or even bathed in fetal blood*

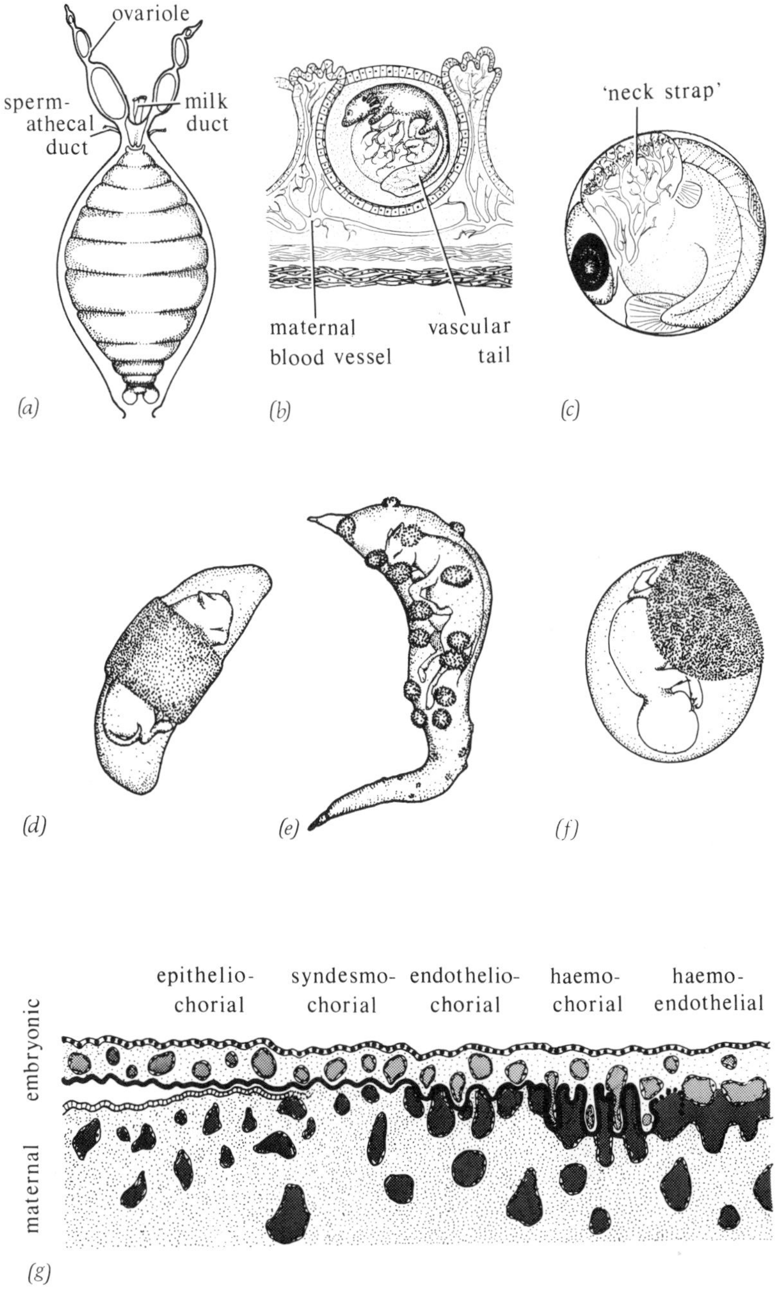
ovariole
sperm-
athecal
duct
milk
duct
(a)
maternal
blood vessel
vascular
tail
(b)
'neck strap'
(c)
(d)
(e)
(f)
epithelio-
chorial
syndesmo-
chorial
endothelio-
chorial
haemo-
chorial
haemo-
endothelial
embryonic
maternal
(g)

mammals contact may be between the outer chorion of the embryo and the endometrial epithelium, as in all yolk-sac placentas, or the superficial tissues may erode in parts of the contact area, exposing the deeper layers to more intimate exchange (*Figure 11.1*). In **haemochorial** placentas like the human, complex blood flow circuitry and regulation are necessary to prevent pooling and to maintain flow in the large blood sinuses.

This intimate contact produces a problem for the mammal, for the embryonic proteins are foreign and would be expected to elicit antibody reactions. Except in rare circumstances like the **rhesus** incompatibility in humans, this does not occur and it is a continuing puzzle why it does not. Embryonic trophoblast (outer layer) is not very antigenic, and the mother is less immunologically reactive when pregnant, but there is almost certainly another protective mechanism as yet undiscovered.

There is a dramatic mode of nutrition for the young, practised by some viviparous insects. Some larvae, like those of *Miastor* and many scale insects, entirely consume the mothers' tissues, and the latter may use mothers' cuticle as a shelter in which to overwinter too. In some tunicates too, the sexual form, or **nurse**, sometimes only carries one young. This soon outgrows the mother, who is reduced to an appendage, although the young one is only fed through the placenta; compare the tunicate life histories shown in *Figures 10.10* and *11.2*.

11.5 Birth

Parturition may be a simple process of escape, as apparently in some sharks, or may involve very complex structures, physiology and prenatal behaviour, as in most mammals. There is room here for only two points, the initiation of birth and the developmental state of the offspring.

It is still unclear whether birth in mammals is initiated by mother or fetus[8]. Gestation periods are often very constant in those species like rabbits and rodents whose offspring vary in number and weight; but there is also evidence for fetal control of the onset of lactation. Certainly there is exchange of hormonal signals, and changes of sensitivity to oxytocin, prolactin and prostaglandins culminating in that cascade of changes which results in expulsion of offspring.

The degree of development at birth may be very different, even between closely related organisms. A series could be constructed, perhaps starting with those birds whose blastoderms are commencing gastrulation on laying (though birth should most properly be considered equivalent to hatching, not mere expulsion from the female). Marsupials, then rodents like hamster, could be followed by 'larval' births like guppies, aphids, daphnia and deer all of whose young resemble parents in morphology but differ in some respects of diet or physiology. Then organisms like gnu and guinea pig, *Glossina* then *Gyrodactylus* could complete the series. These are born very mature and able to lead adult-type lives; in the last-named the offspring when born may already be a grandmother.

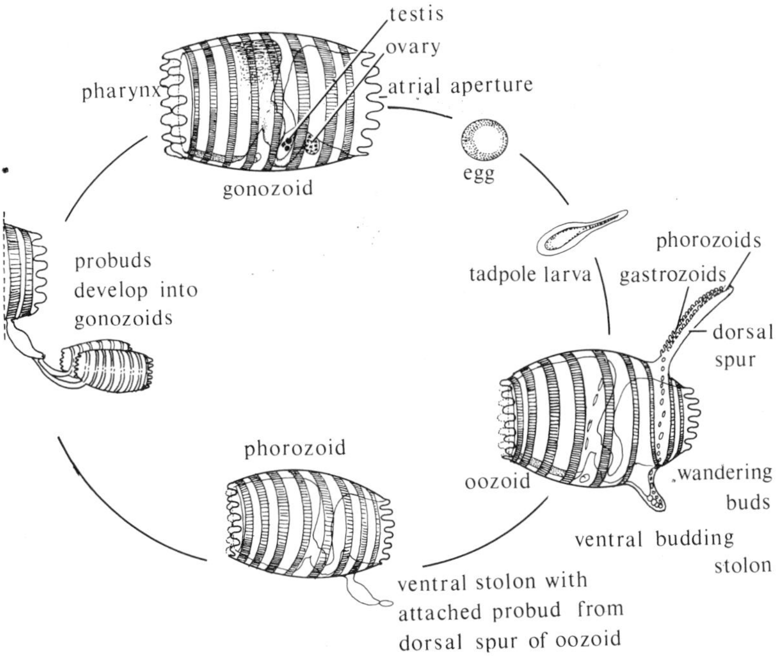

Figure 11.2 THE LIFE CYCLE OF DOLIOLUM *(partly after Berrill, 1950 and 1961). This peculiar and complex life cycle demonstrates the difficulty of using mammalian concepts and terms for other organisms, particularly in connection with the nutrient provision for next generation. Salps are free-swimming tunicates, whose muscle-bands allow them to pump water through the pharynx into the atrium, achieving both nutrition by filtering it and locomotion by expelling it.*

Sexual individuals, gonozoids, are commonly protandrous and produce eggs in later life. These develop into tadpole larvae which metamorphose without settling into oozoids, called 'nurses'. These superficially resemble gonozoids, but have a ventral stolon (compare Figure 4.9*) from which buds migrate around the body on to the dorsal spur. Here some settle as feeding gastrozoids which take over nutrition of the nurse; this atrophies, remaining only as a locomotory appendage to the long spur. Other buds develop into phorozoids on the spur, which leave the spur as soon as they can swim. When they leave they carry a couple of immature pro-buds from the spur attached to their ventral stolons. These develop into gonozoids at the expense of the phorozoid carrying them, and the cycle is completed*

In the non-mammalian viviparous forms, birth seems relatively easy. Often, mother does not stop her vegetative activity, even in those dramatic instances (e.g. some scale insects) where the young devour the mother. Eutherian mammals, however, usually seem to find the event stressful to all concerned. We appreciate the rapid acquisition of locomotion by baby gnus, who must run with the herd within minutes of birth, and feel that rests are proper for baby mammals and their mothers. Some proportions are odd: newborn polar bears are three to six inches long, while the newborn blue whale is about a third of the length of the mother (*Table 16.1*, p. 280).

The human is a special case, not only because of special birth rituals but also because of the common exclusion of parturient women from normal social life. There are several cultures in which, as the time of birth approaches, the woman must subject herself to ritual cleansing or even, in extreme cases, she must build herself a hut for giving birth. Rarely is human birth as 'normal' a procedure as for most mammals, partly because of the large head of the infant. The birth canal of the female bat expands much more than the human, the pubic symphysis expanding to nine times the normal width, but she is not required to walk bipedally at this time.

However, one of the most peculiar birth rituals occurs in western society, where the woman is expected to attend a hospital, traditionally a place where attempts are made to cure the ill. Most of her part in the birth is then taken over by modern medical technology, leaving her at best a bewildered observer and usually denying her the reassurance of her own mother, husband or indeed child. She is then recommended not to endure lactation, despite medical exhortation to the contrary in many publications. This aberration is a good example of the anti-biological trend with regard to human reproduction in western society.

11.6 Numbers

Viviparous forms usually have few young at a time; there is clearly much investment in each of the progeny, and reproductive tactics involving much selection usually show high gametic rather than zygotic numbers, as in mammals. Nevertheless there are many viviparous forms with very large numbers of progeny; these are usually parthenogones; *Daphnia* and aphids are classic examples (*Table 2.1*, p. 16). That the common low number in litters is related to investment rather than to viviparity *per se* can be seen by comparing mammals and birds, or *Daphnia* and *Cyclops*, in *Table 2.1*.

REFERENCES

1. Amoroso, 1960
2. Ivanova-Kasas, 1972
3. Hyman, 1951
4. Asdell, 1964
5. Racey, 1975
6. Ullman, 1976
7. Martinet and Raynaud, 1974
8. Liggins, 1972

11.7 Questions

1. In what ways can development be more complex for an offspring developing *within* a parent?

2. Look again at *Table 6.1*, and consider whether ovoviviparity is a useful category. Indeed, can oviparity and viviparity be usefully distinguished only on a nutrition-route basis?

3. In what ways do birds and marsupials achieve the developmental control found in eutherian mammals?

4. It has been suggested that copulation is a necessary pre-requisite for viviparity. Is this so?

5. In what ways can many of the advantages of viviparity be attained without retention of embryos in the female genital tract?

12

OFFSPRING CARE

12.1 'Yolk'

All organisms more complicated than viruses transmit material and energy to the next generation as well as information, and even viruses make the host cell produce coat protein as well as nucleic acids to give the next generation a start in life (*Figure 16.9*, p. 284). We have already considered the various ways of putting yolk into eggs (pp. 119 and 125); the use of this yolk by the embryo varies, often sequentially in any one embryo. Intracellular digestion is often followed after gastrulation by the development of a special **yolk-sac** to contain and digest the yolk. Although this is commonly associated with the gut, the teleosts enclose the yolk within the peritoneal cavity but outside the gut[1], and some cephalopods and arachnids are also unorthodox (*Figure 12.1*). There is sometimes a special region of the egg which contains the nutrients, but they are usually dispersed in a large proportion of the egg cytoplasm. In teleosts the yolky cytoplasm becomes separated from the clear blastodisc immediately after sperm entry, by a very dramatic streaming of yolk particles away from the animal pole (cf. *Figures 6.1*, p. 116, *8.2*, p. 146 and *Plate 5*, p. 61).

An odd maternal contribution to yolk digestion in some groups is the allowance of a large number of supernumerary sperms into the eggs. These provide nuclei which aggregate beneath the blastodisc of teleosts and birds and may precede the cleavage nuclei into the yolky cytoplasm. There is often contribution of such accessory sperm nuclei to **periblast** 'tissues', which change the yolk and render it suitable for further digestion by embryonic tissue. In teleosts, the periblast may precede the engulfing blastoderm edge extending over the yolk; in the birds it may lie beneath the blastoderm digesting a mass of yolk which, after extraction with histological reagents, shows as the **subgerminal cavity** (*Plate 11f*, p. 157). This is not a cavity in life, but the yolk it contains has been broken down into digestible or absorbable components, and fats, which are extracted by the solvents used in preparing sections.

As well as the intracellular yolk itself, other food materials are commonly presented to the next generation. Hen's egg albumen, trophic embryos, a honey-filled cell, fungus with which to digest elm wood, or a paralysed spider are all contributions by one generation to the nutrition of the next. But the tsetse fly, whose single maggot is laid at the end of its larval life and simply burrows, pupates and emerges as an adult, has fed its young via its own metabolism right up to the adult state; so has *Gyrodactylus*. These viviparous forms transform some of mothers'

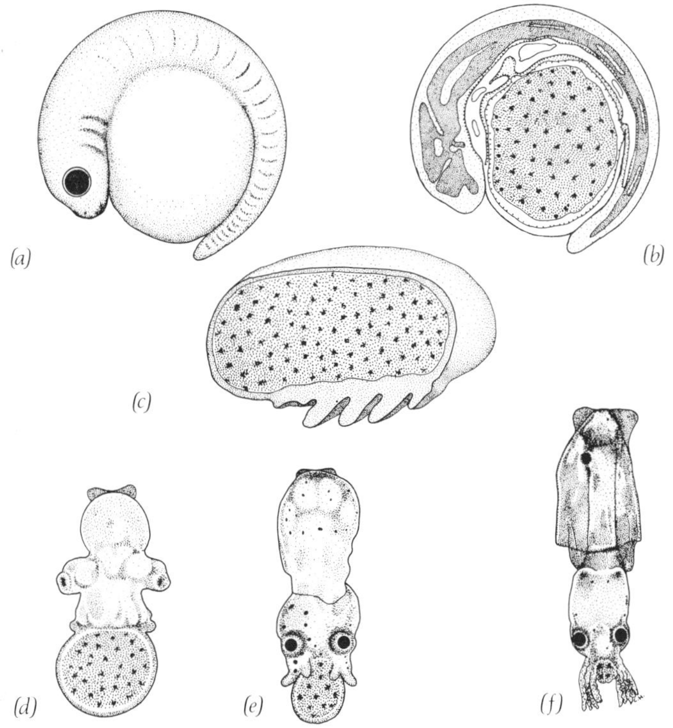

*Figure 12.1 THE PROVISION OF YOLK (a) A teleost embryo, whose yolk lies in the peritoneal cavity, outside the gut, usually invaded by periblast nuclei which assist in its digestion. (b) A sagittal section of (a). (c) A mite embryo cut vertically showing the yolk-filled visceral mass with the appendages developing from the cellular covering. (d) A young squid (*Loligo*) showing the embryo developed on top of the yolk sphere. (e) Later, the arms develop around the yolk which is outside the foregut but inside the epidermis. (f) The yolk is drawn into the foregut and its epidermal coat forms the mouth lining, stomodeum*

current metabolism and material into the next generation; this is not grossly different from the provision of yolk or albumen.

12.2 'Milk'

Many organisms continue feeding the young after they have hatched, or after birth. Sometimes adult food is not suitable, and must be processed to various extents. The nestlings of insectivorous birds can deal with adult food, as can the young of many but not all sea birds. But pigeons produce 'milk' from special crop glands, presumably because their diet is mostly vegetable; seed-eating finches may supply the young with a high-protein diet of insect larvae and worms as well as seeds with broken tests which are in the process of digestion. Mammals, of course, are the specialists in this reproductive strategy, and it has even earned them their name. Uterine secretions are called 'uterine milk' (*Table 12.1*).

The paralysed prey animals left by many *Hymenoptera* for the young maggot to eat are better considered 'yolk' than 'milk' as they are laid down at the beginning of development. The honey in the bee-cell falls into this category too, but the special diet given by the workers to the larvae is comparable with the provision of milk (*see* p. 241). The donation by the parent generation of such a special diet should be distinguished from the deposition of the embryo or larva in a situation where the larval food is expected to be available, as by the alder wood wasp laying her egg, with fungi, in dead alder wood. The essence of the very special 'milk' adaptation is that the nutrient is obtained via the mother's metabolism.

This is especially the case in the mammals, where the mother may be feeding more than her own weight of offspring. The **mammary glands** are related to glands like the apocrine sweat glands[2], both in sensitivity to hormones and in their production of a fatty secretion product and, probably, in their evolutionary origins. The basic structure is of a branching web of ducts leading from a **teat** (with one opening) or **nipple** (with several) outward into subcutaneous tissue, where the ducts end blindly as **acini** whose cells release a complex of cytoplasm, sugars and organelles which make up **milk**. Milk contains sugars, usually lactose (a disaccharide of glucose and galactose) but with variable amounts of other sugars, proteins (**'lactalbumin'**) and fats of various kinds forming a **cream**. Different mammals vary in milk composition (*Table 12.1*). Human milk contains much less fat, and rather less protein, than cows' milk and the curds produced in the infant stomach differ greatly, the cows' milk being of more solid consistency and less digestible by human babies. The milk of marine mammals is extremely rich in fats but also has a high concentration of protein and sugars, as does the milk of the tiniest mammals, shrews and mice.

The teats or nipples are developed at differing places along the **mammary line**, which all mammalian embryos show and which runs from the inguinal region up onto the front of the chest, and sometimes onto the shoulders. Pigs and dogs develop more or less the whole line as a series of glands and this is associated with their **polytoky** (having many offspring). **Monotokous** animals like ungulates, horses and elephants commonly have only one pair of teats, which in horses is at the posterior end of the line but in elephants is thoracic as in the human. Many mammals have a group of mammary glands which feed one or more muscular reservoirs or **udders**. The cow and goat are classic, and useful, examples. In many marine mammals, notably the large whales, the reservoir acts as a pump, forcing the milk into the baby rapidly so that breathing is not complicated by long meals; one has only to see a human baby feeding when it has a cold to realise the problems of breathing and suckling together.

The secretion of mammalian milk (*Figure 16.6*, p. 275) is stimulated by **prolactin** produced at the end of pregnancy, closely associated with changes in level of sex hormones, notably oestrogens, and with prostaglandins and **oxytocin** (which is involved in its release). The initial secretion is **colostrum**, a thick secretion, which is followed by milk after a couple of days in women. Milk not only gives nutrient to the offspring, but is also

Table 12.1 The composition of milk (mostly from Jenness, 1974)

A. Uterine milk (compare *Table 6.1*)

Organism	*Common name*	*Water*	*'Organic substance'*	*Protein*	*Fat*	*Ash*	*Notes*
Lamna cornubica	Porbeagle (shark)	breakdown products of further ova					
Torpedo ocellata	Electric ray	96.3	1.2		0.1	2.5	
Acanthias vulgaris	Spur dog	95.1	2.4		0.3	2.5	
Mustelus vulgaris	Smooth hound	93.6	5.1		0.2	1.3	
Trygon violacea	Sting ray	86.5	13.3		8.2	1.2	
Mustelus laevis	Smooth hound	89.4	9.1		0.1	1.5	
Bos taurus	Cow	87.9		11.0	1.23	0.37	No reducing sugars
Equus caballus	Mare	79.6	19.77	18.0	0.006	0.59	"
Glossina morsitans	Tsetse fly	69.0	15.3		15.0	0.7	"

B. Mammary milk

Organism	*Common name*	*Water*	*Fat*	*Casein*	*Whey protein*	*Lactose*	*Ash*	*Notes*
Tachyglossus	Echidna	9.6	9.6	7.3	5.2	0.9		Sweat gland
Megaleia	Red kangaroo	80.0	3.4	2.3	2.3	6.7	1.4	Early?
Oryctolagus	Rabbit	67.2	18.3	←13.9→		2.1	1.8	
Saimiri sciurus	Squirrel monkey		5.1	←3.5→		6.3	0.3	
Homo sapiens	Woman	87.6	3.8	0.4	0.6	7.0	0.2	
Sciurus carolinensis	Grey squirrel	60.4	24.7	5.0	2.4	3.7	1.0	
Canis familiaris	Dog	76.5	12.9	5.8	2.1	3.1	1.1	
Thalarctos maritimus	Polar bear	52.4	33.1	7.1	3.8	0.3	1.4	
Pagophilus groenlandicus	Harp seal	38.4	52.5	3.8	2.1	0.9	0.5	
Elephas maximus	Indian elephant	78.1	11.6	1.9	3.0	4.7	0.7	
Diceros bicornis	Black rhino	91.9	0.0	1.1	0.3	6.1	0.3	
Bos taurus	Cow	87.3	3.7	2.8	0.6	4.8	0.7	
Capra hircus	Goat	66.8	4.5	2.5	0.4	4.1	0.8	
Tursiops truncatus	Dolphin	58.3	33.0	3.9	2.9	1.1	0.7	
Balaenopterus musculus	Blue whale	42.9	42.3	7.2	3.7	1.3	1.4	

important for the passive transfer of antibodies to the young of some species, including the cow and rat. There is also in some species, e.g. mouse, rabbit and human, a transfer of passive immunity *in utero*, by immunoglobulins transferred from mother to fetus during pregnancy.

12.3 Ventilation

A very important function of the offspring care of aquatic forms is to aerate the eggs as they develop, for a dense clump of eggs can suffocate the innermost. Indeed the masses of frog spawn may use the albumen coats to prevent the eggs clustering so closely that they compete for oxygen. Fish and cephalopods which guard the egg mass take great pains to remove dead (fungused) eggs and to prevent settlement of detritus around the living ones. Substrate-breeding cichlids move the wriggling mass of **alevins** (newly-hatched young with yolk-sacs) to a succession of pits as they grow; the guarding parents continually fan the mass with pectoral fins until they become free-swimming.

There are many species, for example the swan mussel *Anodonta*, in which the young develop protected by the parents' anatomy (but not in the reproductive system) and the parents' own ventilating system is commonly used. In *Anodonta* and many other bivalves the inner aspects of the folded gills hold the developing eggs; the bitterling lays its eggs in the *Anodonta* gills so they benefit by the mussel's respiratory current too. Crustacea also commonly use gills or swimmerets for holding their eggs, while mouth-brooding fishes use the gill chambers too. There are other ways of carrying the young, e.g. in a brood pouch as in seahorses or the toad *Pipa*, but here there must be a special vascular supply and the situation should more properly be seen as quasi-viviparous.

12.4 Sanitation

Excretion by the young, of carbon dioxide and ammonia or urea, is dealt with by adequate ventilation or placentation. The production of solid wastes may be a problem long before solid food is taken, and many kinds of offspring care must include solutions to this problem. These fall into three categories. Cichlid fishes move their young from pit to pit in the substrate, leaving the excreta behind, while mouth-brooders and other 'forced ventilators' presumably remove faeces in the ventilating current. Separation of faeces and young like this is the simplest solution.

Parents may eat the faeces, which are 'packaged' for the purpose by many nestling birds and nest-building mammals. The parents probably find considerable nutrient left in these pellets, which their adult digestive systems can extract. Thirdly, it may not be the actual parents which clean up after the offspring; worker ants and termites constantly 'lick' the larvae and pupae and consume their faeces, and a major function of dogs around the households of primitive human communities may be to look after the messy outputs of the untrained baby. Certainly modern dogs are very willing to assist with babies as they do with their puppies, and actively seek out soiled nappies.

12.5 Protection

Examples of parents protecting young from predators are of course many and diverse. Two good ones are cichlid fish which guard the swimming brood or give them shelter in the parents' mouth, and horned ungulates which protect their calves from hyenas and wild dogs. There are many signals which must pass between parents and offspring for this kind of sophisticated defence against predators; a rapid dorsal fin fluttering, for example, causes cichlid juveniles to find shelter, and a high-pitched squeal seems to be a common distress signal of many young mammals. This may simply be necessitated by the short vocal cords, but is surprisingly alarming to parents of other species.

Parents protect against more than predators. Parental behaviour assists hygiene, i.e. prevention of infection or re-infection with parasites, and parents commonly restrict the movement of the offspring to the parental territory, whose hazards are at least familiar. Food selection, and protection from poisonous animals and plants, may be assisted by imitation and example; transport may be provided between suitable sites, and sexual or other interaction with conspecifics may be avoided by the family situation.

Protection may be more subtle. The mother may provide antibodies which passively immunise the young against common bacterial infections until its own immune system matures, which may be weeks or months after birth. Grooming by parents may prevent infection by ectoparasites which could transmit protozoan or bacterial infection more dangerous to a developing than to a mature organism. And finally, as with viviparity, growing up with care by parents is less traumatic, more predictable, than dealing with all the environmental variation as an immature organism. So a more complex developmental programme can be prescribed, after as well as before birth, as a predictable social sequence is impossible to organisms whose young must fend for themselves. There are fewer constraints on the development of a guarded offspring, and a more complex developmental programme can be reliable.

12.6 Education

There is at least one very important way in which birds and bees do not illumine mammalian reproduction. Both are very programmed organisms, with built-in abilities and responses for a great variety of environmental stimuli. All mammals must learn some of their adult behaviour, although much of the behaviour *is* already built in, as is the ability to learn in certain modes.

Sometimes the attention of a parent to a developing organism can make the developmental acquisition of an ability look like learning. Baby birds do not *learn* to fly, they become able, just as baby humans become able to crawl or to use words meaningfully. Birds become better at flying with practice, as babies do at crawling or talking. Insects, on the other hand, can only refine techniques in limited directions. Some elements of bird-song are like this, others require a model to copy. Cuckoos are said to prefer a certain host species because the calls resemble those of its own parents; so races of cuckoo parasitise different host species. Many birds have a very specific time at which a kind of learning, called **imprinting**, fixes an object as the mother-object of the juvenile and, later, the sex-object of the adult, for example Lorenz and his geese[3]. It is less well-known that parent-association and peer-group association during development have been shown to be necessary, in many mammals, for adequate reproductive functioning of the adult[4]. The Harlows' studies on monkeys[5] deprived of responsive mothers or of other young monkeys has shown that sexual inadequacy engendered in this way cannot be 'cured' in adulthood by even the most persistent or obliging partner. Not only the ability to perform, but the likelihood of achieving reproductive status is determined by parental and peer influence interacting with genetic endowment. Rats brought up by sedated mothers[6] are very easily frightened by new circumstances, to which they react by 'freezing' and defaecating. Rats of the same strain which were either handled or given an electric shock every day were phlegmatic in their reaction to novel situations and reacted rather more successfully than rats reared 'normally' (in laboratory cages). Rats are, in the wild, very versatile exploiters of a great variety of foods and environments, and one might therefore argue that 'security' in the nest is not the best start in the rat race.

Most mammals (and many birds and even insects), especially omnivores like rats and many primates, need to learn about the foods which are available, and most can adapt to the very different diets offered in captivity. They must also learn the local geography and, if they are social, the social geography of their group. Contrast the young baboon in a multi-male troupe, finding his way through the uncharted paths of growing up, with the newly emerged honeybee, with perfect congruence between her internal circuits and the demands made upon her.

Human children also, of course, need to learn their social geography. Many human responses are built in, like the baby's suckling, grasping, smiling, stepping 'reflexes'. Other abilities appear in development, so that children become able to learn language soon after the first year; if they have not learnt a human language by eight or ten years the full ability is lost[7,8]. In most human cultures, parents are the main educators of the babies and young children, and it is repeatedly asserted that this early training is irrevocable, determining many facets of the adult character. Such cultural transmission of particular characters, like language and the courtesies and taboos, gives the child the necessary tools for the next part of its education, tools which it passes on as part of its own reproductive process, and which are as necessary for its human cultural life as the circuitry for the honeybee's dance is necessary for the hive (*Figure 16.9*, p. 284). The later education, often by specially-designated teachers but also by peer-group ritual and adult reaction, is not usually so obviously a part of the reproductive act within the culture concerned. Its means and its apparent objectives usually relate to adult social roles, and often to cultural myths and practices. This overlay, characteristic of each culture, is however a very important element in the continuity of the cultural pattern, and is therefore necessary for reproduction of, and within, each human culture. Those who transmit the culture across the generations, **clericy** in a literate society, may have no children themselves, for example most monks and nuns in medieval Europe, but are a part of the cultural 'germ line' nevertheless. In primitive human societies they often administer the traditions which determine future progeny production by the juveniles; these may be explicit 'puberty rituals' (*see* p. 227) or the arrangement of marriages. In more complex human societies the causal chains of cultural reproduction seem less clear, but the vociferous claims by each generation of juveniles that its life is being constrained by outmoded cultural concepts argue that the chains are still there.

REFERENCES

1. Ballard, 1964
2. Van Tienhoven, 1968
3. Lorenz, 1965
4. Eibl-Eibesfeldt, 1970
5. Harlow and Harlow, 1962
6. Denenberg, 1963
7. Morehead and Morehead, 1974
8. Lennenberg, 1967

12.7 Questions

1. What are the advantages of offspring receiving nutrients ('milk') which have been processed by mother?

2. Consider the provisions made by a cuckoo for her progeny.

3. How do parasites (e.g. cuckoos, mealworms, fleas, parasitic wasps) exploit the offspring care of their hosts?

4. To what extent does domestication of animals involve replacement of natural parental care by a substitute?

5. To what extent do parents, teachers, priests, contribute to offspring care in your culture?

13

LIFE CYCLES

13.1 Seasons and Reproduction

Reproductive activity is nearly always cyclical, usually annual even in the tropics, and temperate examples are very familiar. Events are so timed that the young are growing when food for them (often the young of other organisms) is plentiful. Sometimes a succession of food organisms is exploited, as with freshwater fishes which graduate from ciliates and rotifers through copepods and cladocerans to annelids and other fishes. Individual fish species are often more versatile than expected, at all stages; but some, e.g. the Mbuna (*Pseudotropheus*) of Lake Malawi, are specialised feeders which have such mature hatchlings from few large eggs that diet does not change during life[1]. We will consider two situations, however, in which the life history is still bound to a year-long cycle, but has departed so that a rapidly developing, and temporary, habitat can be exploited. These are the odd cycles of oviparous cyprindontid fish living in temporary ponds, and aphids (greenfly) which have many but different generations in each year (**heterogony**).

There is a great variety of organisms which come to a rapid bloom in ponds formed in the rainy season in arid tropical regions; these include bacteria, algae, protozoa, crustaceans and fishes. The dry spores of the first three do not surprise us, and *Artemia* (brine shrimp) eggs have made us familiar with the idea that branchiopod eggs can survive in a dessicated state. But even some small fishes, closely related to guppies and mollies but oviparous, have eggs whose development can be arrested, by drying, at several stages. The eggs of some *Epiplatys* ('panchax'), *Cynolebias*, and *Aphyosemion* species, common in pet shops which specialise in tropical aquarium fishes, are of this kind. In nature the adults breed in shallow ponds, when they are drying up at the end of the wet season[1], laying the eggs in bottom-mud or anchored on fine leaved plants. The eggs respond to a variety of thermal and osmotic cues by arresting development. Some stop development at blastula stage, others with heart and eye-cups, others with even a complete little fish. They then dry out and can remain quiescent for months or years. When water is added only a proportion usually hatch and more hatch on subsequent wettings, clearly a tactic which avoids premature development of all the eggs in response to the first showers while ensuring that some eggs hatch as early as any rival's. Moss with such eggs attached had a brief popularity as 'Instant Fishes' some years ago, and was sold with a packet of *Artemia* eggs and instructions. The fish develop amazingly quickly; for example *Aphyosemion australe* can mature in four to five weeks as a 4 cm male or 5 cm female on a diet of crustaceans and algae, and will tolerate temperatures from 15 °C to 42 °C. The eggs of *Epiplatys chaperi* and others do not need to be dried out, and several generations can follow each other, but other eggs (*Cynolebias*) will not hatch unless a drying has occurred. Growing up in these fishes, as in other residents of temporary habitats, is greatly abbreviated, but few of the species seem to have adopted the tactic of **paedogenesis**, larval breeding (p. 253); all seem to pass, but quickly, through comparable life histories to their longer-lived relatives. Whether the apparently 'primitive' nature of the crustaceans in these ponds is a secondary simplicity resulting from loss of a previous more sophisticated adult form we cannot know, but such habitats might well foster neoteny of this kind. There may be two or more cycles, of filling and drying out of the ponds, during the year[1]; these cycles, rather than annual ones, are followed by the organisms.

13.2 Heterogony

Quite another route away from one-generation-per-season breeding has been taken by the aphids. In these organisms there are usually several generations in a year, adapted differently and frequently living on different plants. As the crustaceans specialise successive instars in divergent ways, so aphids specialise different generations[2,3] (contrast *Doliolum*, *Figure 11.2*, p. 207). The aphid reproductive cycle is nearly always annual, but several generations are required to complete it (*Figure 13.1*). The fertilised egg, often the over-wintering stage in temperate latitudes, hatches to give the **fundatrix**. In *Acanthochermes quercus*[2], living on oaks in Italy, the fundatrix is hatched in April and lays eggs parthenogenetically, which hatch in May giving a sexual generation (**sexuales**) on the same tree; these mate and the females produce the eggs from which the fundatrices hatch the next April. The production of the two generations takes only two months of the year, the fertilised eggs remaining dormant for 10 months. Parthenogenesis allows rapid exploitation of the growing oak shoots. Such life cycles are called **heterogonic** (with 'different breeding'); aphids usually have much more complex ones, often with two host plants and sometimes taking two years. For example, *Sacchiphantes viridis* lives on spruce (primary host) and larch (secondary host). Fertilised eggs laid by the sexuales on spruce hatch in winter giving fundatrices, which produce winged forms in spring and summer (**alates migrans**) which find larch trees and produce many **virgos** which may overwinter. These produce further virgo generations on larch until the winged **sexuparae** arise the next spring; these are to produce the sexual forms and, although themselves all parthenogenetic females, some (**androparae**) will produce males and others (**gynoparae**) will produce females. But before these sexual forms are produced the sexuparae migrate to spruces. Here the sexuales are produced, which mate, lay eggs and die; the eggs then hatch during the next winter to give fundatrices again (*Figure 13.1*). Some aphids live on different parts of the same, or

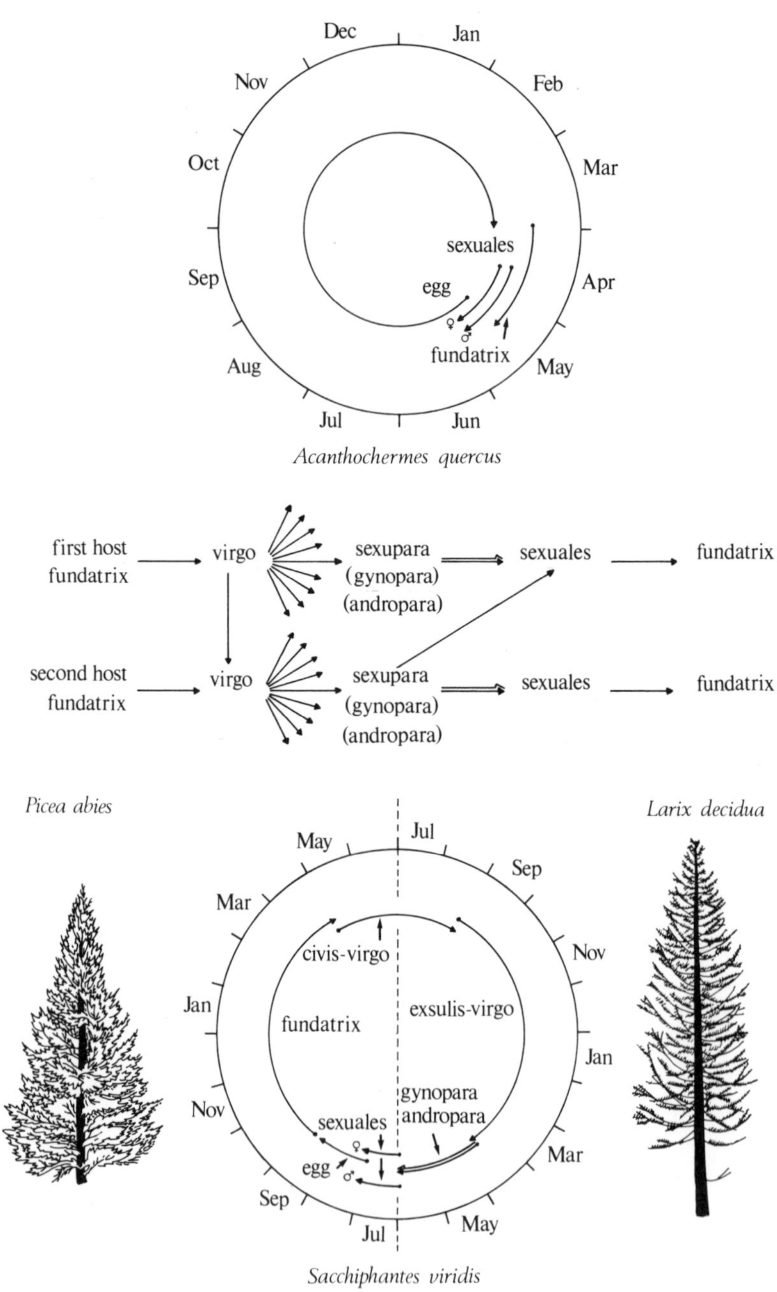

Figure 13.1 THE LIFE HISTORIES OF APHIDS (after Engelmann, 1970). See text for descriptions; the first host is sometimes called civis *and the second* exsulis

different, plants in different phases, e.g. *Pemphigus betae* whose fundatrices, and the virgos produced from them, live on poplar petioles making galls, while the succeeding year's virgos live on beet roots.

Other species, like a variety of the peach aphid in Israel[2], have lost the sexual forms altogether but may retain heterogony. Many species have acquired viviparity, especially of the virgos. The **alate** (winged) forms can often appear in any generation in response to crowding, but it is usually the virgo generations, with their rapid multiplication, in which they appear.

The arithmetic of reproduction of aphids is interesting, because the two-parents-make-two-parents rule is complicated by the question of *which* parents. The same is true, however, of many cyclical forms like lemmings, locusts and especially many social forms, like ants, in which a reproductive cycle may be considerably longer than a breeding cycle or generation. Once again parental numbers are re-produced in the reproductive strategy, but only after some generations.

There are instructive parallels between the successive generations of aphids and the 'alternation of generations' of hydrozoan coelenterates. The **planula** larva hatches and forms the first polyp, which is comparable to the fundatrix, and this produces (asexually) the **polyp** generation (*Figures 1.1*, p. 6, *13.2* and *Plate 1*, p. 3). The latter exploits the vegetative

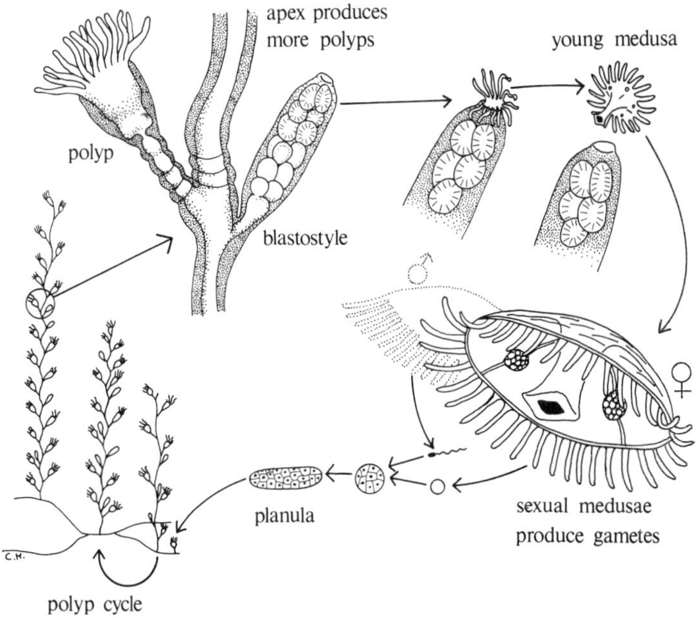

Figure 13.2 THE LIFE HISTORY OF OBELIA *The polyp generation may be considered a secondary larva (after the planula) which multiplies, or as the dominant colony which has specialised certain persons, the medusae, for a distributive sexual phase. Alternatively, the medusae are sexual and are therefore adults. Compare with aphids (*Figure 13.1*) and* Doliolum *(*Figure 11.2*), other examples of 'alternation of generations'. See* Plate 1 *for comparison photographs*

habitat by vegetative production of more polyps, comparable with the production of many virgos by aphids. Some polyps, called **blastostyles,** are like the sexuparae; they produce **medusae** vegetatively. The medusae are sexuales, with gonads of one or other sex (*Figure 13.2*), and fertilisation results in more planulae. Where aphids increase their numbers parthenogenetically, hydrozoans produce polyp colonies vegetatively. Such comparisons can be extended further, into larval multiplication for example as in the *Digenea* (*Figure 4.10*, p. 89). The **miracidium** of the liver-fluke is its fundatrix, which asexually produces (becomes?) a **sporocyst** in a snail and whose progeny, **redia**, are comparable with virgos of *Pemphigus* on poplars; secondary redia produced from them maintain this virgo population, while **cercaria** are like alate sexuales which migrate to a second host. The cercaria secrete cysts and become **metacercaria** on vegetation by the water's edge, and only become adult flukes in mammalian bile ducts.

Such comparisons can easily be pressed too far. Essentially all that they show is that rapid multiplication can be best achieved vegetatively, asexually or parthenogenetically once a genetically appropriate individual has located an exploitable environment, be it growing shoots, a hermit crab shell (*Hydractinia*) or a snail digestive gland. The clone so produced (perhaps with a little genetical diversity after diploid thelytoky with meiosis and crossing-over) represents the vegetative phase of growth of a diffuse colony, of which a few organisms go on to breed sexually. The sexuales generate diversity among the next generation of fundatrices, a few of which locate appropriate niches in which asexual multiplication is again appropriate.

13.3 Growing Up

In Chapter 10 it was emphasised that young and adult must differ, even if only in size. Even the juvenile *Gyrodactylus* and *Glossina* (pp. 200 and 201) are specialised for living inside their mothers, although they resemble them closely as soon as they become independent, even in size, for these are effectively born as adults.

Most organisms grow toward the adult state and reach a metamorphosis, or at least a **puberty**, as adulthood is attained; even if a definitive larva is not present, juveniles differ from adults in size and therefore in physiological proportions. If the juvenile was exactly a miniature adult, with all dimensions scaled down linearly, then physiological processes such as respiration would be easier for the juvenile; but the maintenance of osmotic barriers would be more difficult, simply because its surface area would be greater in proportion to its volume, and diffusion distances in tissues would be shorter. No organism does grow **isometrically**; instead various degrees of **allometry** are commonly found, that is to say proportions change as size changes. Classic examples[4] include the larger head of juvenile (and small species of) mammals, the smaller fins of young fish, and the changes of proportion of arthropod limbs in successive moults. The change of proportion is usually expressed as an exponential equation,

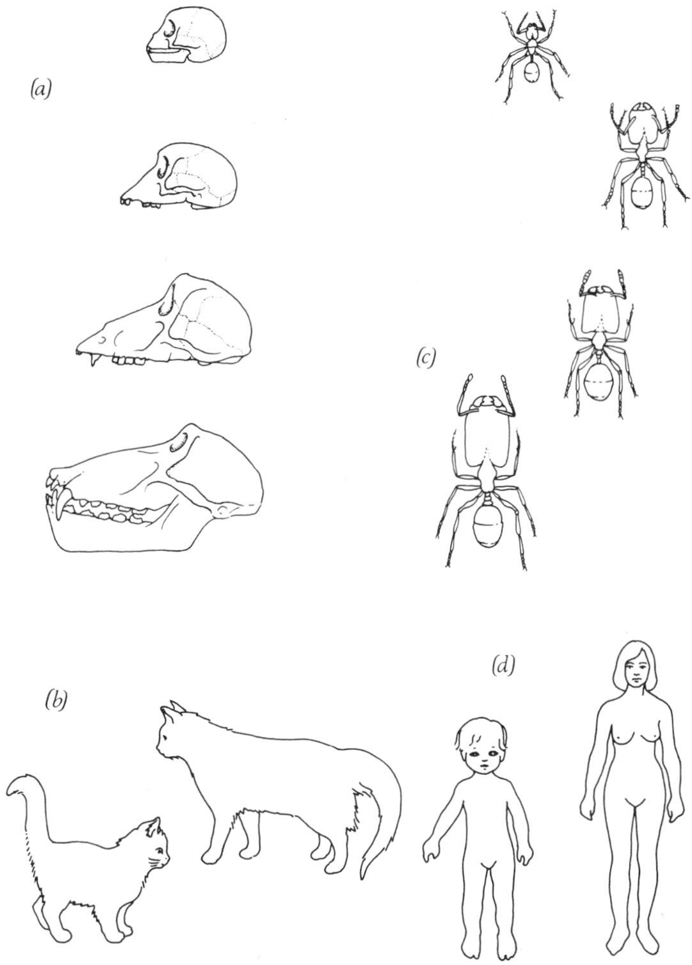

*Figure 13.3 SOME EXAMPLES OF ALLOMETRY (a) The skulls of baboons (*Papio porcarius*) form a series (after Huxley, 1972 from Zuckerman, 1926). Newborn, juvenile with milk dentition, adult female and adult male show a smooth increase in relative face length with increasing size. This emphasises adult specialisation in, for example, the series at the apex of Figure 10.1. (b) Different breeds of dog and cat, of different absolute size, have different proportions reminiscent of stages of development. These cat and kitten outlines show our subtle appreciation of proportion. (c) The ant* Pheidole instabilis *has an allometric relationship of head-and-jaw size to absolute size (after Huxley, 1972). Different functions in the colony can be served by the 'neuters' which have fed differently as larvae. Note that these forms metamorphose to different shapes, unlike baboons and humans which change shape as they grow. (d) Human head/body/limb ratios are markedly different in adult and juvenile, and we have neural circuitry which may have developmentally programmed, as well as learned, recognition of this*

$y = bx^{\alpha}$ where y might be organ size at a certain stage, b a constant for the organ under consideration, x the size of the body at the stage considered and α = 'allometric exponent'. A value $\alpha = 1$ gives isometric growth; the relative sizes of the heads of human beings change from infant to adult and have α about 0.95, and human femur length α is about 1.11[4] (*Figure 13.3*).

Allometric ratios are controlled by genetic systems. Indeed the absolute sizes and shapes of animals are probably always secondary, direct genetic control being related to thresholds, times of onset of effects, allometric ratios, and rates of destruction or production of organ-specific growth-control substances.

At puberty many allometric exponents suddenly change under new hormonal or environmental conditions. The body form then changes, slowly with many slight imbalances as in human adolescence, or rapidly as in the metamorphosis of tadpoles. The hormonal or other controls at puberty which distinguish the sexes often determine changes in ratio rather than presence or absence of organs; clitoris/penis of mammals or the fins of male and female Siamese fighting fish (*Figure 1.1*, p. 6) are good examples of this. Other changes, as in Müllerian/Wolffian duct systems or mammary glands which function in only one sex, become qualitative because of the 'cascade effect', of a sequence of changes which reinforce each other.

13.4 Puberty and Reproductive Status

It rarely, if ever, happens that all organisms reaching sexual maturity are permitted to breed. Even in colonising species, limiting behaviour appropriate to more stable populations is usually retained[5]. Animals with a wide behavioural repertoire (insects, cephalopods, most vertebrates) compete for the mating privilege by 'dances', ritual offerings, and other unusual and complex sequences of behaviour. Such sequences are derived from the sex act, feeding of young, aggression, or even feeding behaviour; and usually this behaviour is related to acquisition of a territory[5]. This territory is sometimes not related to vegetative needs, as in the penguins' patch of stony beach or the stamping ground of the Kob antelope, but is the 'certificate of mate-worthiness' earned by the behaviour associated with limitation of the breeding population. Sometimes the choice seems random, as in honey bees where an unusually high proportion of genes is shared with competitors for the breeding privilege (p. 63). In organisms with less complex behaviour, food or other limitations seem to be the major constraint. Medusae released by *Obelia* colonies, *Daphnia* females at the end of the growing season, *Gyrodactylus* on a heavily parasitised fish, or mealworm beetles (*Tenebrio*) in a squirrel's drey are never all given a breeding chance. Sometimes, as in cercariae finding vertebrates or malaria parasites (*Plasmodium*) sucked up in blood by a feeding female mosquito, the selection is fairly random; but barnacles or mussels must have great intra-colony competition during growth, for the final breeding adults apparently fill the available space.

Many organisms use the acquisition of a breeding territory as the visible sign of breedership. Robins in gardens, rainbow lizard males nodding at each other on their elevated positions, edible crabs in their caves and blennies in their bottles, elephant seal bulls on their strip of beach and bower birds with their elaborate 'front gardens' are all survivors of $R_{puberty}$ (p. 33), which presumably always has *m, n* and *r* components. Breeder selection tactics reduce accident *m*, and *n*, and emphasize *r* selection (*Table 2.2*, p. 30).

A complex tactic of breeder selection exists in many mammals and birds. These advanced organisms produced steroids from their gonads under the influence of pituitary follicle stimulating hormone (FSH) which is itself released by hormones (releasing factors) from the hypothalamus (*Figure 16.3*, p. 267). The hypothalamus is the executive of the whole limbic system, part of the ancestral brain, and contains centres responding to many ancient functions. In territory-holding birds and mammals the size of the territories is not equal or random, but is related to the **aggression**, in a wide sense, of the territory holder. (This is a somewhat tautologous argument, as we usually measure aggression by that behaviour which results in larger territory, but this does not concern the present example.) A large number of experiments have clearly shown that artificial increase of androgen levels[6] in male territorial birds or mammals, or sometimes of androgen *or* oestrogen in females, increases territory size; in the eyeball-to-eyeball confrontations at the boundaries[7] the individual with augmented sex hormones wins more often, and extends his territory. These experiments suggest strongly that, in the original territorial claim, steroid status determined territorial size at least in part. Thereafter, the possession of a territory permits copulation, which itself augments steroid levels[8]. Meanwhile those organisms which have failed to acquire territory usually fail to copulate and their steroid level rests or declines.

Steroid level is not physiologically *useful* to the organisms above a certain minimum required to maintain the reproductive organs adequately, but it is used as the internal parameter of the external symbol, territory, whose possession permits breeding. Because individuals who could accept small territories often failed to rear offspring, territory now ensures adequate rearing of the young. Variation in steroid level presumably results from a multitude of enzymic, perhaps nutritional, social interactive, and feedback threshold phenomena; so that the organism which comes to a territorial dispute is the result of a very complex gene–environment interaction. The environment (often including parents) acts upon the juvenile organism, and elicits from it programmed responses (e.g. when days lengthen, FSH is secreted) whose extent determines whether that organism will succeed or fail in a later territorial confrontation; a major final parameter by which the outcome may be judged is steroid level. From one viewpoint this is only one ancestral tactic to improve feedback circuitry of secondary sexual characters, but from our present viewpoint it is a measure of very many genetic–environmental endowments of that organism, of its history of success. If variation in several characteristics affects the result of a confrontation, ambiguity can arise, whose result is wounding, not victory/defeat; so the representation of many characters on one linear scale, steroid level, produces decisive confrontations. Such decisive selection of breeders uses a metabolically trivial parameter which represents a whole array of variables, its genetic and environmental endowments. Females may not be territorial, but their attractiveness to territorial males is certainly steroid-dependent, so steroid-level selection is constantly fostered in both sexes, once 'chosen' evolutionarily as the **status character**. Steroid level is symbolic of integration.

This delay in the testing of integration is a new strategy, formally comparable to yolk production: the young benefit from mother's abilities, nutritional or territorial, whatever their own genetics, before their delayed individual tests. This removes old necessities from the early life of the organism, just as egg architecture, then yolk, shell and membranes, then viviparity, then parental protection and nutrition (milk) all show use of parental abilities to give advantages (**privileges**) to *their* young (*Figure 16.9*, p. 284). Non-human social mammals find some difficulty in arranging this inheritance of privilege. In a social situation like a herd or a troupe, the position is not as simple as a stable physical or social geography. There is no private territory in which the mother can give her offspring special advantages in a herd; and we usually find that herd animals keep the young in the only place where mother's endowments can protect it for as long as possible, in the uterus. In a large troupe of baboons or a group of rhesus monkeys, however, there is such a 'territory' but it is notional; it is up with the aristocracy, with the high-status animals who lead the troupe. All females with newborns get a chance at this position, just as all females get a chance to mate with the top-status males when they are at peak oestrus. Females nevertheless return quickly to their old status, *with* their babies; so, independently of the baby's genetics, low-status mothers have low-status offspring and high-status mothers have offspring who are more often the winners in status games with their peers[9].

We must not consider this as a nature/nurture or genetics/environment argument but as a question of the total inheritance. The hereditary genetics of a young bird includes its yolk, its shell and its nest as well as its genes. The hereditary genetics of a young baboon is not only its genes, but that egg structure which makes trophoblast and prevents immunological rejection, its uterine nutrition, its milk, its mother's fur *and* her status.

In human societies the metamorphosis to adult status is very varied. In some cultures it is timed by reproductive status; **menarche** in girls or semen production in boys may mark the change, or it may be a pubertal birthday event.

Few, if any, cultures permit breeding, or indeed full adult status, to all who qualify by age or even by a puberty ritual. This is seen most clearly in polygamous societies, especially in polygynous cultures where a few men have many wives and much status, so wives are denied to other mature men. Status among adults is normally a prerequisite for breeding, as by achieving a kill as a warrior, acquiring a boat as a fisherman, or finding the money for a **dowry** to buy a bride. In **polyandrous** cultures (e.g. the Toda) it is often only the eldest girl in a family who is polyandrous, marrying the younger brothers of her chosen husband too. **Concubinage**, the customary liaison of young girls (or occasionally sexual adepts) with married high-status men, is practised in many cultures, and is difficult to distinguish from polygyny where one wife usually has senior status, except by the legal status of the resultant progeny. Children of concubines usually do not inherit anything in the paternal line, not even their names.

Puberty rituals are a very common phenomenon, often organised independently by the several subcultures a child lives in; so adults may test him publicly, parents may fete him, but his peers may play extended games which test his behavioural status most severely. There is a good case which can be made for puberty rituals as a very important factor in man's evolution[10]. In very many present primitive societies, and presumably in our comparable ancestors, these rituals have a very specific form. The applicants are tortured or mutilated for mystical or traditional reasons, and the constraints on the freedom to escape are symbolic. These symbolic restraints may be wisps of straw over wrists and ankles of the supine young man, which he is told are sacred ropes preventing his movement while he is circumcised; or they may be the limits of the area consecrated to the ritual, and the boys are forbidden to escape the homosexual attentions of older men by crossing the notional line; or the cheeks may be slashed or scarred for the symbolic prize of admission to the society of men. Nearly all the puberty rites of modern cultures can be seen as a trial of symbolic effect against 'instinctive' escape reactions. It is quite remarkable that all these tests do *not* test simple strength, or endurance, but they match strength of symbolic imagery, or persistence of mystical compulsions, against the 'infant' or 'animal' escape reactions. Perhaps, in our ancestors' culture, those who passed these tests, and so were permitted to breed, were the more 'human', symbol-oriented. Those who failed were the more 'brutish', the less subject to the rule of higher brain centres.

This view of puberty rituals sees man as a self-evolving creature, using societal, or at very least learning, behaviour to discriminate between the breeders and the failures. Present ritual practice in most cultures would pass nearly every applicant, for the system has clearly worked, producing the only intelligent organism ruled by tradition, by religion and by the imagery which has now become complex language. Human society forms a flexible, but protective because predictable, homeostasis around each individual into adulthood. *Glossina*, gnus and *Gyrodactylus* (p. 201) use the maternal abilities to protect the offspring until they are adult, but man uses the ability of many previous generations to protect the present generation throughout their lives (*Figure 16.9*).

13.5 Senility

No discussion of life cycles, or indeed reproduction, can be complete without consideration of the mechanisms and varieties of programmed senility. First the incidence will be considered, and varieties of senility, then the mechanisms, evolutionary origins and consequences. (The author is indebted to a thought-provoking article by Medawar[11] for many of the ideas in this section, and to Hamilton's article on senescence[12].

A distinction must be made first. Senility is an *increase* in likelihood of dying, in an adult animal, and must be distinguished from continuous low survival and from cumulative accident. Many small songbirds suffer more than 50% mortality per year and this is moderately age-independent, so birds in their eighth year would be less than a thousandth of such a population. These may not be senile even though they have outlasted very nearly all their peers, and may go on to live tens of years more in a cage or even in the wild. The same seems to be true of many fish, reptiles, and marine molluscs and crustaceans; their numbers apparently show fairly constant attrition and no sudden change in mortality patterns[13] after adulthood is reached.

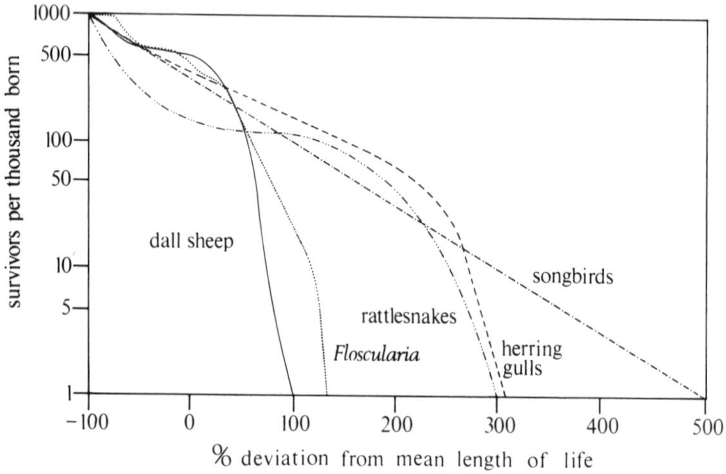

Figure 13.4 SURVIVAL IN SOME NATURAL POPULATIONS (after Deevey, 1947 and Porter, 1972). The form of this graph is intended to allow comparison of disparate species. Floscularia *only lives for months, songbirds for years, and rattlesnakes and sheep for tens of years. The 'songbird' line has been averaged from the survivorships of various species in Deevey, from Lack (1943). The figures for blackbird, songthrush, British robin, lapwing and starling are not more than 7% from this line, and American robin not more than 10%; most deviation is at the right-hand end*

Some other organisms do not show senility in the sense in which it is defined above; nevertheless they become more likely to die as they get older, because of accumulated accident without perfect repair. Adult insects, which do not moult again, would show this form of attrition even if they did not show senility as well. Each accident suffered, for example loss of a limb, makes another accident more likely, so the curve steepens with increasing age as the organisms have had more accidents. Senility is not a continuing process like this, but a sudden change of slope at a particular part of the life history, usually signalled by the cessation of reproductive effort.

The most dramatic senility is shown by terminal breeders, who usually have no provision for continuing metabolism after breeding. The adult mayfly, for example, cannot feed, and the same is true of many *Lepidoptera* (e.g. silk moths) and most parasitic *Hymenoptera*. After one or two days of frenzied reproductive activity these imagos cease activity, and are eaten or simply die. There are many organisms which, when kept in the laboratory, demonstrate senility; nevertheless, it is most unusual to find senile individuals in the wild. Perhaps the pathologies which cause death in the 'soft' conditions of laboratory or zoo culture are antedated in the normal life history by less obvious changes which nearly always cause death in the more competitive wild situation[11]. We are familiar with a great number of normal human senile conditions, ranging in severity from arthritis, loss of sight and hearing and malignant tumours, to greying of the hair and brown spots on the backs of the hands. In the western world we survive to experience an ever-increasing range of such conditions, but our ancestors saw many fewer. Probably the senile changes we observe in animals are only the visible tip of a large iceberg of changes affecting all the vital functions. We recognise some of these in other mammals, like tumours, 'dental and ocular insufficiency', and greying of hair, but the symptoms in many animals elude us. Some fishes seem to change shape as they age, finally producing grotesque bumps on the forehead (many cichlids), deformed jaws (salmon, some cyprinodonts), or peculiar thickened fins (gouramis, some cyprinids).

What are the mechanisms of this sudden incidence of a variety of pathologies? Can any general points be made, or does each kind of organism need separate treatment? There are, perhaps surprisingly, several possible generalisations.

Firstly, senility is not simply a wearing-out or accidental phenomenon. Many of its characteristic pathologies occur at different times in the life histories of related forms. For example, the collagen of rats becomes cross-linked and inelastic at about 18 months[13], whereas the very similar protein in man shows comparable changes only at 40 to 50 years. This pathology is clearly under the control of another metabolic clock, or rats have 'chosen' a collagen molecule of shorter effective life than man's; whichever is the case, the timing is precise and fits the life cycle.

Secondly, there is a spectrum of mortality for somatic cells. Some organisms, like the eutelic nematodes, only have somatic mitoses during development, and mitosis seems not to be possible for the adult soma (except perhaps in the epidermis). On the other hand, vegetative and asexual reproducers like the tunicates and coelenterates seem to be able to continue mitosis almost indefinitely. But most metazoans come between these extremes, and probably have a 'reserve' of mitoses in most tissues, as discovered by Hayflick[14] (p. 173) for man. Whatever the mechanism, many kinds of mammalian somatic cell seem not to be capable of an indefinite number of divisions but to be restricted to about 80 at most, even in long-lived species like man. Man has some cells, like those of striated muscle and most neurones, which never divide and whose number therefore slowly falls as accident befalls them. Perhaps we show the onset of senility when other tissues begin to 'run out' of replacements; this could happen fairly suddenly compared with the steady loss of muscle cells or neurones.

Thirdly, while this explanation might serve for many of the minor pathologies and possibly even for the high incidence of tumours in the senile, it cannot explain the synchrony of all these. This synchrony is the most obvious reason for considering senility as one programmed process rather than a coincidental breakdown of a variety of metabolic systems. In rats senile changes occur from about 12 to 18 months in tissues very similar to man's, who takes 40 to 60 years to reach his senility. This programming is emphasised by two anomalous conditions in man. In the **Loraine–Levy syndrome** (also called 'Peter Pan' syndrome) the characteristic changes of maturity and of early senility are postponed, while in **progeria** they occur much earlier than usual, so that a girl of only eight may resemble an aged crone[15]. Skin is wrinkled and 'liver-spotted', face, shape and stance are all characteristic; according to medical textbooks gum recession and falling hair occur in just the pattern expected at about 60 years. If this is not a 'textbook myth' then the whole complex of (at least dermal) senile changes may be moved as a group. In both the Loraine–Levy syndrome and progeria the anterior pituitary has been implicated in this timing change, but not on very sound evidence. So in one species, as well as between comparable species, senile changes form a characteristic association which is modified as a whole, and not an accidental assemblage.

There are many questions about senility which impinge directly on reproductive strategy, but we will consider only three. They concern its function in the reproductive cycle, its origin as a group of unrelated pathologies, and the lack of a defined senility in several kinds of animals.

Senility removes post-mature, usually post-reproductive, individuals and therefore occurs in those organisms which are most able to deal with the environment, the adult breeders who have survived their peers and still have enough profit on their metabolic transactions to produce eggs or sperms. Unlike predators, senility attacks the healthiest. This is a clue to its major 'function'; it removes the competition faced by the next generation of juveniles as they mature, competition which would always lead to defeat of these juveniles by the selected, experienced, and larger adults who already possess all the territories. Senility in these adults provides a set of compensating disadvantages which permit juveniles to grow up in competition with them, and therefore to grow up at all[11,12]. If adults remained as competent organisms, juveniles could only colonise new and therefore marginal territory, and few would have appropriate genetics; if they did, they themselves would prevent further juveniles from succeeding. Replacement of parents by offspring can only occur regularly if senility occurs, for if all individuals above a certain age have the same chance of dying, as perhaps in small songbirds, very slight advantages of some alleles might result in small inbred populations taking over an entire territory, with consequent loss of versatility. Can we envisage a genetic mechanism for this acquisition of obsolescence by the older organisms, preferably without appeal to 'group selection' (p. 63)? *Each* organism may be competition for its own offspring, so its demise will increase their chances of breeding in their turn.

The temporal linkage of such developmental time-bombs as a composite degeneration of mature organisms is very interesting, as there seem to be at least two distinct kinds of pathology concerned. One is apparently the result of continuing processes which operate throughout development but which reach crisis in late maturity, like the change in surface/volume relationships as the animal grows[16], or the increase of chromosomally abnormal cells[17]. The other kind of pathology seems to be the 'other side of the coin' from an early developmental innovation: the progressive cross-linking of the beautifully designed and elastic collagen of the juvenile and early adult mammal; problems deriving from continued growth of the vertebrate eye lens which has adopted this way of remaining transparent (*Figure 13.5*); or the loss rather than continuous replacement of teeth in old age, which seems to be associated with the modification of the second set for specialised diets by the early mammals.

It should be appreciated that different parts of the life history, like different organisms, can be compared in 'relative reproductive fitness'. *Figure 13.6* illustrates the breeding of a hypothetical parthenogenetic organism, producing one young each year and breeding each year, with no senility. This organism produces, in its *n*th year, one offspring compared with a potential 2^{n-1} offspring from its first breeding. So, in the 10th year, the offspring produced in that year faces an environment colonised by 512 potential siblings from the *first year only* of its parent's breeding. To put it another way: viewed from posterity, the 10th year of breeding contributed only 0.2% of the first year, and the 'relative

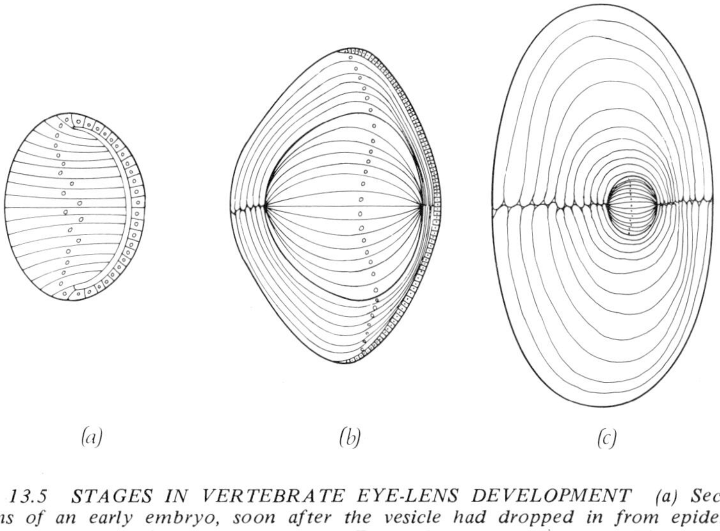

Figure 13.5 STAGES IN VERTEBRATE EYE-LENS DEVELOPMENT (a) Section of the lens of an early embryo, soon after the vesicle had dropped in from epidermis into the outer ring of the optic vesicle. External (lateral) is to the right, optic cup (medial) to the left. Cells of the rear wall have elongated, forming 'lens fibres' mostly containing crystallins. (b) More cells have become lens fibres in an older embryo (e.g. hatching tadpole, 90-hour chick embryo); more will be recruited from the edge of the anterior epithelium. (c) An older lens in section (e.g. hatching chick, internal gill tadpole). The lens 'nucleus' is the original lens, and on this scale the epithelium is represented by the thicker line around the edge

genetic fitness' of those years of the life history is 1 : 512. If a mutation appears which improves relative fitness in the first year by 1%, but decreases fitness of the 10th year by 90%, it will still be selectively advantageous. It will spread because of the greater contribution of offspring to the future from those individuals possessing it. Even if the mutation killed the organisms in their 10th year, it could still spread if it gave 1.01 the chance of producing an offspring in the first year. The simple 1:1 breeder we have been considering is representative of all the more complex cases too (*Figure 13.6*), as a few calculations will show early advantage given by an innovation is always 'preferred' to longer term advantage.

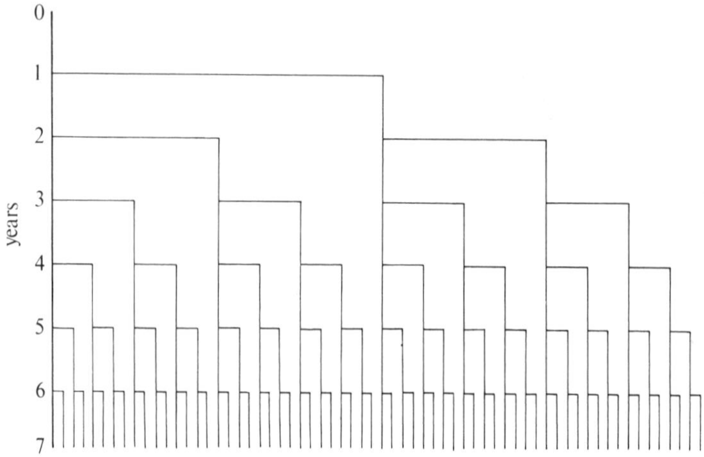

Figure 13.6 THE BREEDING OF A HYPOTHETICAL ASEXUAL ORGANISM PRODUCING ONE PROGENY PER YEAR In the text, ten-year figures are suggested; it is said that 512 potential progeny from the year one breeding compete with year 10's progeny. Here, at the 7th year, 32 1st-year, 63 total, compete with this year's progeny

Many ancestral genetic innovations must have had this property (e.g. lenses, collagen, specialised teeth), so the later part of an organism's life becomes the repository of the side-effects of all the short-term gains his ancestors invented, thereby becoming ancestors[11]. As soon as the sum of these (or any one of them) prevents reproduction after a certain age or stage the period subsequent to that moves out of the scope of ordinary natural selection, for nothing that happens after reproduction has finished can affect an organism's numerical contribution of offspring (though much can happen that affects the offspring's chances). In social situations old individuals may have value, but often even they compete with their own offspring and so reduce their likelihood of evolutionary survival. So the removal of post-reproductives is fostered, as is the continued accumulation of any accidental side-effects of developmental innovations which have spread through the population because they are useful to the young adult.

As soon as the sum of this **developmental detritus**, or any individual aberration, kills the bearer at a predictable time (e.g. 18 months in the rat or 70 years in the human) then new programming resulting in adverse effects *after* this time cannot be selected against. Human eye lenses, which continue to grow, would at about 250 years be larger than the eyeball, but this cannot be considered a real disadvantage of this tactic for achieving good eyesight when young. Equally, our collagen cross-linking resulting in loss of suppleness, our atherosclerosis, our loss of teeth, are all programmed to occur *after* death from some other cause, in the 'wild' at about 40 one supposes. Such an accumulation of programmed pathology in the post-reproductive period is not simply neutral, as we saw above; it compensates for the larger size, experience and territory-holding of the older adults and allows their juveniles to compete with and supplant them.

There are, nevertheless, several kinds of organism in which this programmed senility seems not to occur. Some animals are so heavily predated anyway, like some songbirds at 50% per season[18] or marine colonies like oysters at 70% per year[19], that external forces prevent attainment of senility and presumably therefore do not allow the temporal clustering of pathology we have considered above. Some organisms have radically different environments for juveniles and adults, like many marine molluscs and crustaceans, and here also attrition of the adults does not seem to change slope downwards as they age. This may be effectively true also of some freshwater fish like carp. But the apparent lack of senility in some snakes and tortoises, songbirds and parrots, lobsters and anemones, and its presence in lizards, owls and pigeons, crabs and medusae, is very puzzling whatever its biological basis as part of the reproductive cycle.

REFERENCES

1. Lowe-McConnell, 1975
2. Engelmann, 1970
3. Chapman, 1969
4. Huxley, 1972
5. Brown, 1975
6. Watson and Moss, 1970
7. Wynne-Edwards, 1962
8. Macrides *et al.*, 1975
9. Eibl-Eibesfeldt, 1970
10. Campbell, 1961
11. Medawar, 1957 (pp.44–70)
12. Hamilton, 1966
13. Comfort, 1964
14. Hayflick, 1965
15. Cooke, 1954
16. Thompson, 1961
17. Burnett, 1974
18. Lack, 1954
19. Walne, 1961

13.6 Questions

1. Compare colonial coelenterates (*Plate 1*, and corals) with aphid colonies.

2. How can allometry allow radical changes of life style as an organism matures?

3. Imagine a pubertal breeder-selection *not* using sex-hormone-related structures as the arbiters of competition. What other systems could be, or are, used?

4. What are the puberty-associated rituals in your society? Do they have reproductive sequels?

5. What is the 'kin-selection' argument for programmed senility?

14

REGULATION OF NUMBERS

14.1 Numbers and the Life Cycle

It should be unnecessary to emphasise again that the number of surviving progeny is less the later the sample is taken, until in a stable population only the parental number survives to breed (*Table 2.2*, p. 30). The deaths may occur mostly among adults, as in the hyperviviparous *Glossina* and *Gyrodactylus* which are born nearly adult. More usual, however, is a **'critical period'**[1] when offspring care by the parents declines and they must fend for themselves. In fish this critical period comes when the yolk has been used up and the alevins should begin to feed; many do not, even in *Anabantidae* or *Cichlidae* where they are being guarded by a parent. In roach (*Rutilus*) there has been investigation[2] of these deaths and it has been found that, of each 10 000 eggs laid, only about 150 survive to three months, in the wild. In laboratory tanks, the surprising discovery was made that about the same number survived, from a variety of starting densities. Most of these, like the larvae of many marine fish, do fail at this 'critical period' when the yolk has been used. Some survival figures for fishes are shown in *Table 14.1*. It will be seen that most of these deaths occur very early, but that there is further, apparently equally unavoidable, death later in development too. It is clear, however, that provision of ample food together with an absence of predators may permit many more to survive than do so in nature; soft as well as hard selection occurs (p. 19).

Losses do not begin at this critical period. The eggs themselves are eaten by many organisms, including the parents, and at first sight this looks totally accidental and unaffected by the genetic properties of the eggs. The density of fish eggs, however, is usually modified by oil droplets in the yolk and their visibility depends mostly upon yolk pigments. Both of these, in turn, depend upon the mother's ovary and upon her diet and her genetics. It would be imprudent to exclude selection from egg predation, but its significance cannot be large.

Many of these fish fail to hatch, while their siblings do so. Here hard and soft selection (p. 19) are clearly operative, because a proportion regularly survives, but accidental factors are also involved.

Many fish have a metamorphosis, but few are as dramatic as those of the flatfishes (*Pleuronectidae*). *Figure 14.1* shows the four most apparent changes: one eye moves around the dorsal side of the head, 'trailing' its optic nerve; the bones around the jaws rearrange so that the mouth

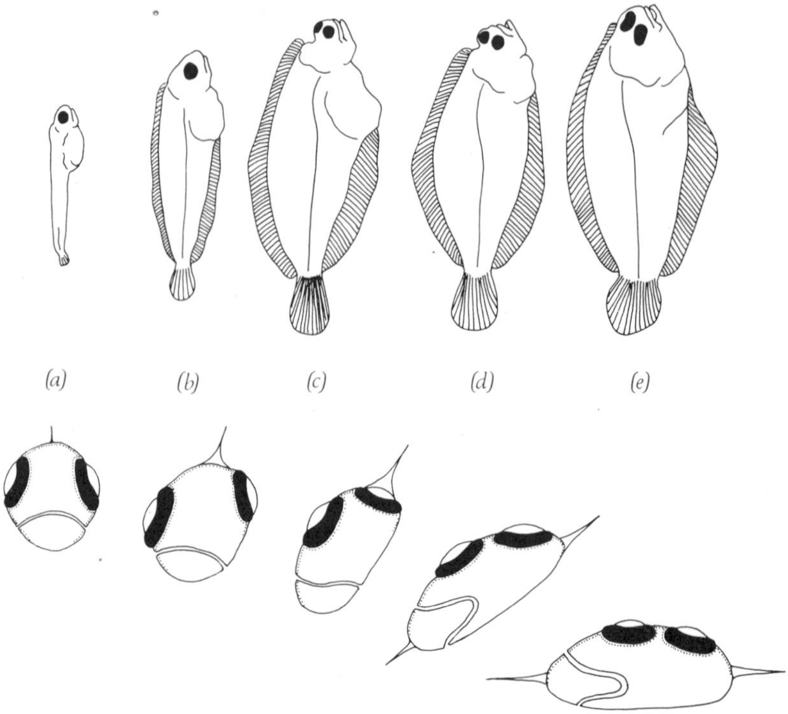

Figure 14.1 STAGES IN FLATFISH METAMORPHOSIS (compare Plate 9*).*
(a) The early alevin, symmetrical. (b, c, d) Stages in eye and mouth rotation.
(e) The metamorphosed flatfish. Upper figures show dorsal views, lower figures show very schematic front profiles. For ventral views see Plate 9f

twists; one side of the body adopts a very versatile camouflaging pigmentation while the other loses its larval pigment and becomes white; the air bladder loses its function and the fish settles to the sea bottom on its side. Only a proportion succeeds in this metamorphosis, however; some of the juveniles have 'wrong' eye movement, some have pigment on both or neither side, some have blind guts or other deformities so that only some 1 to 5% achieve metamorphosis[3] (*Plate 9*). In culture there is little loss after this stage, but of course in the wild numbers are reduced dramatically even after the final form has been achieved.

It is these late juvenile and early adult stages which often exhibit a new phenomenon, intraspecific aggression. Intraspecific competition has been occurring to some extent from the earliest stages of development; even the oocytes may have competed for ribosomal or yolk precursors, and competition for oxygen and food among siblings is usual. The behaviour of the progeny and parents, however, usually minimises the severity of this competition, as when a brood of cichlid larvae in a pit combine their tail-wriggling to provide respiratory currents. Young adults, however, do not usually cooperate, but separate as they acquire territories or fail to

Table 14.1 Mortality of fishes. (Data from Mathews, 1971; Bannister *et al.*, 1974; Cushing, 1974)

Baltic Cod	*Gadus morrhua*	
0–10 days		90%
0–50 days		99%
1–3 years		5–11% per annum
North Sea Plaice	*Pleuronectes platessa*	
Eggs I–V		6.93% per day
Larvae 1–4		4.57% per day
0-group		44.9% per month
0–1 group		14.0% per month
2 years		*9.5%
2–9 years		7.0% per annum
Roach (Thames)	*Rutilus rutilus*	
0–3 months		98.1–99.9%
3 months–1 year		53.5–92.3%
0–1 year		*99.94%
1–9 year		58.75% per annum
Bleak (Thames)	*Alburnus alburnus*	
0–1 year		*99.83%
2–6 year		79.59% per annum
Dace (Thames)	*Leuciscus leuciscus*	
0–1 year		*99.90%
		26.24% per annum
Gudgeon (Thames)	*Gobio gobio*	
0–1 year		*99.97%
1–5 year		63.27% per annum

*Calculated from 'instantaneous mortality rate'

do so; in this process what Wynne-Edwards[4] has described as 'eyeball-to-eyeball confrontation' is usual. It occurs between established adults at territory boundaries, and between an established adult breeder and challenging young adults hoping to usurp the territory.

Territory is a common prerequisite for an adult to become a breeder, because in most territorial species breeding behaviour can only appear in territorial context. Territorial behaviour of diverse birds, for example of herring gulls or robins, or of antelopes, is familiar from film and vivid accounts[5].

There is, however, one crucial issue relating territory to reproductive ability. This is that the territory is *not* directly a measure of the food requirements of parents and offspring. It may be tiny relative to this, as in breeding gulls or penguins, or very much larger than even the most pessimistic would suppose necessary, as in grouse on heather moors. This is because territory size is **symbolic**[4], and relates to internal demands of neural circuitry which have been selected for a plethora of reasons. Some of these reasons may have related to successions of very unproductive years, or to parasite transmission, or to cyclical predator or competitor numbers. It is important to relate territory size to the animal's demand in evolutionary time, and not to its present need.

Watson has performed a study of grouse[6] on heather moors which amply justifies this belief, and which demonstrates further that those birds which cannot achieve territories are those which die over winter. In the grouse, territory is largely irrelevant as they often feed in groups, but those animals excluded from satisfaction of the ritual demand are nevertheless those which succumb to starvation, to wild predators, to disease and even to the hunter's gun. The established breeders, with their notional but rigidly defined territories, seem to have charmed lives and go on from year to year.

This final intraspecific competition for breeding 'places' occurs in some form in nearly all animals, and should be regarded as the 'fine adjustment' of numbers after the 'coarse adjustment' of accident, predator and parasite have interacted with the reproductive strategy.

14.2 Selection or Accident?

Until we have longitudinal studies of gene variations in developing organisms in the wild, it is probably hopeless to attempt distinction between accidental and selective mortality. Experimental evidence for larval selection is accumulating fast[7,8,9], but so also are alternative explanations of these same experimental results, which deny selection. In this book I have attempted to emphasise selection, partly because accident has been the chosen explanation in the past but also because I believe most deaths to be at least in part selective. I believe that we are moving, in biology, from a paradigm of 'like begets like' toward a new dogma whose central tenet will more closely resemble 'like begets mostly unlike, but among these it is mostly the like which beget in their turn'. In this paradigm both selection and accident are important parameters of reproduction.

14.3 Inter- and Intraspecific Forces

It is nearly always misleading to use a term from one discipline in the argument of another. So 'force' should be used warily in biological reasoning. It is probably permissible, though, to divide selective 'forces' which determine survival into those which are engendered mostly from the same species, and those from outside the species, usually from other organisms.

An evolving species constantly adapts to outside pressures and, in so doing, changes the relationships between its own members. Selection for dark colour (melanism) in moths, for example, is said to be related to predation in industrial areas[10]. But the melanistic moth must adapt, perhaps, to new visual recognition cues for mating, to new catechol metabolism, to new thermal effects of incident radiation, and so on. There is clearly intraspecific adjustment associated with interspecific selection.

Nevertheless, it is possible to list phases of reproduction in which inter- or intraspecific effects dominate.

Competition among growing oocytes for limited yolk resources, competition among sperms for few available eggs, and competition among embryos in a mass for available oxygen are all possible intraspecific situations which could be selective. In these three cases, especially the first two, selection is unlikely to distinguish between genotypes. But oligopyrrhene sperms are destined to fail, as are trophic embryos; they run a race which has been 'fixed'. Competition among young fishes, or *Daphnia*, for inadequate food or oxygen exposes each individual to 'honest' competition from its peers. They are the creatures best adapted to the way of life concerned, so they make the most effective competitors for each other. When the young of one species dominates in an ecological niche, this intraspecific competition must usually be the dominant cause of juvenile death. In some cases, as in the African wild dogs studied by Jane Goodall[11], competition among the puppies firstly for milk and later on for parental attention and food results in selection, for there is not enough of either to feed all the puppies born. This juvenile intraspecific competition for inadequate resources is very close to Wallace's soft selection[12], and is probably the major cull in the life history of most species. There may, apparently, be an evolutionary route away from such direct competition of siblings by specialising and diversifying the brood into a social unit (*see below*), and this may be a major force towards colony formation.

After the offspring leave parental care, or after metamorphosis, the juvenile usually resembles an adult in shape but is much smaller. This is true of *Suctoria* (e.g. *Dendrocommeres, Figure 10.10*, p. 195), crabs (megalopa, *Figure 10.8*, p. 189), some tapeworms (plerocercoid of *Schistocephalus, Figure 4.11*, p. 90 and *Plate 2*), and of nearly all animals with the major exception of the holometabolous insects. This is only somewhat tautologous in that what we call the juvenile is the 'small adult' form; before that it is a larva. The holometabolous insects, however, show that this nearly universal assumption of adult form before adult size is not a necessary part of development; nor is it a necessity for evolutionary success.

The small adults are usually the most solitary phase of the life history, between the crowded and competitive larval or family group and the adult world of direct confrontation for symbolic gain. It is during this phase of the life history that the classical 'survival of the fittest' occurs, and it is surely unnecessary to give examples. It is during this phase too that the organism comes up against the adaptations of other species, whether they be predators or food, and his competence is tested in interspecific encounters. This solitary phase may be avoided in a few social circumstances like shoals, flocks and herds where only a few animals leave the group to return later. Even in some group animals a sojourn away from the group in young adulthood seems mandatory, especially for young male vertebrates[5].

When sexual maturity occurs, usually synchronised in virgin adults and experienced breeders alike by seasonal changes, the organisms again show intraspecific competition. Only those which can attain a place in the breeding group, usually by direct confrontation, breed in their first mature year. Some of these may survive to try again the succeeding year, but this is probably rare[6] except in large mammals. Senility and accident ensures that there are always places among the breeders. Some organisms avoid this final confrontation of young adults with mature breeders by terminal breeding, for here the breeding niche is empty when the adolescents come to it and they fill it by competition among themselves.

14.4 Speciation

It may be that a species becomes adapted to a particular niche, so that its siblings form its major competition in each generation. Then polymorphisms may arise so that different kinds of organism can exploit the various facets of the niche, and these may diverge. Then the organisms with their several ways of making a living in the niche will each specialise, and this leads to the classical kind of speciation when barriers to interbreeding occur. Whether this simple story[10] ever occurs is debatable[9], but it serves to show how numbers may rise when diversity broadens the demands made on the environment and dilutes the competition for each growing organism. Polymorphisms are very common in successful species, and this is one possible explanation: they reduce the pressure of soft selection by allowing greater numbers of organisms to live through each stage of the life history. The situation may be unstable and temporary, but there are many values to the parameters which give stable polymorphisms rather than speciation[10,12]. Or, if speciation occurs, there are some parameters that permit co-existence rather than extinction of one of the kinds of organism (*Figure 14.2*).

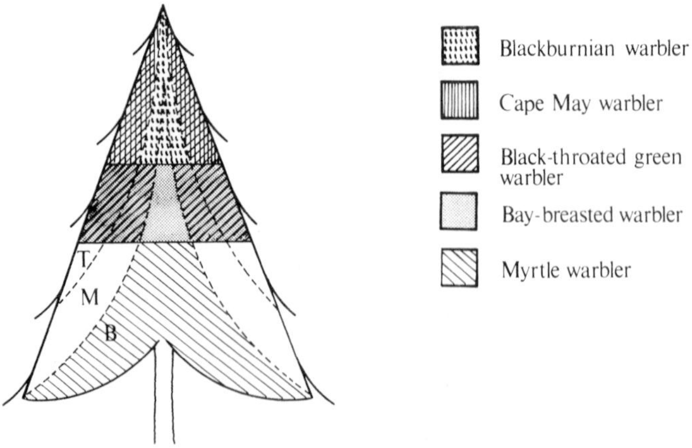

*Figure 14.2 THE SPECIALISATION OF WARBLERS (*DENDROICA*) IN SPRUCE TREES (from Salthe after MacArthur, 1958). All the warblers feed mostly on insects, but their preferred sites range from the older, lichen-covered lower limbs (myrtle) to the waving tips of the youngest branches (Blackburnian). Bay-breasted and Cape May warblers only occur in some numbers on bud-worm infested trees, giving a temporal as well as a spatial distribution pattern for these species. B, bases of old branches; M, middle zone of older needles; T, terminals of new branches, with buds and young needles*

14.5 Animal Societies

Much of the intraspecific competition we have been considering can be regulated by the formation of heterogeneous groups *within* the species, with the different ages or different **castes** fulfilling different roles in each group. The group then regulates its own numbers by internal controls and presents a 'super-organism' to the environment, a **society**. Within the society the environment is monitored and regulated so that homeostasis of animal interaction is achieved in the social milieu, as well as homeostasis of the internal milieu of each of the organismal units. Three kinds of animal society will exemplify this; insects, wolves, and baboons with multi-male troupes.

Two orders of insect, the *Hymenoptera* and the *Isoptera*, have social species. In both there is considerable variation in the pattern; the honey bee and a termite will be considered only briefly, as many excellent accounts are given elsewhere[13]. A **queen** bee, produced late in the season in a well-fed colony from a specially treated larva, receives spermatozoa from one or several males **(drones)** during the **nuptual flight**. These drones are themselves the products of unfertilised eggs (i.e. haploid). The queen is then effectively bisexual and can serve as the **foundress** of a colony. Queens take a group of **workers** with them **(a swarm)** when they leave the parent colony, in most races of the honey bee. After they have constructed a small **comb** the foundress commences egg-laying. The workers fill the cells of the comb with honey, pollen and other food for the developing larvae, which superficially resemble fly maggots. 'Bee milk', a secretion of worker mandibular glands, is fed briefly to all larvae but for the full 21 days to the future queens. When the new workers emerge from their cells they have attained the imaginal stage and appear female, but have only 8 to 15 ovarioles compared with a queen's 100 to 150. They pass through a fairly well-defined sequence of duties[14] in an established nest, ventilating, feeding larvae, then foraging when they are older; in a new nest, particularly, the sequence is more variable and one worker may perform a variety of tasks. In some bee varieties workers occasionally (e.g. *Apis mellifera capensis*[15]) lay eggs which hatch into more **diploid** workers without fertilisation, but usually any eggs laid by worker females produce drones if they survive. The workers of other kinds of honey bees will lay eggs if the queen is removed, but in a **queenright colony** rarely do so. This is because[13] the queen secretes 9-oxydec-2-enoic acid and 9-hydroxydecanoic acid; the latter inhibits ovarian development in the workers, and the former inhibits the production of more queens. These **pheromones** are synergic, i.e. act together more strongly than separately, and are spread through the colony passively or actively on the bees. The comparison with hormones in a single animal's body is useful but should not be extended. When the queen dies or is removed workers begin to lay eggs, but more importantly they provision a few of the new larval cells with **royal jelly**, which contains more lipid, vitamins B and E, biopterin, neopterin and pantothenic acid than the ordinary food fed to those larvae which are to become workers. Surprisingly it is not these substances which cause the **trophogenic** (food-produced) change to a queen, but a small molecular weight organic substance which passes dialysis membranes. Workers continue to feed queens the 'bee milk' diet. Poorly fed colonies, or those with a high larva/worker ratio, do not produce queens, but well-fed colonies may even produce a crop of young queens in autumn while still queenright. These will found new colonies, often by **sociotomy**[15], the splitting of the old colony and swarming of part of it.

Some species of bee have more than one queen per colony instead of the single queen of the honey bee, and the reproductive structure of these colonies has close parallels with the vital and trophic embryos produced by other species. In this case, however, the trophic individuals continue to supply food, shelter and even warmth to the reproductives. Competition between siblings has been bypassed and cooperation, except in egg-laying, has become the rule. Colonies, each the product and work force of a few reproductives who are usually siblings, do compete with other colonies. Ants, with a similar system, sometimes develop specific soldier-like workers for this inter-colony warfare, or for interspecific warfare, and here the reproductives avoid involvement in this dangerous, competitive but essentially vegetative activity.

The soldier system has been perfected in many termites. These *Isoptera* have males and females in all castes, including the reproductives, and it is almost certain that caste determination is again trophogenic. Unlike bees, whose maggots are only sensitive to royal jelly briefly during an early instar, the five larval instars of termites (at least of *Kalotermes*) are pluripotent, and they become either **soldiers** (via a **pre-soldier** instar) or **'pseudergates'**, which can become **sexuals** or soldiers (there are no workers in this genus).

The numbers and ratios of individuals are interesting in these insect colonies[15]. The sexuals lay many more eggs than do related non-social species, but fewer of these eggs, probably, develop as sexually competent individuals. The common wasp queen (*Vespula vulgaris*) lays about 22 000 eggs per year; honey bee queens lay about 120 000 eggs per year of which at most hundreds normally become queens. Small meadow ant (*Formica rufa rufa pratensis minor*) colonies may contain 25 000 queens and 1 500 000 workers. The African Driver Ant queen (*Anomma wilverthi*) may lay a brood of three to four million eggs every 25 days, or 40 million per year, a rate of egg-laying which averages out faster than one per second. Some termites do lay continuously; *Bellicositermes natalensis* lays 36 000 eggs daily, i.e. about 13 million per year probably for 6 to 10 years. But this is not comparable to the egg-laying of non-social insects, just as metazoan developmental mitoses are only in a trivial sense comparable with protozoan multiplication. The massive production is of the 'social soma', and only few of the eggs will be part of the germ line.

Wolves, and other wild canids, have only a few animals in a pack, and these are usually closely related. Only one pair, or the dominant female with some polyandry, breeds. The puppies as they grow up do compete a little, but they do so within the social framework of bluff and ritual[11].

An animal may starve to death without 'real' fighting for food. The appeasement gestures completely turn off attack, and so avoid nearly all intra-pack violence. Subordinate males and females have been observed to copulate, but they rarely produce offspring and these may not be suckled if they are actually born. These subordinates assist in obtaining food, in grooming and in defence, but they are banished from the pack if they attempt to challenge the leaders repeatedly and are defeated. Sooner or later, however, a challenger will be successful. Senility, or a wound, will guarantee the succession. So a wolf pack, unlike insect colonies, may have a social tradition which continues through a succession of leaders. One might presume that the habits of different packs would be different, and this has been described[11]. The young growing up in any one pack will learn slightly different patterns of behaviour which distinguish its own group from other packs. There is **cultural inheritance**, only made possible by the sheltered but interactive infancy of the puppies. Insect larvae are sheltered but are only chemically, not behaviourally, interactive; so behavioural or cultural traits have no continuity between generations in insects.

A multi-male baboon **troupe** has many more members than a wolf pack, perhaps as many as 70; chimpanzees and rhesus monkeys also live in large multi-male groups[13]. Here there is no restriction of breeding only to one pair, though a group of dominant animals does produce most of the offspring. Any female at the height of oestrus will be attractive to the dominant male, but will revert to her previous status as her hormonal state changes after ovulation. These baboon males (e.g. *Papio anubis*) form two groups. Several subdominant males determine, with the dominant male, which way the troupe will travel. Other males, lower in dominance, lurk at the periphery of the troupe but engage in some social interaction with it. This is related to their developmental status in the troupe. Babies born to low-status females seem unlikely to achieve high status, whereas those of a more dominant mother usually outrank others[16]. There is, of course, controversy over whether such dominance is genetic or environmental. In the view of such societies proposed here, the genetic determination of hormone levels and so on interacts with the hereditary determinants passed through the social milieu at the level at which the mother interacts with it. There is little determination, but interactions (like those of oocyte structure and zygote genetics) determine the later social and hormonal responses of each organism which develops in the society.

Life in these primate societies is much richer than in the wolf pack. Animals move up and down in **status**, there are peripheral animals, and offspring born at all levels of dominance grow up and respond to the social milieu which itself varies at these levels. So this society has brought back intraspecific competition, but it is now at a social level and uses social symbols instead of trophic symbols like territory. Among the social insects there is mechanical cooperation, among the wolves there is one dominant pair and lower-status non-reproductives, but in baboons there is change in status of individual animals within a stable social framework. The major predator of baboons is the leopard, but it is probable that more die from accident, disease, and starvation (especially the young of low-status females) than from predators. Banishment from the troupe has some effect on numbers, but it seems that adolescents may change troupes – there may be gain as well as loss, and it is very difficult to understand the population restriction of baboons. There is territorial activity, but its relevance to population control is difficult to detect[13]. This complex (and dubious) kind of control is common in mammals, especially in primates, and our ignorance of its mechanisms has implications for our study of human reproduction.

14.6 The Regulation of Human Reproduction

This subject has been discussed so widely in the last few years that little remains that needs emphasis. However, some baselines need to be drawn, because for many readers the 'planned' family of two or three children, and use of a contraceptive method such as the Pill or condoms, defines the ideal of human reproduction. This attitude may be seen in surveys to determine the additional morbidity or mortality associated with the Pill; the control population does not consist of naturally pregnant or lactating women, but of women in the same society who use other methods of birth control. This is a statistically proper comparison, but leads us to view pregnancy and lactation as the exceptional state for a woman. If no contraceptive precautions are taken, for a fertile couple, pregnancies and lactations alternate producing babies about every two years.

Except for societies like the Trobrianders, where it is said that they do not relate copulation to babies[17], human cultures have a tradition of separating the procreative, reproductive function of copulation from the recreational, pleasure-giving functions. It must be emphasised, however, that even without any attempt to divorce these functions by contraceptive methods nearly all acts of human copulation can have no direct reproductive function. This is because the woman does not ovulate while pregnant or, usually, while lactating; and coitus generally continues for most of this time. Even in the couple considered above, producing a child every two years, very few of the copulations have any chance at all of being fertile. If two copulations occur per week, then because fertility is possible during only one week each month and only three cycles can be fertile per two years in our example, a more realistic estimate is that only some 3 per 100 acts of intercourse can be fertile. Human sexual intercourse does not only have reproductive function directly; in most couples in monogamous cultures, this activity is enjoyed by both sexes and helps to keep the pair bond a close and loving one. Indeed, these reproductive and sensual functions may be physiologically distinct (p. 71).

Human beings of most cultures limit their reproductive potential in some way, usually because their life style involves disease, starvation and neonatal death. There is an apocryphal story of a fishing village which regulated its reproduction perfectly. A couple generally had only about two children, for the fishermen went to the local village to sell their catches as soon as they had their own boats, staying the night and bringing back gonorrhoea from the village prostitutes, which sealed their wives' oviducts and brought a speedy end to the fertile period. The story continues that a UN investigating team, who discovered this situation, brought in antibiotics to cure the prostitutes and their customers; now this village starves like all the others. Most human ways of limiting fertility are not so natural, and involve many possible kinds of action: abstinence, avoidance of coitus by other sexual practices, withdrawal, barrier methods, interference with the physiology of sperm transport, prevention of ovulation or implantation; early abortion and sterilisation are all practised by many people[18].

All these methods are in common use in western society, but research is continuing into other methods of fertility control. Immunological methods of several kinds have been suggested, including the making of antibodies against sex hormones, against human chorionic gonadotrophin (HCG) or other specifically placental antigens (without cross-reactivity to somatic hormones or tissues), or against sperm enzymes or other sperm antigens. Such antibodies could be induced from the woman's own immunological system leaving her effectively sterile until countermeasures are taken, or antibodies could be passively injected leaving the woman sterile for some months until they are finally lost from her system.

Some prostitutes, and a few other women, may already be infertile for immunological reasons[19]. Some of these women, despite previous fertility, no gonorrhoea which might seal the oviducts, and many acts of normal intercourse without precautions, still do not become pregnant. A large proportion of these can be shown to have circulating antibodies against sperms. So do many fertile women, however, and it remains to be shown that these antibodies of sterile women are different in a significant way. Do they perhaps coat those few sperms which fertile women permit to the oviducts, or do they have special lysing or immobilising, as well as agglutinating activity? If it should prove possible to use such a 'natural experiment' method of infertility induction, this would be preferred to other more radical methods because the preliminary test population has already shown its feasibility and safety.

There is an asymmetry in most contraceptive practice which relates to its voluntary, and somewhat technical nature. Some women *do* insert the oral contraceptive pill into the vagina, and there are stories of them being taken by men 'for a change'. Any so-called 'accident' results in a child, unless abortion is sought successfully. *Failure* produces some children in western society; a recent guess suggests that, as about a tenth of births are illegitimate, about a fifth are unintentional. This is borne out by a study of **rhesus factor** heredity using families re-housed in high-rise apartments in a large English city[20]; a large proportion of children have blood groups incompatible with their legal paternity[21] so, because compatibility is not proof, it was argued that 20 to 30% of births were from other than the legal father. Some, as some other illegitimate births, may be intentional; but it seems clear that if we could make *not* having babies easier than having them, we might get nearer to a rational reproductive arithmetic.

A more worrying question concerns not 'how many?' but 'which?'. The UK presently has approximately **zero population growth**; slightly more people died than were born in the first half of 1975 so, because we have a voluntary contraceptive service, this means that those people who are using contraceptives successfully are not replacing themselves. Clearly those who fail, forget or are feckless contribute more offspring to our future. This may not matter, for the inheritance (genetic or extragenic) of the children may not include fecklessness; it may also be important to promote lovingness, spontaneity and fecundity as well as dull responsibility. But I shall be much happier when a contraceptive method appears which is easily applied – once; thereafter people should be contraceptive until they take action to become receptive instead of, as now, taking action to become contraceptive. Only in this way can the elements of human reproductive strategy be combined to maximise loving sexuality, strengthening the pair bond and allowing the responsible production of wanted children from couples who opt into reproductive status.

Some discussion about human population statistics and predictions has used a very misleading figure, **death rate**. This, the number of deaths per annum per 1000 population, is used as equivalent to birth rate; indeed the two are usually compared to give an indication whether the population is growing, or more usually whether the rate of growth is changing. Birth rate and death rate are not equivalent – they are equal; 'the death rate is the same for us as for everybody – one person, one death, sooner or later'[22] or, in our terms, one birth, one death, sooner or later. All that crude death rate measures is gain or loss of population since those deaths were born; if gain, their death comes to less per 1000, not because less die but because there are more thousands. At best, death rate is a

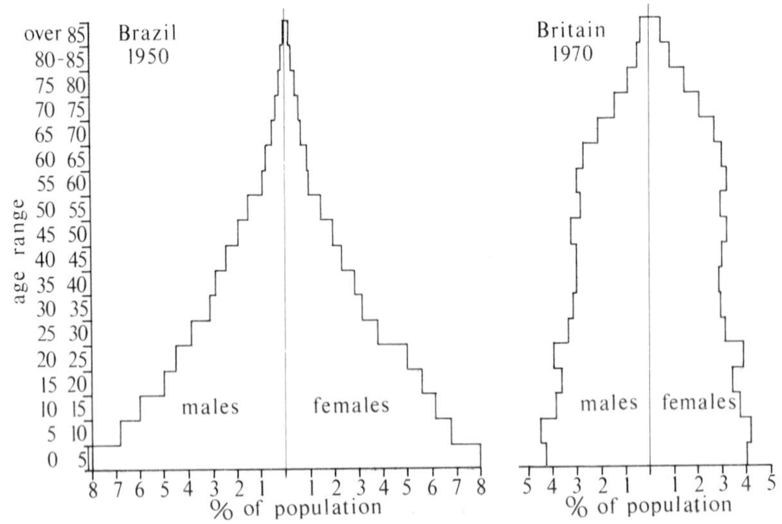

Figure 14.3 POPULATION PYRAMIDS (after Fremlin, 1972). See text for discussion

measure of population change; in practice, it measures recent birth rate, immigration, emigration, variation in survival, and the peculiarities of different cohorts of the population, with all their errors, in one useless figure. A population pyramid (*Figure 14.3*) is much more useful. It includes a prophecy, for it shows the size of the breeding population of the future. For most developing countries, this prophecy is a dire threat which crude death rate per birth rate doesn't hint at. When half of a population is under 15 years old, and effectively all are expected to breed, there is something of locust or lemming about the future, but without the possibility of control measures such as would be used for these pests. Public health measures, and some changes in medical practice, are supposed to account for such population explosions. Clearly, food production and the unsuitability of contraceptive philosophies or methods also contribute. If the public health measures can be maintained in the new population, there is hope that fewer babies will be born, as affluence, and particularly the obvious survival of most babies, become fulfilled political promises. If not, then infant survival will again drop, and juvenile misery and death will reward adult fecklessness as in all of human history. We would then show that we have learnt little of value since Malthus.

REFERENCES

1. May, 1974
2. Orwin, 1975
3. Jones, Alderson and Howell, 1974
4. Wynne-Edwards, 1962
5. Ardrey, 1967
6. Watson and Miller, 1971
7. Williams, Koehn and Mitton, 1973
8. Beardmore, 1976
9. Lewontin, 1974
10. Shepherd, 1975
11. Van Lawick-Goodall and Van Lawick-Goodall, 1971
12. Wallace, 1970
13. Brown, 1975
14. Lindauer, quoted in Brown, 1975
15. Engelmann, 1970
16. Koford and Beach, 1963
17. Ford and Beach, 1953
18. Potts, 1972
19. Shulman, 1971
20. McLaren, 1975
21. Phillip, 1973
22. Heinlein, 1955

14.7 Questions

1. Relate 'hard' and 'soft' selection (p. 19) to K- and *r*-selection (Glossary, pp. 313 and 316).

2. Attempt to distinguish intra- and interspecific competition in the life cycle of a wild species you know well.

3. To what extent do you think that sociality is a reproductive tactic, employing the 'reject' organisms as 'slaves' to the breeders?

4. Is there a 'natural' reproductive strategy for human beings, against which contemporary practice should be measured?

5. Relate territoriality to the reproduction of parental number. Is there a similar process in domestic animals or human beings?

15

REPRODUCTION AND EVOLUTION

15.1 Embryos and Evolution

The increase of complexity during development has its parallels in the sequence of living things in geological time, and in the scale of complexity to be found in nature now, the **scala naturae** of the natural philosophers. The word 'evolution' used to mean 'development' and we still speak of the evolution of an idea, or of a gas from a liquid. There is still confusion because of these parallels between the sequence of developmental stages in one organism's lifespan **(ontogeny)** and the sequence of ancestral structures during its evolutionary history (**phylogeny**). Confusion about relationship of ontogeny and phylogeny has been at the root of many of the great biological arguments of the past[1]. There are obviously some parallels between the stages of development and of fossil history, but embryology during the nineteenth century was stultified by Haeckel's belief[2] that phylogeny caused ontogeny. This, his **principle of recapitulation**, stated that organisms had to pass through all their phylogeny during each life history.

The final dissolution of this belief in the historical causality of development was produced by de Beer's complete, penetrating and scholarly analysis[1] in *Embryos and Ancestors*. Embryos do not recapitulate their sequence of adult ancestors, 'climb their own family tree'. Instead, because the instructions for development which they have inherited resemble those of their ancestors to a greater or lesser extent, the embryonic structures and stages formed in response to those instructions will be similar. The embryo of the descendant resembles the embryo of the ancestor because both have similar sets of developmental prescriptions. One does not cause the other, any more than a man can cause his cousin. This concept frees us from the tyranny of the historical causation idea, and we now look at embryology in a different way. Woodger gave a useful comparison, according to de Beer:

> 'Rockets go up on November 5th (a) because of an historic tradition in virtue of which the practice of firing rockets is repeated each year on this day, (b) because the rockets contain charges of a substance, gunpowder, whose properties are to undergo rapid combustion and to produce powerful gaseous expansion resulting in the exertion of force on the rockets. The details of the properties of gunpowder are not deducible from a study of the biography of Guido Fawkes; nor are those of a fertilised egg from a study of the phylogenetic series of

adults in the evolutionary history of its species. Evolution therefore does not explain embryology.'

Nevertheless, as de Beer shows in full measure, the similarities and differences in the developmental instructions of ancestor and descendant, or of two different descendants from the same ancestor, are a fascinating study. Throughout our study of reproduction we must remain aware that reproduction in the short term produces variety, and in the long term produces either evolutionary change or extinction. Exceptions, like *Lingula* which appears to have lasted unchanged for some 300 million years, we treasure as examples to test our generalisations.

15.2 Evolution and the Life Cycle

Evolutionary changes can make their appearance at any stage of the life history. It cannot be emphasised too strongly that adaptive changes produced by mutation can affect any organ at any pre- or post-phyletic stage of the life history. If recapitulation *did* occur universally, then eggs, blastulae, gastrulae would be similar in all vertebrates. As we saw when we considered larvae (Chapter 10) there is often more diversity at an early stage in the life history of a group than among the adults, showing that some earlier stages have changed further from the ancestral early stage than the adults have from the ancestral adult. This is not common, because adults usually retain some larval changes to their own advantage, or perhaps, as in gastropod torsion, disadvantage. The ways in which life histories may change, as innovations are acquired by a species, are shown in *Figure 15.1*.

Adult variation is the first picture to come to mind. After a very similar development, new morphological characters appear in the adult form. It is most obvious, of course, where metamorphosis to the imago occurs, as in insects. Very nearly all of the gene changes studied in *Drosophila* are of this kind. It probably has little or no importance in evolution above the varietal level, except for one major kind of variation, in breeding habits. Different breeding habits are often found in similar species, and this is a very common kind of variation. *Creagrutus beni*[3] is a South American characin (like *Astyanax*, *see below*) looking rather like a sardine. Most characins breed in small groups, of 2 to 20, and scatter slightly adhesive eggs among fine-leafed water plants, the milt emitted by the males fertilising them. But *Creagrutus* associates in pairs (or more) and the milt is emitted but no eggs; a day or more later the female deposits fertile, developing eggs on water plants, having retained them in her oviducts. Perhaps she draws in water containing the sperms, for the male has no intromittent organ. This behaviour, like mouth-breeding in cichlids (p. 55), can be regarded as an adult variation, often called a **'specialisation'**; it can clearly be seen as a 'stage on the way to' a more advanced sexual congress involving copulation.

Deviation is the acquisition, by an early stage in the life history, of an innovation which directs the later part of the life history along a new

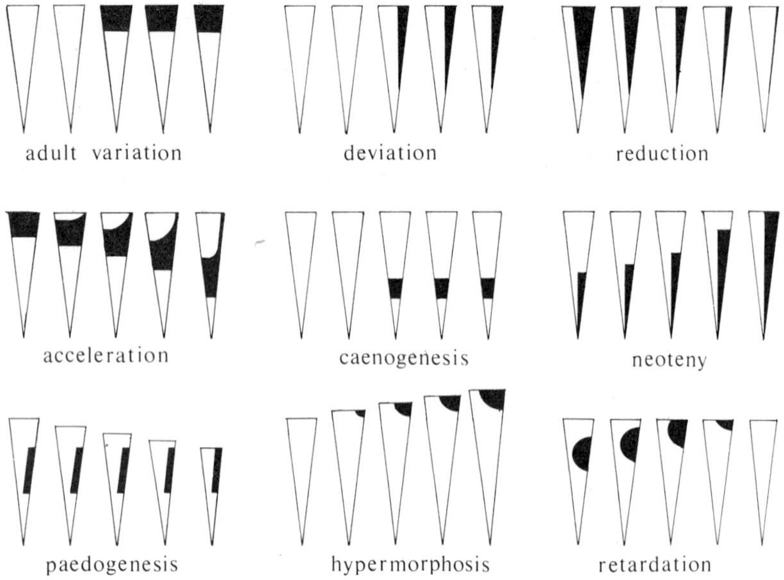

Figure 15.1 ***EVOLUTIONARY RELATIONSHIPS OF ONTOGENY AND PHYLOGENY (mostly after de Beer). Each diagram is of five life histories (ontogenies) arranged as ancestors (left) to descendants (right) i.e. as a phylogeny. Individual development time reads up the page, and geological time reads left-to-right for each sequence. The black part of each life history refers to the interesting character. For example, adult variation reads: two successive ancestors had uninteresting adults, but the three descendants show innovation in their adult stage. See text for the other examples***

route. It would be expected that few of these routes would be viable, but is is surprising how many dramatic cases of deviation can be found. Many of them may, however, have been achieved in three steps: a neoteny/paedogenesis, followed by a period of 'consolidation' of the old larva as new adult, followed by hypermorphosis (*see below*). The barnacles, which have a sessile adult but 'normal' crustacean larvae (*Figure 10.5*, p. 184), presumably had a more conventional crustacean adult ancestor. *Sacculina* (p. 186 and *Figure 10.6*, p. 185) presumably had an ancestor whose adult state was a barnacle, but the present adult resembles a fungal mycelium more than a crustacean. Sometimes it is surprising what extreme larval modifications can be made *without* apparently affecting the adult, as in dipteran flies (*Figure 10.2*, p. 181) or frogs.

Sometimes a dramatic larval innovation, like **torsion** in gastropod molluscs, causes adult modifications of extreme kinds too. The torsional twist brings the anus over the head, and later development has been modified in all gastropods to avoid the unfortunate results for the adult. Many untwist to a greater or lesser extent and *Haliotis*, the ormer, puts its anus out of a succession of holes in its shell.

Many kinds of species differences, which seem at first sight to be simply deviational, can on further examination be seen to be a **series of deviations** affecting juvenile and adult, which have diverged as a result.

No single juvenile difference remains to distinguish the adults. For example, *Astyanax* is a genus of fairly common small river fish (characins), with many species common in Central America[3]. *A. fasciatus* is probably the most generally distributed, with many varieties including *A. fasciatus macrophthalmus*, with large eyes (*Figure 15.2*). Near San Luis Potosi, in Mexico, *A. fasciatus mexicanus* is common, and there is a series of subterranean rivers and pools in which occurs *Anoptichthys jordani*, very like *A. f. mexicanus* in shape but with no skin pigment and with no apparent eyes. This fish, the 'blind cave fish' of tropical aquarists[3], is clearly a simple deviation from the silver, eyed species in the open rivers – until a breeding analysis is performed. Sadoglu[4] has crossed blind cave fishes from several different caves, and she has found that the first generation are all commonly both silver and eyed, i.e. that the genetic reasons for lack of pigment and lack of eyes are different in different cave fish populations, and that the deficiencies are complemented in the first cross. Deviation has, then, occurred in different genetic ways to give phenotypically identical (or at least very similar) phenotypes. By rearing the second generation (in which the differences segregate) in dark or light and in various nutritional states, she has been able to determine that eyed fish survive less well in the dark, perhaps because the eyes get bruised and infected, and possibly that albino fish have a more efficient use of food protein, perhaps because they waste less tyrosine and phenylalanine on making melanin. This case of deviation, then, has at least two differences between the new form and the 'wild type', and these differences are themselves different for the different new races.

Reduction is a common evolutionary trick. A structure present in the ancestral juvenile and adult is lost from the descendant adult, or is not present in effective form. The lizard family *Scincidae* shows a great variety of locomotory forms; there are many burrowers, which have lost

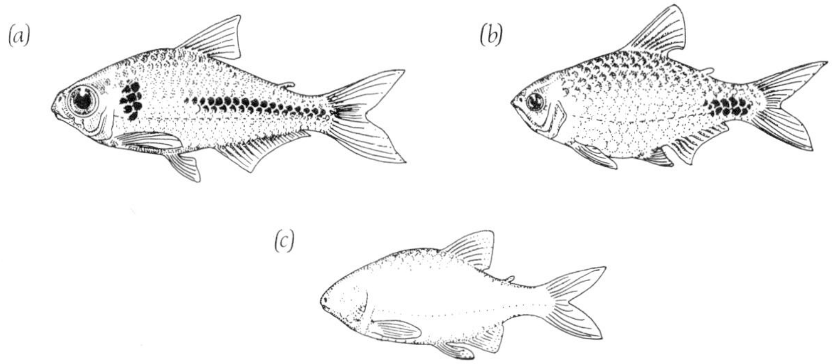

Figure 15.2 THREE CENTRAL AMERICAN CHARACINS (after Sterba). (a) Astyanax fasciatus macrophthalmus. *Note the large eyes. (b)* Astyanax mexicanus. *The eyes are of normal size. (c)* Anoptichthys jordani *(cave form of* Astyanax mexicanus*). This fish has no apparent eyes, and is barely pigmented (white or pink). See text for consideration of the genetic relationships between the cave and river forms*

their limbs (like the slow worms) or only retain vestiges (*Figure 15.3*); others have small (*Tiliqua*) or normal legs. The new-hatched or newborn (many are ovoviviparous or viviparous) usually have reasonably proportioned limbs, but these 'fall behind' as the animal grows, i.e. they show a fractional allometric exponent (p. 222).

Vestigial organs are said to be derived by reduction, and this explains their presence in some cases. The human appendix is an example, as are the muscles for moving the ears, and the coccyx. However, some structures which appear briefly during development, and which may have been useful in adult ancestors, for example the notochord and the pronephros of vertebrates, still have a crucial function for each embryo. The **notochord** is that part of the embryo which our kind of prescription uses to induce neural tissue from the overlying ectoderm, and which later serves as the basis of the spinal column. The **pronephros**, the anterior or head-kidney, functions to initiate kidney tubules in the nephrotomic mesoderm, and passes the stimulus back to the functional **mesonephros** which in turn sends back the **Wolffian duct**, whose outgrowth induces the adult kidney of amniotes, the **metanephros**. The mesonephros itself is 'vestigial' in female amniotes, but very necessary in their males, where it forms the **vasa efferentia**, the testis ducts. Invention of a new way of making old organs does occur, but it is rare; as long as the old bit of prescription works, it seems to be retained even if structures seem to be 'wasted'.

Some of this 'morphogenetic waste' is involved in male–female difference, like the mesonephros quoted above; so that a structure which in one sex is useful is often a vestige in the other sex. Nipples in the male are the obvious example, but **uterus masculinus** and **epi-oophoron** are others. Vertebrates generally use a dual-purpose prescription but exaggerate some characters hormonally to distinguish the sexes, unlike the insects; however,

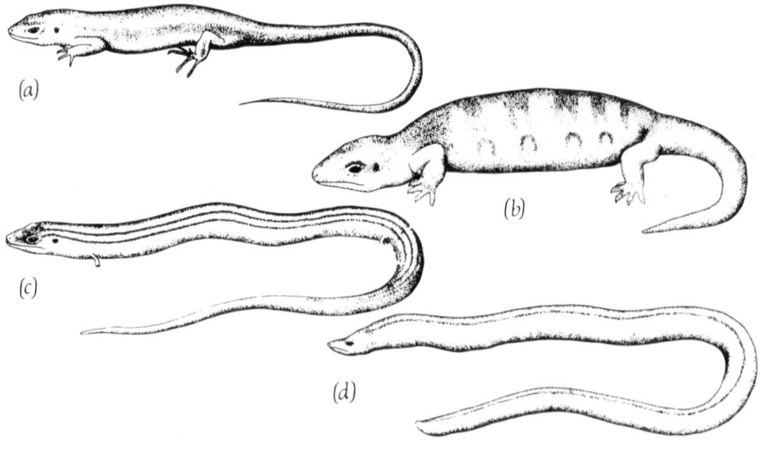

Figure 15.3 FOUR LIZARDS IN THE FAMILY SCINCIDAE *(a)* Eumeces, *a common skink. (b)* Tiliqua scincoides *(Australian blue-tongued skink). (c)* Chalcides chalcides, *a burrowing skink. (d)* Feylinea currori, *the limbless skink. This series shows reduction of relative leg size in the adults; all the embryos have apparently normal limb-buds*

even the insects have male vestiges in the female and the converse. This emphasises the function of reduction but not complete loss of an organ. The prescription may require its presence for another organ to develop; or there may be a quite different necessity, for example the enzyme prescribed by that genetic instruction may be required for quite another necessary function, i.e. the gene may be **pleiotropic**.

Acceleration of an organ system may attain the same end as reduction, and will give the appearance of recapitulation. A structure developed in the adult ancestor is developed earlier and earlier in the descendants until it appears in late embryology; it may or may not be retained into the adulthood of the descendant. *Care must be taken to exclude the obvious recapitulatory examples, like gill pouches, from this category*[1]*; they are examples of embryonic resemblance to ancestral embryo because of similar prescription.* 'The 'grasp reflex' of babies is a marginal case whose provenance is nearly in this category, and even de Beer finds few and poor examples of acceleration. The development of tubes from coelom to exterior was probably originally an adult device to release sex products; it now occurs in most coelomate embryos. The most dramatic example of apparent recapitulation concerns the teeth of sirenians (manatees and dugongs). The embryo first shows unworn cusps on its teeth, then long before it can grind them it produces flat, as if worn, cusps presumed to be like those of its adult ancestor[1].

Caenogenesis is the acquisition, by an early developmental stage, of a morphological adaptation for its needs at that time, unrelated to adult needs as far as we can tell. Examples include most larval adaptations like ciliary bands on planktonic larvae, the tail of the cercaria, torsion in gastropod mollusc embryos, the embryonic contribution to the placenta, yolk sacs, and so on. The importance of caenogenetic adaptations to evolution is usually not great when they appear, but they may later be 'exposed' as an adult adaptation if the later part of the life history is lost by paedogenesis or neoteny. So the axolotl must live with permanent, not temporary, external gills which were originally caenogenetic, for larval use only.

Neoteny is a greatly abused term, and should be restricted to the delay, appearing later in the life history, of a character shown only by the larval stage of the ancestor but now present in the adult descendant.

Paedogenesis is its converse, the earlier attainment of sexual maturity by the descendant, so that the larva breeds. Characters may be neotenous, but the organism is paedogenetic. It is often practically difficult to distinguish these two processes, even by comparing maturation time with related forms, and both are commonly subsumed under **paedomorphosis**. Examples abound, and paedomorphosis has doubtless provided the major mechanism for evolutionary innovation on a grand scale[1]. A Cambrian echinoderm, with a larva like modern echinoderms, lost its sessile adult phase as the planktonic larva acquired precocious sexuality, we are told, and hence the chordates (*Figure 10.1*, p. 179). There are some grounds, too, for supposing many of man's most human characters, including flat face, large brain, relative hairlessness, unspecialised hands, and continuing curiosity into adulthood, to be juvenile characters of our ancestors. They are all juvenile characters of our cousins the great apes, and it is tempting to see them as having persisted longer in each human life cycle until we reached our present happy state. But the evidence is, by and large, against this romantic view of man as the baby of the primates. The common ancestor of modern apes and men, at the time of *Ramapithecus* and the dryopithecines some 20 million years ago, had at least as many human traits as pongoid (ape-like). The three great apes may indeed have overlain the curiosity of this common ancestor with a morose (gorilla), manic (chimpanzee) or depressive (orang) adult outlook. Dobell[5], writing in 1911 of evolutionary argument, declared that it was just as logical to speak of the man-like ancestor of the apes as the ape-like ancestor of man, the same creature. Fossil evidence is now turning this into more than a philosophical point, and makes man look deviant rather than simply paedomorphic, and the great apes all a bit hypermorphic (*see below*).

Hypermorphosis is the opposite tactic from paedogenesis, and is characterised by a longer-lived descendant who may grow bigger and stronger before breeding. Organs which were small in the adult ancestor can be given time to grow in the descendant juvenile before it becomes an adult. There can be additional advantage if the character is positively allometric with body size (p. 222) for then a small increase of body size gives a more than proportional advantage of organ size. Classical examples are the horns of titanotheres and the canines of sabre-tooth tigers; as these animals increased in size during geological time their horns and teeth enlarged faster, until some of the later forms seemed ludicrously overdone (*Figure 15.4*). The suggestion

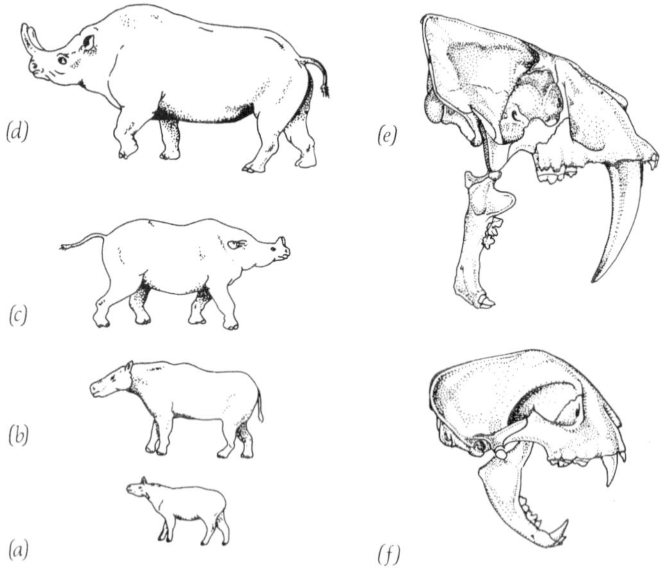

Figure 15.4 ALLOMETRY IN TITANOTHERES AND CATS (a) An early titanothere, Eotitanops. *(b) A later form,* Dolichorhinos. *(c) Later and larger,* Brontotherium leidyi. *(d) Still larger,* Brontotherium platyceros, *(from Piveteau, 1958, after Osborne). (e) Skull of* Smilodon, *a sabre-tooth tiger (after Romer, 1959). (f) Cat skull, for comparison*

that each maintained selection pressure on the other to get bigger, so titanothere could beat off sabre-tooth and sabre-tooth might kill titanothere, looks plausible until one looks up their actual distributions; the titanotheres probably did evolve slowly, but fairly extreme sabre-teeth are found in the earliest machaerodonts. The principle, however, might apply elsewhere. A pair of species might each exert selective pressure on the other until both are forced into an 'evolutionary blind alley'; mastodon/sabre-tooth has been suggested by Romer. I do not believe such to exist but many biologists do, and they also include male display structures like peacock's tail, bowers, and even antlers like those of the Irish elk. For they argue that if sexual selection always chooses the more exaggerated tail, or comb, or whatever, there is no way back to moderation. On the other hand, of course, one could argue that those too grotesque to survive will not be able to breed. So it is difficult to see a whole species, with all its variation, driven over this evolutionary precipice by the relentless force of its reproductive necessity.

Retardation is another of de Beer's categories of evolutionary change during ontogeny, and involves delay in initiation of an organ in the descendant compared with the ancestor, or a slower rate of growth of the organ. The 'wisdom' teeth of humans are the best examples of this. They are third molars which only just erupt before growth stops in the twenties, or fail to erupt at all; trouble is caused when they attempt eruption in a jaw too old to make adjustments, so that they 'impact' between the second molar and the angle of the mandible. Our ancestors had three molars and we seem to be caught half-way to losing them. Whether we will lose them cannot be predicted, as there is no **'evolutionary momentum'**; each generation interacts with its own environment, and there is no historical imperative to continue a trend set by ancestors. Some proponents of this 'momentum' or **orthogenesis** (evolution along straight lines) pointed to evolution of the horse as an example of striving upward and onward along a path started by *Eohippus*. G.G. Simpson[6] has demonstrated, however, that the appearance of simple increase in horsiness is an artefact created by palaeontologists from a great plethora of fossils of this group.

It must be emphasised that the above categories and examples refer mostly to morphological characters. Some behavioural examples are given, but no biochemical, for example. The reader may decide for himself whether the ammonia, then urea, then uric acid excreted by the chick *in ovo* represents its phylogenetic history or present necessities, and consider whether the haemoglobin ϵ chain is caenogenetic.

15.3 Polymorphism

There is another question not fully considered by de Beer and yet very relevant to the evolution of reproduction, and this concerns the divergent specialisations of alternating or succeeding generations, as in hydrozoans, or aphids. The hydrozoan example is best considered as a larval multiplication but the aphids are clearly a divergence from a *Daphnia*-like situation, with sexually produced eggs overwintering and hatching into a time of plenty, capitalised by parthenogenesis. Divergence of such species as *Pemphigus betae* on to two hosts enables the two generations not to compete directly, as in species with larvae, but here either generation can increase its numbers and so be a reservoir for the other, should one or other plant have a bad year.

It is possible to regard the assortment of related species which divide a niche, as for example the warblers in *Figure 14.2*, p. 240, as a divergent super-species. If any of the species becomes extinct, its niche will surely be filled by divergence from one of the other species. So the species group has an adaptability not possible to any single species. Although no appeal is made to group selection, such a diverse colonisation of a niche probably gives a more stable exploitation than would a single versatile species, for periods of intraspecific competition alternate with interspecific. This will only be marginal if the species overlap as little as the warblers in *Figure 14.2*; of course, we have no idea whether there is intense pressure on any of these species by the others. **'Species flocks'** of cichlids in the African Rift Valley lakes seem to possess about the same spectrum of types in each lake, including floating algae feeders, attached algae feeders, carnivores, specialists which eat scales and finrays from the caudal peduncle and fin of other fishes, and **paedophagous** forms which specialise in sucking the young from the mouths of mouth-brooding relatives[7,8]! Such a species flock exploits many of the lake's food chains, and when the fish are caught from the wild their stomach contents confirm the diet supposed from examination of their dentition. But in aquarium tanks all seem to take commercial brands of fish food, and only to 'top up' with their speciality. This suggests that conditions in the wild are far more stringent, driving each species to 'do its own thing' so that little effective competition exists. The existence of a similar spectrum in each lake, from a *different* founder species[7] in each case, argues for the super-species coloniser stratagem I have been describing. The variety engendered during the over-reproduction of the early generations persists as adult variety and is fixed as species types, instead of failing to survive in each generation.

The aphids with their different generations, the social insects with their different castes, the hydrozoans with planula, polyps and medusae, and the digeneans with reproductive stages both in snails and vertebrates all contrive to have two or more kinds of interaction with their environment. Sometimes male and female differ, as in mosquitoes where the male is a nectar feeder and the female requires blood meals to lay eggs, but this situation is not comparable to the above list because *both* **morphs**, not *either*, are required for maintenance of the species. The same is true of the larval form without larval multiplication. Larvae can only make adults and adults can only make larvae; neither can 'mark time' waiting for the *other* niche to open again. The polymorphs can, whether they are spatial or temporal polymorphs, provided they can maintain a population in more than one niche.

Population geneticists see genetic polymorphisms in single populations as serving a similar end, and adding to the versatility of the species[9]. The reproduction of the variety is necessary, however, if the versatility is to be maintained. This is difficult to imagine in a normal situation of stabilising selection, trimming the ends of population heterogeneities in each generation. Heterosis, where few offspring are produced and *s* and *t* are both less than 1 (*see* p. 25), does maintain variety, but heterosis makes for *less* variety among breeders if large numbers of homozygous offspring form the recombinational genetic load. Different offspring may survive when conditions change, however, so this still increases versatility for each breeder.

Some odd selections have been shown to foster polymorphism[9]. For example, populations of snails with several shell marking morphs are less likely to be decimated by thrushes, because of the 'search image' retention of the predator. The thrush, finding a striped snail and being rewarded, will then search for striped shells and take no notice of other morphs, e.g. plain or blotched shells. It is important to realise that this population will produce all the variety again in the next generation, for **variety** is being selected for. It is easy to see, however, how this polymorphism can lead to speciation. The species flock would then maintain the apparent polymorphism, but on a much longer time scale; if striped snails become extinct in a species flock, many generations may elapse before that morph is regenerated. Cases can easily be invented where this would be useful, for example in long cycles of temperature fluctuation, like ice ages.

15.4 Variation and Redundancy

The expressed variation of the polymorphisms of caste, stage, generation or breeder morph considered above are all reproductive stratagems which minimise the ecological pressure of the species, by spreading it over a wider area. They only secondarily have the occasional benefit of allowing evolution into the various avenues of exploitation taken by the various morphs.

In many species there is **cryptic polymorphism**, in the sense that juveniles possessing the variations do not survive to adulthood, so breeders with certain morphological characters do not normally appear. If a tetrameric enzyme (like lactic dehydrogenase (LDH) and many others) requires AAAB or AABB to work in the larva, and the gene locus is not (yet?) duplicated[10], only AB heterozygotes can survive to adulthood, because BB and AA larvae die. A change of temperature may permit homozygotes to survive marginally, allowing AAAA (genetically AA) adults to appear suddenly a whole new class in the adult population. This kind of change probably occurs frequently, but with no dramatic effect. Some cases will occur, however, with compensating defects in the AABB larva at the new temperature, allowing survival of a morph very different from the normal yet from the same reproductive event. For many generations these might be produced as 'sports' among the breeders, and it might make little difference whether they breed. A change of circumstance, new marginal niches or change in food supply, might foster the sport as a breeder; then there are a variety of evolutionary tricks to **fix** the odd allelic combination uniquely required to survive both larval and adult life. Judging by *Drosophila*, chromosome **translocations** and **inversions** are among the common methods used to avoid crossing-over and so to retain heterozygosity or homozygosity in a particular chromsome region[11]. **Gene duplication** has fixed allelic combinations in vertebrate haemoglobin and LDH[10], and according to Ohno this is the important evolutionary stratagem; inversions/translocations seem to be a temporary, varietal, tactic.

In organisms with a vast spectrum of zygotic types, of course, no such fixation would be required because cryptic variation occurs in every set of progeny. Such fixation would in fact destroy the value of the large progeny variety. A rather different population of breeders survives, perhaps, when developmental conditions vary, but they generate effectively the same zygotic spectrum because they are still highly heterozygotic at many loci. In Chapter 2 we used flatfishes as our example of such a reproductive tactic. So cryptic polymorphism, normally lost during development, may provide evolutionary material for speciation by fixing the gene combination which allows rare survival to adult; or, on the other hand, a profligate strategy of reproduction may allow a species to be versatile without fixation, by selection of different kinds of breeders in different circumstances without loss of alleles.

It is interesting in this connection that both *Drosophila* and flatfishes,

Table 15.1 DNA weights in nuclei of various species. (Data from Rees and Jones, and Ohno)

Organism	*Common name*	*DNA wt/nucleus*
Phage *T4*		0.000025
Escherichia coli	Bacterium	0.009
Aspergillus nidulans	Fungus	0.044
Cucurbita pepo	Pumpkin	2.6
Helianthus	Sunflower	9.8
Tradescantia virginiana	Tradescantia	70.8
Gryllus	Cricket	12.0
Lytechinus	Sea urchin	1.8
	'Tube sponge'	0.1
Drosophila	Fruit fly	0.2
Ciona	Sea squirt	0.21
Eptatretus	Hagfish	2.80
Scaphirhynchus	Sturgeon	1.75
Lepidosiren	South American lungfish	123.90
Clupea	Herring	1.00
Corydoras	Catfish	4.40
Anguilla	Eel	1.40
Xiphophorus	Swordtail	0.80
Pleuronichthys	Turbot	0.63
Xystreurys	Sole	0.80
Hippocampus	Sea horse	0.60
Ambystoma mexicanum	Axolotl	c.61.0
Amphiuma means	Congo eel	168.0
Triturus alpestris	Alpine newt	77.6
Bufo bufo	Common toad	15.8
	'Alligator'	5.0
	'Black racer snake'	2.9
	'Domestic fowl'	2.3
	'Sparrow'	1.9
	'Kangaroo'	6.2
Homo	Man	6.0
Mus	Mouse	5.0

perhaps using these two opposite ways of maintaining cryptic polymorphism without gametic selection, have very small amounts of DNA in their nuclei[10] (*Table 15.1*). Organisms which show great gametic redundancy, and therefore possibly selection, have much larger amounts of DNA. This may be because the genome has been duplicated to form 'permanent heterozygotes', as in LDH with its M and H loci and haemoglobin with ϵ, ϕ, α and β loci, which were but one locus in our presumed agnathan ancestor (as they are in modern lampreys)[10]. This is a method of 'fixing' heterozygosity which, unlike the translocations/inversions apparently employed by *Drosophila*, allows the maintenance of a larger pool of cryptic polymorphism by gene rearrangements. It also provides a supply of cryptic genetic possibilities 'at one level down', as the unused duplicated alleles mutate and occasionally produce something useful. Ohno[10] has documented many examples of such 'invention' of new enzymes by unused duplicate genes.

Reproductive tactics which employ frank polyploidy are common among angiosperms, rare among animals. There are a few animals which employ comparable tactics, including some oligochaetes whose soma is effectively triploid, probably **allotriploid** in origin (from hybridisation between different species). These worms duplicate the chromosomes of the germ cells by a strange endomitosis just before meiosis[12], so that the spermatogonia and oogonia are hexaploid briefly, giving ample opportunity for assortment of the loci, and so for the cryptic polymorphisms associated with such gene duplications.

Redundancy, with the possibility of a cryptic polymorphism, may therefore appear at four levels. It may be zygotic, in plaice and oysters (and bees?); it may be gametic in man and locust; or it may be cytogenetic, as in axolotl (possibly) or some earthworms. Or it may be genomic: tandem duplication, gene repetition, short sequence reiteration or even nonsense sequences all provide material for evolutionary experiment without loss of a working prescription.

15.5 The Evolution of Reproduction

Little remains to be added to what has already been said, but some points may be repeated for emphasis. That breeding tactics and reproductive strategies may evolve, like all other facets of an organism's life, is obvious. Distinction has been made here between reproductive **strategies** like parthenogenesis, polymorphism and polyploidy and **tactics** like viviparity, copulation or lactation. These latter are minor solutions to species-specific problems, whereas the strategies are seen as grand plans fitting in with the whole ecology of that kind of organism. Tactics will vary between similar forms, but they will use the same overall strategy. Therefore we might expect to find tactics to be mutable during the course of evolution, but strategies to remain characteristic of large groups for all of their existence. By and large we do find that, for example, urochordates rely on budding for increase in numbers, but maintain sexuality, presumably as a diversity generator because they may not increase their numbers this way (*Figures 4.9*, p. 88 and *11.2*, p. 207). Hydrozoans have alternation of generations and even when, as in trachyline medusae, polyps are absent the medusa often makes up for this by indulging in budding itself[13]. Echinoderms use the larval stratagem, as do insects. But vertebrates seem to have adopted a 'yolk' system, and few of them have true larvae and metamorphosis. On the other hand, viviparity turns up sporadically everywhere, as do spermatophores and true copulation, egg jellies or males looking after the eggs.

Within each strategy, however, can be distinguished primitive and advanced ways of adopting it. Advanced methods of reproduction commonly employ many obvious tactics, and result in a very predictable cycle of environments for the growing organism, which in its turn allows developmental complexity.

Attention has classically been drawn to the distinctions between external and internal fertilisation and between oviparity and viviparity. These are, however, tactical solutions to specific problems and we must use distinctions more subtle than these if we are to judge the degree of sophistication of reproductive methods. We might, for example, consider the 'wastefulness' of the system. Nearly all of the elephant's reproductive effort seems to produce more elephants, but nearly all of an oyster's reproductive products seem simply to feed other organisms. We might ask about the degree, and nature, of the intraspecific control systems which regulate reproductive behaviour and numbers; palolo worms, aphids and mice might all be judged advanced by this criterion. Or we might consider complication of the system to be a sign of sophistication, and give the highest marks to *Doliolum*, *Heteropeza* and the flea (*Figures 3.3*, p. 58, *10.4*, p. 183 and *11.2*, p. 207).

If the words **'primitive'** and **'advanced'** have any meaning in biology, then primitive organisms are considerably constrained by environmental factors, while advanced organisms control their internal milieu and choose their environments to suit their physiologies. Primitive animals are **poikilosmotic** and **poikilothermic**, i.e. they regulate neither internal salt concentration nor internal temperature, and in consequence they are 'at the mercy of' environmental change in these variables. Advanced animals, like mammals for example, regulate many of their internal processes by internal feedback mechanisms. Reptiles and insects seek temperatures which suit them, mammals and birds tolerate wider extremes of temperatures because they regulate their internal environments independently of fairly wide excursions in outside temperature.

There are many animals which have 'internalised' reproductive processes so that they are controlled by the parent organisms' physiology rather than by the vicissitudes of the outside environment. *Glossina* and *Gyrodactylus* are obvious examples, in which an adult reproduces an adult directly, the entire development toward that adult occurring within the parent. So parental behaviour and physiology regulate the entire sequence of developmental milieu. Clearly this is more effective, because the adult is better able to regulate its internal milieu.

The mammals really do seem to be those organisms on this planet most independent of environmental change, the most advanced physiologically. By using this internal constancy to protect the developing offspring from outside variation, by the complex workings of parents' metabolism, the mammal allows itself just that predictable environment in which such a complex organism can make itself. *Its physiological complexity ensures developmental stability, but would not be possible without it.* This 'pulling oneself up by one's own bootstraps' is a common evolutionary trick, illustrating the 'to him that hath shall be given..........' principle so common in life.

Among the mammals there are two major trends. One is towards the *Glossina* situation, the young remaining inside the parent until their own physiology has matured, that is to say until they are self-supporting. Guinea pig and gnu are obvious examples, as is the kangaroo because the change of abode is unimportant in this context. The other trend is towards an organisation of the outside world (a **nest** or **den**) in which the young may grow while **learning** by inter-organism behavioural interaction not possible while *in utero*. These latter organisms emphasise behavioural development, and get the young out into the world of relationships as soon as possible; most of them have very immature babies. These are clearly two ways of contriving a consistent milieu for development: inside the uterus of a homoiotherm, or outside in the social space made safe and predictable by parents' actions. The first option emphasises developmental circuitry, **instinct**, and produces a young animal able to respond appropriately to a great variety of stimuli soon after birth. But the second option emphasises **behavioural versatility** and relationships with other members of the species, and puts a premium on intelligence rather than instinct. The social insects contrive a completely controlled environment for the development of their larvae, and so show a very advanced reproductive mode. But they have used this consistency of developmental milieu to produce the most exquisitely designed machines, beautifully engineered automata. There is barely instinct, only complex circuitry usually coming in several models, various soldiers and workers. The trick of using the reliability of development to foster versatility of behaviour is the mammals' great invention. The birds nearly got it right, but like the insects they took the instinctive route, and few birds show the versatility of even a mouse or hedgehog. The social mammals have taken the more exciting route, and they will, I am sure, 'inherit the earth'.

REFERENCES

1. De Beer, 1958
2. Haeckel, 1874
3. Sterba, 1962
4. Sadoglu, 1967
5. Dobell, 1911
6. Simpson, 1950
7. Fryer and Iles, 1972
8. Greenwood, 1974
9. Shepherd, 1974
10. Ohno, 1970
11. Lewontin, 1974
12. White, 1954
13. Hyman, 1940

15.6 Questions

1. Animal development is conservative; to what extent does this simple concept relate phylogeny and ontogeny?

2. Relate specifically human characters to the diagrams in *Figure 15.1*, and criticise *Figure 10.1*.

3. What changes in selection/survival occur in animals which colonise caves? List adaptations which occur, and consider whether 'disuse atrophy is important.

4. How can neoteny of some characters be distinguished from paedogenesis?

5. Consider the reproductive stratagem of two (or more) multiplying morphs (e.g. polyp/medusa or the gonozoids/nurses in *Figure 11.2*).

6. What criteria can be used to distinguish advanced methods of reproduction?

16

REPRODUCTION IN MAMMALS

16.1 The Gonads as Endocrine Glands

The mammals are, by almost any sensible criterion, the most advanced organisms on this planet. This has been achieved by development of a whole series of control systems for the internal milieu. The 'symmetry' argument, that any two organisms with a common ancestor have probably diverged equally from it[1], becomes untenable when one considers the evolution of mammals. Asymmetry in evolution is quite common, with one group remaining primitive and continuing to exploit the ancestral niche, while the other innovates and creates a new niche. Those changes which led to mammals from the earliest vertebrates have mostly involved simplifications of gross morphology, and very probably physiological and certainly biochemical specialisation by gene duplication of several kinds. This is well shown by the skull bones[2], for example, and to some extent also by the reproductive system. The reproductive systems may have become more morphologically complex, but the major advance is physiological[3].

The gonads not only respond to internal cues by making sperms or eggs in season, they also produce signals for many other tissues and organs. Whereas in some fishes, at least some accessory sexual characters depend directly on pituitary hormones[4], most of the secondary sexual characters of mammals depend upon steroid hormones from the gonads. Indeed, parts of the brain itself are responsive to these hormones, not only those where steroid feedback loops return, but also in many diverse centres some or all of which have tracts ending in the hypothalamus, from which control is exerted on the pituitary (*Figure 16.3*).

Steroid hormones from the gonads have another effect too, much earlier in the life of the organism. Newborn baby rodents, if they have only oestrogens, retain a cyclic pituitary control of the gonad after puberty; but if they have androgens, or if they are injected with androgens or a high dose of oestrogens, this cyclicity is lost[5], the pubertal males showing no overt cycling.

The steroid hormones (*Figure 16.1*) are produced in cells which are not of the germ line, and may continue to be produced in the male after the germ cells have been destroyed, for example by aspermatogenic orchitis following mumps or vasectomy. These cells, called **Leydig cells**, are found between the testis tubules in association with blood vessels. Some of the testosterone and androstenedione they secrete probably passes directly into the tubules, unlike large protein molecules which seem unable to pass the **blood–testis barrier** around the tubules. The situation is complicated by

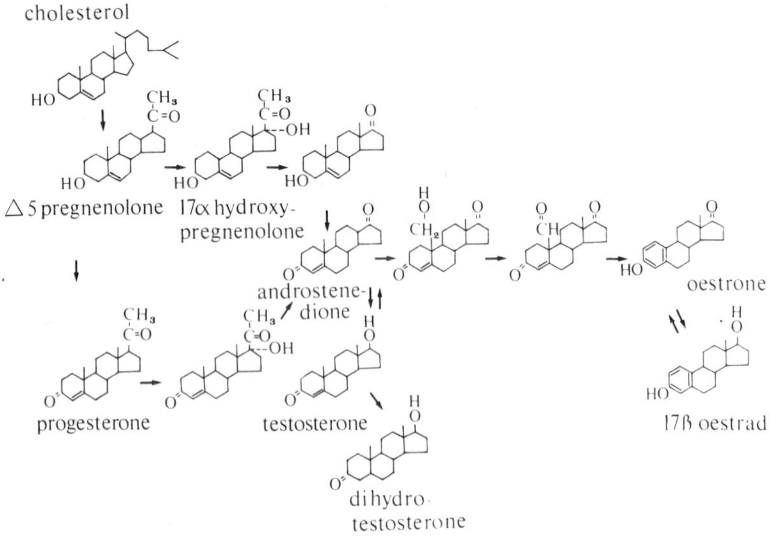

Figure 16.1 SOME STEROID HORMONES OF VERTEBRATES The synthetic pathways of progesterone, and the 'male' and 'female' hormones are available in many cells of the embryonic kidney system (e.g. adrenal cortex) but probably not of the germ cell line. (Follicle cells of the ovary are not germ-cell line, and it is probable that Sertoli cells are not either.) The similarities between maleness-eliciting androgens *(androstene-dione, testosterone and the more active dihydrotestosterone) and the femaleness-eliciting* oestrogens *(oestrone and oestradiols) is much more obvious than their differences. This is reminiscent of the organs affected, which also diverge slightly from a common pattern (*Figure 16.5*)*

the involvement of **Sertoli cells** in steroid metabolism; they are concerned with cholesterol from degenerating surplus cytoplasm of spermatids (*Figure 5.4*, p. 97).

The ovary uses a variety of cells, associated with the growing oocyte in the maturing Graafian follicle, to produce oestrogens, especially cells of the **theca interna**, which can be shown to have receptors for FSH (*see below*). No oestrogen is produced when no more oocytes are left, after the **menopause**, so steroid production in the female, unlike that in the male, depends upon the presence of germ cells.

The cells remaining in the Graafian follicle after ovulation reorganise, with many blood vessels, to produce a **corpus luteum** which secretes another steroid, **progesterone** (*Figure 16.1*). It is the alternation of phases dominated by the effects of oestrogen and progesterone which produces characteristic oestrous or menstrual cycles in eutherian mammals (*Figure 16.3* and *16.4*), when they are not pregnant or lactating.

In the mammalian embryo, the gonads arise as longitudinal thickenings of the dorsal peritoneal wall parallel with the dorsal mesentery and to either side of it, by or over the developing mesonephric kidney (*Figure 16.2*). The germ cells wander up the dorsal mesentery, or along blood vessels, and colonise them. If they are to become ovaries, the germ cells divide in the **cortex** to produce a large population of oogonia. If the

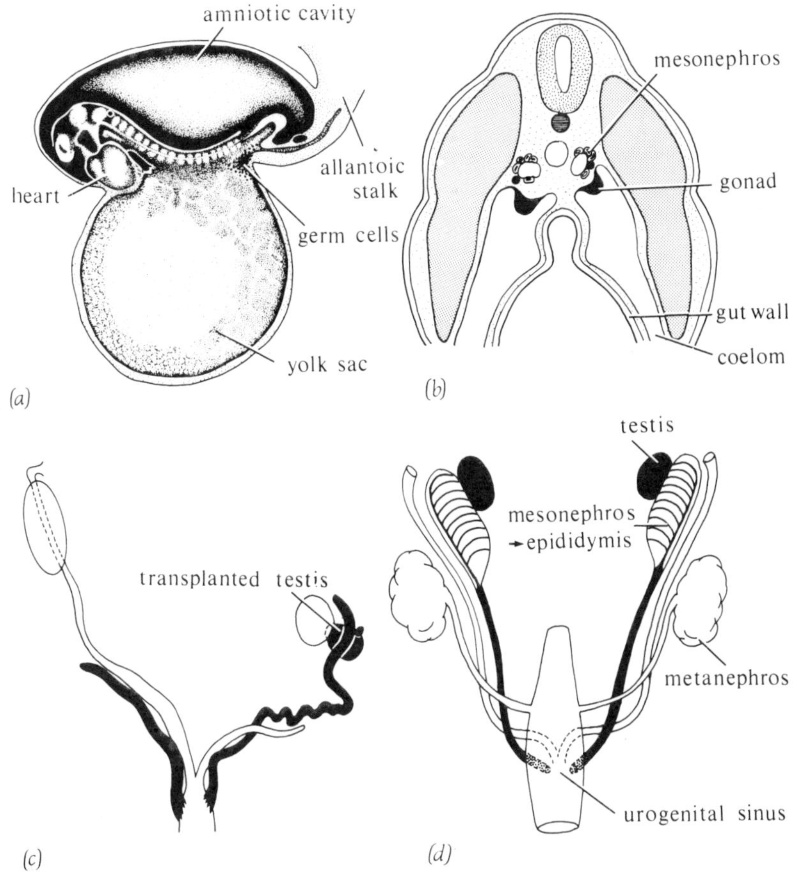

Figure 16.2 EARLY DEVELOPMENT OF THE GENITAL SYSTEM OF MAMMALS *(a) A four-week-old human embryo, drawn in semi-transparency to show the germ cells remote from the developing embryonic axis (from various sources, after Witschi, 1948). (b) Diagrammatic transverse section of the dorsal part of a somewhat later embryo; the gonads are developing in the dorsal coelomic wall by the forming mesonephric kidney. Germ cells migrate up the gut wall, into the dorsal mesentery supporting the gut, and by the mesonephros into the gonads. (c) Part of a female rabbit fetus genital tract which has received a testis graft (from Short, 1972, after Jost). Müllerian ducts (oviducts) are shown in outline, Wolffian ducts (vas deferens) are shown in black. On the side with an ovary only, oviduct is complete and Wolffian duct degenerates; on the side with ovary (outline) and testis, oviduct degenerates and Wolffian duct is fostered. This demonstrates* local *effects of the gonads (see text and* Figure 16.5*). (d) The early duct system of a normal male mammalian embryo; note that Müllerian system is present till quite late. Compare* Figure 16.5 i. *The mesonephric kidney tubules form the rete testis and the collecting ducts, while the mesonephric duct (black) becomes the vas deferens. When the testes descend the Wolffian ducts will loop over the ducts from the metanephros, the ureters (***Figure 16.5***)*

gonad cells are XY, then it becomes a testis and tissue of the medulla develops as a series of **sex cords** whose walls contain mostly germ cells; these become testis tubules. At this stage the young mammalian testis[6] secretes at least two substances which affect growth of the accessory sex organs, the **Wolffian** and **Müllerian duct** systems (*Figure 16.2*). The effect of the testis on the Wolffian system is stimulatory and probably via steroid androgens; but the inhibitory effect of the testis on Müllerian tissue is not via steroids but uses an unknown substance, locally acting for the effect may be unilateral. The experimental cases enumerated by Short[6] assist understanding of the effects of this early testis. Following castration of an embryo the Müllerian system develops somewhat like a normal female and the same happens in the XO humans (those with Turner's syndrome), who have no effective gonads. If androgens are supplied after castration, the Wolffian system develops into the complete male duct system and glands, but the complete female system is developed too. If the anti-androgen cyproterone is administered to the intact male embryo (preventing the action of androgens but not their secretion) the Wolffian system fails to develop, but so does the Müllerian because the testes inhibit it.

Because these effects are local an animal may come to have an ovary, oviduct, uterus etc. on one side and a male system on the other. The effects on the midline organs like penis/clitoris and perhaps urogenital sinus/vagina/vestibule are, however, produced by circulating hormones and these organs only very rarely show functional hermaphroditism. It is usual to find differences in these organs to be minimal in most very young mammals (*Figure 16.5*), the sexual differences only becoming gross as the testis secretes copiously when the animal enters puberty. In some species, like the hyena, the external genitalia are naturally very similar even in the adults, the penis resembling the clitoris and the scrotum being divided into two lateral sacs like labia. Secretion of oestrogens by the juvenile mammalian ovary at puberty also causes some changes in the genitalia, including enlargement of uterus, but these are rather less than those shown by the male and which are caused by androgens.

So much is known of the hormones and their interactions in mammals that any account here can only be derisory. Attempts to summarise some of the interactions between gonads, pituitary and placenta are made in *Figure 16.3* for a rodent and *Figure 16.4* for the human, but the reader is referred to Book 3 of the Austin and Short *Reproduction in Mammals* for a good short account.

16.2 The Receptive Tissues

Figure 16.5 shows the effects of the embryonic gonads on development of the genital tract in the human. In mammals it is usual to find that testosterone and other androgens are male-producing and that lack produces a neutral condition somewhat resembling the female, whereas oestrogens cause less pronounced changes. In humans, the XO condition (Turner's syndrome) lacks functional gonads but resembles the female and so these babies are brought up as girls. In birds on the other hand, where the female is heterogamic (WZ, ♂ WW) the castrate of either sex resembles the male and rather surprisingly will often show bright colours, long feathers and even a rudimentary intromittent organ.

It is a general rule that sex differences in mammals are steroid-dependent, unlike those of most of the animal kingdom. In insects, for example, the

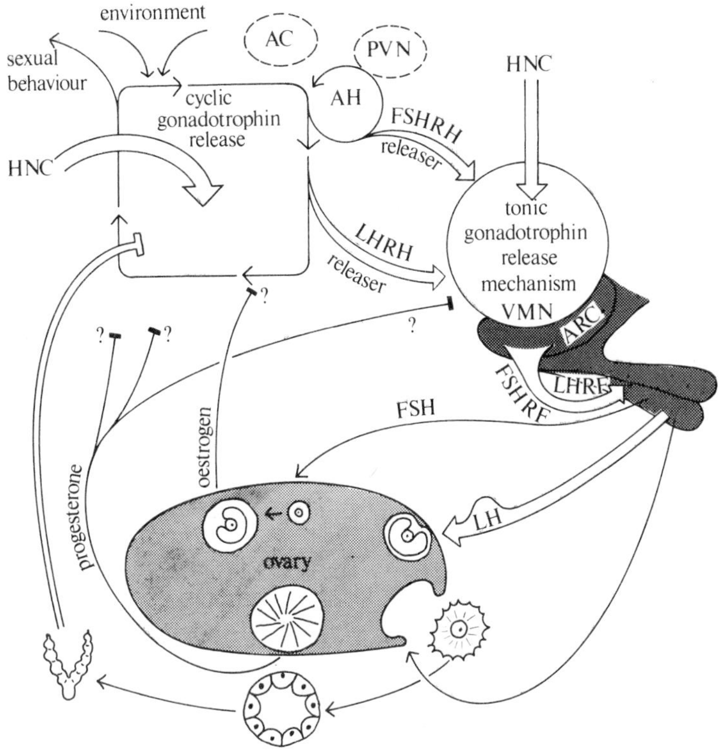

Figure 16.3 A FEW OF THE CONTROLS OF THE OESTROUS CYCLE OF RODENTS (modified, after Donovan, 1970, from Gorski) A cyclical system in the pre-optic area of the brain is sensitive to a variety of influences: i. There is 'higher nervous control' (HNC). ii. Environmental factors (day-length, temperature) can depress the cycle or change its manifestations. These include: i. Sexual behaviour, related to oestrogen (and progesterone) levels. ii. LH (Luteinising hormone) surges occur every day between 2 pm and 4 pm, which on every fourth or (fifth) day exceeds the level required to cause ovulation (see below). From this cyclical system, and controlled and mediated by paraventricular nucleus (PVN) and anterior commissure, comes neurosecretion down the nerve fibres from the anterior hypothalamus (AH), which results in secretion of releasing hormones (FSHRH and LHRH) in the hypothalamus, into small blood vessels (the portal system) from where they reach the pituitary gland, which secretes FSH and LH in response. This mechanism is 'tonic' i.e. is responsive, on a moderately long-term basis; it occurs in the median eminence area, specifically in the ventro-medial nucleus (VMN) and the neighbouring arcuate nucleus (ARC). The FSH secretion affects the young follicles in the ovary, some (about 20–40) of which enlarge, and secrete oestrogen which feeds back to moderate FSH secretion. The LH surge (shown as lumpy arrow) causes ovulation of few enlarged follicles and continued LH causes the remaining follicle (granulosa) cells to become a corpus luteum, which secretes progesterone; this has diverse effects on the brain, affecting thresholds in many brain centres. If the eggs are fertilised, the implanting blastocysts cause suppression of the overt ovulatory cycle, probably by a complex effect on higher centres as well as suppression at tonic centres

sex constitution of each tissue, sometimes of each cell, determines whether it shows male or female character. In mammals, cells of either genotype can respond in either way except for the cells of the gonad itself, which misbehave variously and produce sterility or intersexuality when hormonal

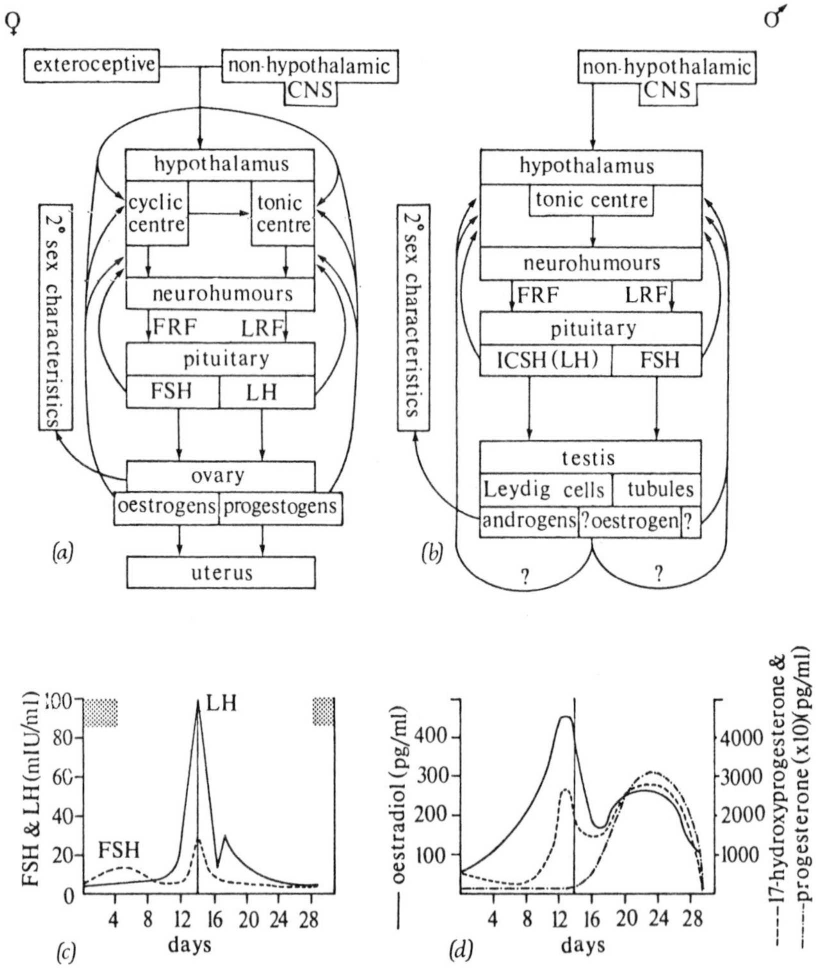

*Figure 16.4 SOME OF THE CONTROLS OF HUMAN REPRODUCTIVE CYCLES AND ORGANS (partly after Odell and Moyer, 1971). (a) Interactions between the female central nervous system (CNS) and pituitary, as in female rodents (*Figure 16.3*); the cyclical system has about a 28-day cycle, and the uterus reacts to lack of pregnancy or lactation by monthly menses. (b) The male appears to lack a cyclic system and to that extent the controls seem simpler. The testis produces androgens from the tissue between tubules (interstitial tissue containing Leydig cells) and within tubules, probably from the Sertoli cells. Some oestrogens are produced in testes (as well as in adrenals) and other hormones (see* Figure 16.2*) may act locally. (c) The changes in gonadotrophin (LH and FSH) levels during the menstrual cycle (menses shown as dotted areas). The LH surge can be seen, with increased FSH too, and the luteinising increase after ovulation, which causes the corpus luteum to organise and secrete progesterone. (d) Steroid levels during the menstrual cycle. Oestradiol, from the growing follicles, dominates the first half of the cycle, but progesterone rises after ovulation. Low levels of both cause endometrial shedding, the menses ('periods'). 17-hydroxyprogesterone shows a pre-ovulatory burst; it is made by granulosa cells before and after luteinisation. Some contraceptive pills act by providing negative feedback and suppressing gonadotrophin cycles; others act by raising progesterone-like activity throughout the cycle*

milieu and chromosomal complement are at odds[7]. The suggestion has been made[6] that this fixedness of sex of mammalian germ cells, compared with their plasticity in other vertebrates, where they conform to gonadal sex, is necessary to prevent their conversion by the mother's hormones.

Once the embryonic road towards maleness or femaleness has commenced (*Figure 16.5*), the young gonad produces little steroid until the onset of puberty, except for an occasional birth episode in which mother's hormones may 'spill over' into the fetal circulation producing, for example, 'witches milk' from babies' mammary glands, or a burst of steroid secretion.

As the young male mammal matures, FSH and luteinising hormone (LH) (*see below*) from the pituitary stimulate growth of the testis and its testosterone production, until some spermatogenesis commences in preparation for the first breeding cycle. All the tissues of the young male respond to the steroids and he usually becomes more muscular and heavier-boned than his female siblings. His genitalia and their associated glands grow phenomenally, and specifically male dermal glands and other structures appear. **Musk glands** of cats and mustelids, thoracic marker glands and **preputial glands** of rodents, **antlers** of deer and so on all depend upon steroid action.

The presence of these male hormones in human blood results in growth of the penis, the **beard**, axillary and **pubic hair** follicles make coarse hair instead of the pre-pubertal down, the larynx elongates and the growth pattern becomes typically 'masculine'. Some of these differences may depend upon the chromosome complement of the cells but it cannot be many, for those with 'testicular feminisation syndrome' (*see below*) look feminine even though all cells are XY. In other mammals that have been investigated the secondary sex characters nearly all depend upon male sex hormones, with exceptions like sebaceous glands which react to the pituitary hormone[8] as well. There is a great variety of human pathologies, some of which have increased our understanding of the basic mechanisms of hormone action. **'Testicular feminisation'** must be mentioned in this context. In this condition, found in goat and mouse as well as human, testosterone is produced normally, and the pituitary–testis relationship is fairly normal despite the abdominal or inguinal position of the testes and their consequent high temperature. But the tissues do not respond to the male hormone, and the Müllerian tissue is often incompletely suppressed; female-appearing external genitalia result, usually very immature-looking with no pubic or axillary hair, but there is some development of the breasts. The vagina is usually blind-ending and the uterus tiny. This syndrome results in slightly epicene, often very sexually attractive, young ladies who attend infertility clinics because they fail to menstruate (primary **amenorrhoea**). It is very difficult for the physician[6] to inform such a patient that 'she' is male, and that her gonads are testes, which must be removed for they are likely to become tumourous because they have not descended.

As the juvenile female mammal matures, there are small changes in external genitalia, and the mammary glands enlarge, but internal changes are more dramatic. The uterus enlarges, follicles may be seen maturing on the ovarian surface, and accessory glands of the vulva, and other sex glands,

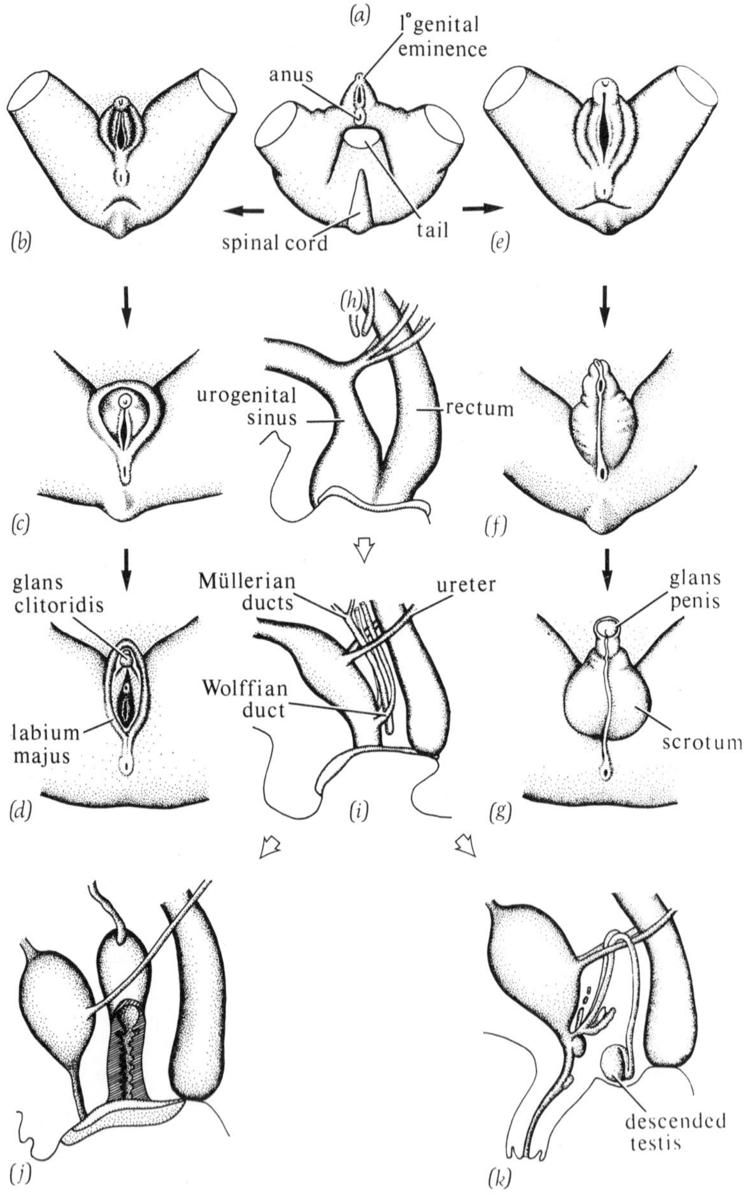
(a)
1° genital
eminence
anus
spinal cord
tail
(b)
(e)
(h)
urogenital
sinus
rectum
(c)
(f)
glans
clitoridis
Müllerian
ducts
ureter
glans
penis
Wolffian
duct
labium
majus
scrotum
(d)
(i)
(g)
descended
testis
(j)
(k)

may discharge their characteristic secretions. In the human female coarse pubic and axillary hair appears, replacing the previous fine bloom, and the breasts fill out, often asymmetrically as they start. The hips widen, subcutaneous fat deposition rounds off the adolescent angles, and the womanly shape appears. Equivalent changes probably occur in all mammals, but we lack the neural circuitry to respond to them as we do to the changes in our own species. Horse, dog or even cat breeders can usually tell mature males from females at a distance, without, for example, such obvious clues as the neck bulge (**dewlap**) of female rabbits.

There is good reason, then, to suppose that the steroid hormones affect development of many other tissues and organs than those used directly in sexual intercourse. Many of these other effects relate directly to the animal's position in social dominance systems, and so indirectly determine the sexual opportunities presented to a young mammal. Although the attractiveness of females to males is usually correlated closely with their oestrogen levels, varying through the cycle (*see below*), her status and the number of copulatory attempts which she initiates seem to depend more upon her androgen levels, at least in some monkeys[9]. Such behavioural responses, like the changes in female behaviour during the cycle, show short-term responses of the nervous system to circulating hormones, as well as the longer-term morphological effects.

We now know something of the cellular mode of action of some hormones, or at least of the first steps in such action[10]. Cells of the receptive tissues, for example the uterine endometrium, contain special protein molecules on their surface, probably integrated into the cell membrane, to which oestrogen attaches. These **oestrogen receptors**, once oestrogen is attached, are detached from the membrane into the cell where, probably after interaction with another substance, they enter the nucleus and may attach to a chromosomal site. This may be a **releaser site** for a genetic operon (p. 169) so that inhibition is released and m-RNA synthesis at the structural locus can begin; the enzyme or protein produced would presumably be part of the spectrum elicited in that tissue by oestrogens. It

---

*Figure 16.5 THE DEVELOPMENT OF SEXUAL DIFFERENCES IN HUMAN GENITAL ORGANS (modified from Nelsen, 1953). The male and female sexual organs are derived from the same set of embryonic rudiments by different emphasis caused by ovarian or testicular secretions. These secretions (*Figure 16.1*) are also variations on a theme. (a) The posterior end of a 7-week embryo. (The tail and legs are shown as cut, for clarity). (b) A female embryo, 10 weeks. (c) A female embryo, about three months. (d) Female, at birth. (e) Male, about 10 weeks. (f) Male, about three months. (g) Male, at birth. (h) A side view of the internal organs of 6–7 week embryo (like a). The Wolffian ducts have met the urogenital sinus (which has separated from rectum almost completely) and the ureter is growing dorsally from the junction to induce metanephros. No oviduct (Müllerian) system is to be seen. (i) Side view of the internal organs of an eight-week embryo (compare *Figure 16.2d*). The twin ducts of* h *have been spread on to urogenital sinus wall so that their connections to the sinus have separated; ureters now connect anteriorly to the future bladder part, while Wolffian ducts connect to future urethra. The Müllerian ducts have fused in the mid-line to give the uterovaginal rudiment. (j) The female internal organs at birth; no Wolffian ducts remain. (k) The male internal organs at birth; nearly all the Müllerian system has disappeared. A little remains, in the prostate, to give trouble in post-mature life* should be borne in mind that oestrogen alone is not sufficient for stimulation; at least nutrients, probably trace factors such as vitamins, and other hormones will be required for tissue responsiveness. Also the tissue itself must have differentiated appropriately so that its cells are competent to respond. Not only triggering, but loading, should be considered and perhaps even cleaning, ordering of ammunition, racking, sighting and the effects of a communal hunt if we can stretch the fire-arm metaphor. Single cells may not respond, where tissues do; and even tissues may need complex **priming** hours, days or months before the acute stimulus is given[11].

16.3 The Pituitary Gland and Hypothalamus

Twenty years ago it was fashionable to call the pituitary gland the 'conductor of the endocrine orchestra' with the implication that it exercised ultimate control over the function of the other endocrine glands. It was thought that it monitored their functions and secreted thyroid stimulating hormone (TSH) adrenocorticotrophic hormone (ACTH) and the **gonadotrophic hormones, follicle stimulating hormone** (FSH) and **luteinising hormone** (LH) in response to low blood levels of thyroxine, corticosteroids or sex hormones. While this is still regarded as largely true, the gonadotrophic activity of the pituitary is now known to be under more proximate control by the **hypothalamic region** of the brain, via **releasing hormones** (RH) and that these pass via a very short blood link, a **portal system.** The hypothalamus itself is probably mostly another relay system too, itself responding to other brain centres (*Figure 16.3*).

It seems that LHRH (luteinising hormone releasing hormone) and FSHRH (follicle stimulating hormone releasing hormone) may be the same short peptide, composed only of 10 amino acids. It may be that sugars are attached *in vivo*, modifying the action towards LH or FSH release, or that the LH or FSH secreting cells themselves respond to other components of oestrous or menstrual cycle and that this determines which gonadotrophin shall be released in response to RH in the female. However, both are usually released together, the ratio varying at different times; even the so-called 'LH surge' also has FSH, but the two peaks probably do not coincide.

The non-cycling hypothalamus of male mammals is produced by the fetal surge of androgen, around birth in myomorph rodents, but during uterine life in mammals which are born older, guinea pigs and probably humans. The difference between male and female pituitaries, non-cycling or cycling, is complex and only the rodent story is beginning to be understood. It already seems clear that the primate situation is very different and it is possible that no such fetal surge occurs. Rodent ovaries implanted into castrated males go into a permanent oestrous state[6], and this may correlate with the continuous sexiness of many male mammals. It is uncertain, however, whether the female rodent pituitary gland cycles independently and the ovary is simply responsive, or whether the pituitary and ovary 'converse'

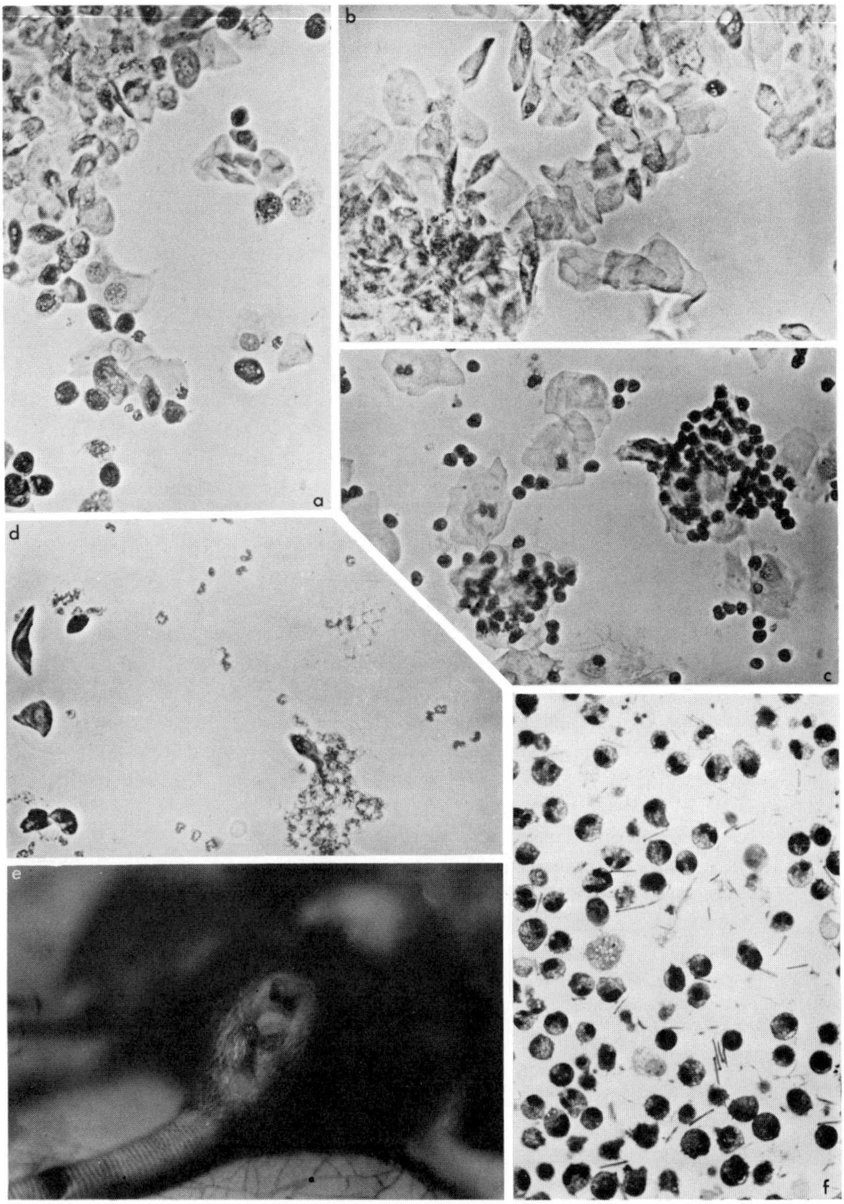

Plate 17 OESTROUS CYCLE OF THE MOUSE (see p.276 and Figure 16.7*) (a) A vaginal smear, pro-oestrus. (b) A vaginal smear, oestrus. (c) A vaginal smear, met-oestrus. (d) A vaginal smear, dioestrus. (e) External* **genitalia** *of female mouse. The two dark spots are on the ventral aspect of the clitoris; below them can be seen the external end of a vaginal plug. (f) A thin resin section of (rabbit) sperms and leucocytes in uterus.* Plate 7 (f) *shows one interaction. The leucocytes outnumber the sperms and few sperms survive*

producing the oscillatory cycle between them. The hypothalamus/pituitary axis may be involved, too, in presence or absence of the characteristic cycling female system.

16.4 Other Endocrine Controls and Lactation

Because mammals are such complex creatures, with many of their short-term as well as long-term responses mediated by hormones, it is not surprising to find that reproductive responses are mediated by many nervous and hormonal effects. The best example we can use is that most characteristic of mammals, **lactation**. Most experimental work has been done on mice and rats, but clinical experience and tissue culture studies have made the **lactogenic tissue** of the human among the most-studied tissues[12]

The embryonic development of the nipple and duct system is similar in both sexes, but differs early in juvenile life. Growth is depressed by testosterone, and may be released experimentally in male embryos by cyproterone, the androgen inhibitor. Isometric growth of ducts continues in females until puberty, when the glands increase allometrically (p. 222); a is said frequently to be as high as 1.5 or even 2 for a short time, but this is the period of transition from one characteristic breast/body ratio to another, and is probably best thought of as stimulation and not as allometry proper.

This stimulation requires oestrogen and anterior pituitary hormones, probably including **prolactin**. In pregnant females, the duct and alveolar system develop further, under the influence of a protein hormone, **lactogen**, produced in the placenta, also called **chorionic somatomammotrophin**. The actual production of milk, **lactogenesis**, requires the special cooperative activity of several hormones: prolactin, with oestrogen and ACTH in the rabbit, plus adrenocorticosteroids in rodents, growth hormone too in goats, while sheep require all these plus thyroxine[12]. High progesterone inhibits lactation and oestrogen assists, perhaps because of interaction with pituitary secretion of prolactin. On the other hand, high oestrogen administration after lactation has started will cause failure of lactation. In tissue culture[13] both insulin and cortisol are necessary in the culture medium before prolactin will induce the enzyme **lactose synthetase**, and progesterone may also inhibit α-lactalbumen synthesis directly. Even the parathyroid hormone has been implicated. In the intact pregnant female, however, all these other controls, like insulin and parathormone, will normally fall in the physiological range and can be 'taken for granted'. *Figure 16.6* shows some of the major controls of lactogenesis.

However, the mere secretion of milk is not sufficient, except in marsupials where the very young joey is permanently attached to its teat; there must also be release in response to suckling. The **milk-ejection reflex (let-down** or **draught)** is produced by a combination of hormonal control and direct nervous control, arising from the suckling infant's mouth contact on the teats or nipples, which affects the hypothalamus, causing stimulation of the posterior pituitary lobe which releases **oxytocin**. This causes contractile elements (myoepithelial cells) around the alveoli to act, expressing milk into the ducts and udders. In rats, whose infants may continue to suckle for

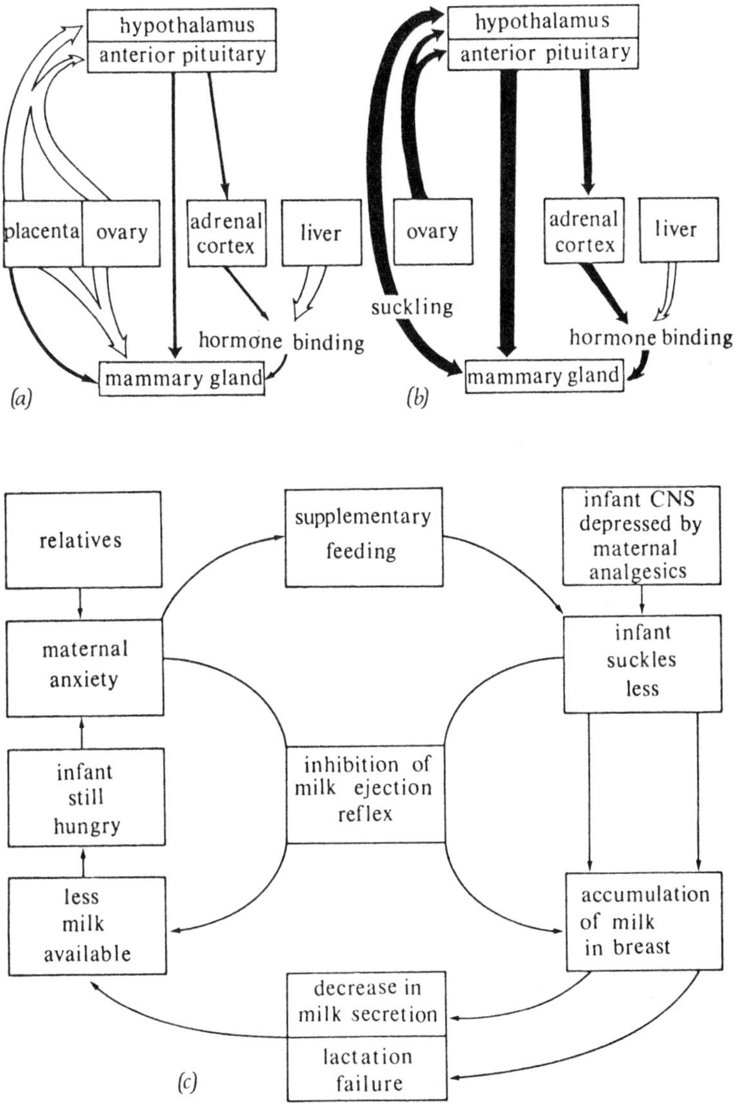

Figure 16.6 THE CONTROL OF LACTATION (modified from Findlay). (a) Before the birth, the mammary gland grows and begins secretion. Anterior pituitary, responsive to hypothalamus, secretes prolactin which, with placental lactogens (placental somatomammotrophin) stimulates this activity. The adrenal cortex maintains corticosteroids under pituitary control, and these permit lactogenesis in alveolar cells (see text). Liver cells may compete for, or destroy, these necessary corticosteroids. Low oestrogen and high progesterone levels (ovary and placenta) prevent full turning on of lactogenesis before the birth. (b) After the birth, progesterone drops (no placental source) and ovarian oestrogen rises. These affect the pituitary controls as well as lactational biochemistry directly; liver cells cease competing for corticosteroids and lactogenesis moves into high gear, maintained by suckling. (c) The maintenance of lactation may be upset by dietary, hormonal, social or iatrogenic factors. Commonly, in 1976 England, babies are sedated when born and a deprivation/anxiety cycle ensues, resulting in lactation failure and considerable maternal and infant stress

long periods, there is a periodic milk-ejection reflex every four hours or so which forces milk into the infants, who do not wake; time-lapse film of this is very amusing, as the baby rats 'erect' on the nipples as let-down occurs[14]. Presumably the milk-pump of dolphins and whales has its evolutionary origins in this oxytocin effect, but the udder or **cistern** is now extremely muscular and capacious (500 ml in *Tursiops*). Lactating women may experience 'let-down reflex' in response to a variety of stimuli, from hearing a baby cry or laugh to the pre-orgasmic departure from the plateau phase of sexual arousal.

Suckling and milk ejection are required for the continued release of ACTH and prolactin from the pituitary, and for the continuation of synthesis in the alveoli. This need for chronic stimulation to maintain lactation is presumably related to a high incidence of neonatal mortality in wild mammals; lactation must cease rapidly if there are no more infants to feed. The sensitivity of the system produces many problems for lactating human mothers, because a chronically high level of suckling is needed to maintain lactation, and this often conflicts with other demands (*Figure 16.6*). The use of narcotic analgesics to ease the birth pains for human mothers is often accompanied by depression of the infant reflexes, so that lactation may be turned off as if by infant mortality; at least a cycle of disuse may be set up.

16.5 The Oestrous Cycle

Although most male mammals are ready to mate at almost any time during the breeding season, it is common for female mammals to be receptive, and attractive, relatively rarely and briefly. Allen[15], in the 1920s, described the cyclical changes. Female mice were receptive about every fourth day if they were not pregnant; he called this state **oestrus**, and found a characteristic vaginal cytology at this time with many keratinised cells (*Figure 16.7* and *Plate 17*). The other stages, appearing on the other three days, he called **metoestrus** (after oestrus), **dioestrus** (between oestrus) and **pro-oestrus** (before oestrus).

Metoestrus is characterised in the mouse by the presence of increasing numbers of leucocytes among the squamous cells, until all the squames have disappeared and only the leucocytes are seen, in a thick mucus. These then disappear too, and vaginal smears two days after oestrus, in dioestrus, show very few cells. Then, on the day before oestrus, nucleated cells appear; these are often elongated comma shapes, and are characteristic of pro-oestrus. Late on the pro-oestrous day, squames lacking nuclei appear, and ovulation usually occurs early the following morning. With experience, these stages may be recognised and graded as in *Figure 16.7*, so that the state of the internal organs can be judged by the vaginal smears, to four hours in some cases. We now understand the events of such an oestrous cycle in terms of their hormonal regulation, although many puzzles still remain (*Figure 16.3*). FSH secretion begins the **follicular phase**, and a group of small follicles begin to grow in response. If only one ovary is present, the same number of follicles (30 to 50 in the mouse) grow in

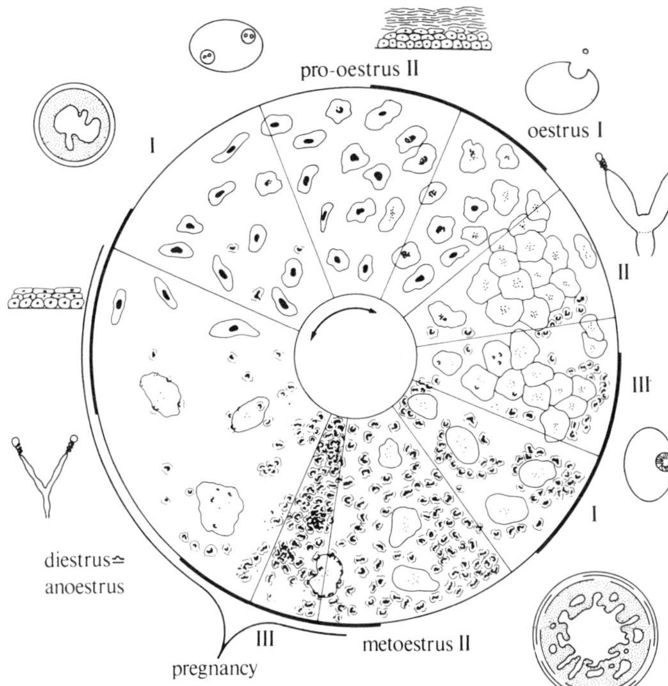

Figure 16.7 THE OESTROUS CYCLE OF THE MOUSE (compare with Plate 17, *and see* Figure 16.3 *for the endocrine controls). The central 'clock' shows four days and four nights (thickened rim) of the vaginal smear cycle of a non-pregnant female. At the top, evening of the first day, nucleated keratinised cells are found. Increasing numbers lack nuclei as they desquamate from a thicker vaginal epithelium (inset). Early next morning ovulation occurs (inset) and the morning smear shows masses of squamous cells. The uterus is swollen with 'oestrous fluid' (inset). By evening leucocytes appear in the smear and during the night the corpus luteum becomes fully organised (inset). Leucocytes clear squames from the vagina during the next day, and the uterine glands grow in preparation for implantation of embryos (inset). If no fertilisation occurs, the uterus regresses (inset) and the vaginal epithelium reduces to two to three cells thick (inset). As a group of follicles begin growth and oestrogen secretion, the uterine wall thickens (inset) and by the time the Graafian follicles have achieved an antrum (inset) nucleated cells are again found in the smear. Female mice will, not very reliably, mate in the time of the central circle arrow; if pregnancy supervenes the vaginal smear shows any of the conditions from Met III to Pro I*

one ovary, so there is clearly quantitative feedback, probably via the oestrogen they produce, acting upon both the hypothalamus and upon the pituitary directly. The oestrogens cause the uterine endometrium to thicken and its cells to become more basophilic as RNA synthesis occurs; exfoliation of vaginal epithelium results first in nucleated then in denucleated squamous cells in the smear. A sudden increase of LH and FSH, the **LH surge**, precedes the ovulation of these follicles by about a day, and probably causes both the final maturation and ovulation of some oocytes and the final degeneration of the others, giving **corpora albicantia (corpora atretica)**.

The LH, perhaps only in the high oestrogen milieu immediately post-ovulation, now causes **luteinisation** of the remains of the Graafian follicles, forming **corpora lutea** which commence to secrete **progesterone**. The uterine endometrium has become highly glandular, and is nearly ready for the cleaving eggs which might be rolling down the oviducts towards it. If eggs implant a new hormonal system of controls takes over. In many mammals, trophoblast cells produce another hormone, **chorionic gonadotrophin**, which substitutes for pituitary LH and maintains the corpus luteum. If no eggs implant, the corpora lutea regress (probably because of **luteolysins** produced by uterus and oviducts), progesterone level falls, and the pituitary is permitted to commence the next cycle with a rise in FSH.

As the progesterone level falls the uterine endometrium regresses, but apparently without much cell loss, and the vaginal wall also thins down to two or three cells thick (*Figure 16.7*). Such an oestrous cycle may occur only once per season (**monoestrus**), as in dogs and foxes whose oestrous phase lasts 9 days and 2½ days respectively. The eggs of these animals can be fertilised during the first meiotic division, and there is some evidence that ovulation occurs in the middle of oestrus; sperms of the dog are long lived. Most mammals, however, are **polyoestrous** and have several cycles in each breeding season unless the first is fertile. Even then some small mammals, with short gestations, may manage to fit in more cycles and more pregnancies; this is usual in wild mice and voles. *Table 16.1* shows oestrous patterns and gestation lengths for some mammals. There are many variations, however, like the mare which accumulates unfertilised eggs from infertile cycles in the oviducts, but any fertilised egg passes by them and into the uterus. The interested reader is referred to the excellent works by Austin and Short for more information, and to Asdell[16], or more recent works, for more details.

16.6 The Menstrual Cycle

The higher primates also show **spontaneous ovulation**, like the mouse and unlike the rabbit (*see below*), but the cycles are much longer (*Table 16.1*). The most notable difference is that the uterine endometrium is shed at the stage corresponding to dioestrus, and regenerated during the following **proliferative phase** (pro-oestrus) under the influence of oestrogen. This is produced by the ovary under the influence of pituitary FSH, as in the mouse. The 'LH surge' (really both FSH and LH) which induces maturation of the ovulating eggs also causes changes in uterine glands and in cervical mucous secretions so that the latter become penetrable by spermatozoa (this may be an indirect, steroid, effect). The **luteal phase** of the cycle (metoestrus) may, in some primates including the human, be maintained by chorionic gonadotrophin if a blastocyst implants; if not, the corpus luteum regresses and the uterine lining, now deprived of steroids, sloughs and is shed through the cervix, which relaxes slightly at this time (**menses**). Behavioural changes occur during the menstrual cycle of most primates, both receptivity and copulatory initiations being highest about the time of ovulation, and often premenstrually as well. **Premenstrual**

tension is evident in some zoo animals by increased touchiness and reluctance in social contacts. In consequence of these behavioural changes, status of female monkeys changes during the cycle, and such social mobility of our female ancestors may have contributed to our peculiar human social system (*see* p. 288).

Women do not usually show such a pronounced cycle of sexual aggressiveness and receptivity, except insofar as the **menstrual period** is associated with guilt, 'uncleanliness', prohibitions or other social pressures and so may define a period of abstinence. Copulation may occur at any time during the cycle, and any preferences are individual rather than universal. On a population basis, however, there seem to be slight preferences for mid-cycle, but the data are contradictory.

16.7 Induced Ovulators

There are many mammals in which copulation is not timed by the female's cycle, but by the availability of the male. Such animals, like rabbits or ferrets, have a prolonged receptive period sometimes called oestrus, but this should not be confused with the brief ovulation-associated receptivity of spontaneous ovulators like the mouse. Female rabbits in the wild in Britain are usually pregnant or lactating or both, but if not will mate readily at any time from a mild February to early September. Laboratory or meat-line does kept under a 16-hour daylight regime will mate readily throughout the year, although even under 'constant' conditions there is often a drop in receptivity in November–December. About 13 hours after mating ovulation occurs, and the mating stimulus can be mimicked by injection of 50 iu LH or **HCG (human chorionic gonadotrophin**, which has LH activity). If for any reason a mating is infertile does often become **pseudopregnant**, plucking their breast fur and 'carting hay' at 17 to 20 days instead of the 28 days characteristic of a true pregnancy, which lasts 30 to 31 days. After a pseudopregnancy receptivity reappears, usually about 20 days after the first mating stimulus. After a normal pregnancy and delivery there is a **post-partum ovulation**, very common in eutherian mammals, at which some wild rabbits may be inseminated. Some breeders try to 'catch' the doe at this time so that she will be pregnant and lactating together, requiring more food but less space than two does. While lactating the doe is not very receptive, and will usually only accept the buck again at three to four weeks, when the litter is taking some food in addition to her milk, i.e. when **weaning** has started. The post-partum oestrus is not shown in crowded or otherwise poor conditions.

Ferrets and many other mustelids also have induced ovulation. In the mink and its relative marten and sable, the stimulus for ovulation is not copulation but the male biting the neck of the female[17], lacerating the skin and incidentally spoiling the fur. Some breeders pierce the neck skin with special forceps, and either artificially inseminate or use a muzzled male to avoid this laceration. Other mammals with induced ovulation are the cat, the camel and some voles[18], a motley collection; this is clearly a tactic

Table 16.1 Breeding of mammals (from Atmann and Dittmer, and Asdell)

Organism	*Common name*	*Oestrous (O)/ menstrual (M) cycle days, (P if poly-oestrous, A if annual)*	*Age at puberty*	*Gestation period (days)*	*Young/ litter*	*Wt of litter/ wt of mother*
Homo	Woman	M 24–33 P	11–16 years	278 (253–303)	1	3.4 kg/60 kg
Gorilla		M 45 P	5 years	257–259	1	
Macaca mulatta	Rhesus	M 28 P	2–3 years	168 (144–194)	1	0.49/8.00 kg
Galago senegalensis	Bush baby	M 42 P	c.2 years	120	1–2	
Bison bison		O 21 P	2 years	300	1	
Bos taurus	Cattle	O 14–23 P	6–10 months	284 (210–335)	1	36.8/500 kg
Capra hircus	Goat	O 21 P	8 months	151 (135–160)	1–5	3.3 X 3/70 kg
Ovis aries	Sheep	O 14–20 P	7–8 months	151 (144–152)	1–4	3.8 X 2/38 kg
Hippopotamus		O 30 P	3–5 years	237 (210–250)	1	
Sus scrofa	Pig	O 18–24 P	5–8 months	114 (101–130)	9	1.23 X 9/100 kg
Rhinoceros		O 40–50 A (?)	4, 5 years	488	1	
Equus caballus	Horse	O 10–37 P	1 year	336 (264–420)	1	52/408 kg
Loxodonta africana	Elephant	O 42 A (?)	9–12 years	630–660	1	160/1600 kg
Phoca vitullina	Seal	O ? A	5–6 years	270 (330 in most other seals)	1	
Felis catus	Cat	O 15–28 P	6–15 months	63 (52–69)	4	100 X 4/2445 g
Panthera leo	Lion	O 21 P	2 years	105–113	1–6	
Lutra lutra	Otter	O 26 P		61–63	2–5	
Meles meles	Badger	O ? A		210	1–4	
Canis familiaris	Dog	O 9,2 per A	6–8 months	63 (53–71)	7 (1–22)	0.25/13 kg
Ursus arctos	Bear	O ? A	6 years	210	1–2	
Balaenoptera borealis	Sei whale	O ? A	2 years	360 (as all baleen whales)	1	750 kg/13 000 kg
Physeter	Sperm whale	O ? A	3–5 years??	510	1	750 kg/13 000 kg
Tursiops	Dolphin	O ? A	4 years	360	1	
Chinchilla		O 28 P	4 months	105–115	1–4	
Cavia	Guinea pig	O 16–19 P	55–70 days	68 (58–75)	3 (1–8)	94 X 3/400–800 g
Apodemus	Field mouse	O 6 P	88–90 days	23–29	5–6	

Organism	*Common name*	*Oestrous (O)/ menstrual (M) cycle days, (P if poly-oestrous, A if annual)*	*Age at puberty*	*Gestation period (days)*	*Young/ litter*	*Wt of litter/ wt of mother*
Mus musculus	Mouse	O 4 P	35 days	19–31	6 (1–12)	1.38 × 6/33 g
Rattus norvegicus	Rat	O 4–5 P	40–60 days	21	6–9	5.63 × 7/350 g
Mesocricetus auratus	Hamster	O 4 P	35–56 days	15–18	1–12	3.3 × 6/100 g
Lepus europaeus	Hare	O ? A ?	8 months	42	3	
Oryctolagus	Rabbit	O (induced ovulator)	5–8 months	31	8 (1–13)	0.065 × 8/4 kg
Macropus	Kangaroo	*See* p. 202	18 months	38–40	1 + 1 (*see* p. 202)	
Dasyurus viverrinus	**Native cat**	**O 4–12 A**	?	**8–14**	**20–35 (!)**	

easily derived from spontaneous cyclic ovulation by a shift of stimulus of the hypothalamus from endogenous hormone levels; modulated by daylength, to the stimuli associated with copulation. This may involve only a shift in emphasis, for the various brain centres controlling the hypothalamus are already sensitive to different stimuli in the 'spontaneous' ovulator. It only requires one of them to become 'permissive' for that factor to 'induce' ovulation.

16.8 Social Interactions and Reproduction

The potency of external, particularly sexual, stimuli in the control of the events surrounding fertilisation is well shown by the **'Bruce effect'**. Mice which have been fertilised within three or four days will fail to implant if they smell the secretions of a strange male[19]; particularly the urine and probably the secretion of the **preputial glands** in it. Some will even abort implanted embryos of up to seven days. Although this appears to be a tactic to favour outbreeding, in wild mouse colonies it seems much more likely to act as a curb on reproduction when density is high.

It was shown by Calhoun[20] in the early 1960s that there was a variety of reproductive problems which appeared as a breeding rat colony became more crowded. He set up colonies in cages connected as in *Figure 16.8*, and gave water, food and nest-building materials *ad libitum*. Dominant males soon took up residence in cages 1 and 4, which were defensible, and kept harems, juvenile or otherwise subordinate males, and growing offspring in these territories but excluded any other adult males. These males would sleep across the entrance, which ensured that cages 1 and 4 were never invaded by the degenerating conditions in cages 2 and 3 and so served as control situations for disease, food quality, etc. As population increased in cages 2 and 3, which could not be the territory of one dominant male for each had two entrances, a number of behavioural changes occurred. Recruitment to the 'middle' cages was from the 'end' cages as well as from their own breeders, so even when breeding in the 'middle' cages fell to below replacement levels, numbers still increased.

The first change noticed was the lack of careful nest-building by females; they would even give birth out of the nest, and the young would be eaten by other animals. As density went up this latter pathology increased till very few young survived even to weaning. Some, hyperactive, males also invaded such sanctuary as the nests provided, and attempted to copulate with the nursing mothers; this never occurs with normal rats. Some males were passive, well groomed, but totally uninterested in sex, even with the normally irresistible oestrous females.

The lessons drawn from this, and applied to man's population problems in cities, are probably inappropriate. Nevertheless these experiments show some of the internal feedback mechanisms potentially available to rodent colonies, but not used until they are elicited by social interaction. Experiments similar to Calhoun's have been performed with rabbits and a variety of semi-wild and domestic animals, with varying results. Many apparently regulatory mechanisms have been described[21], often

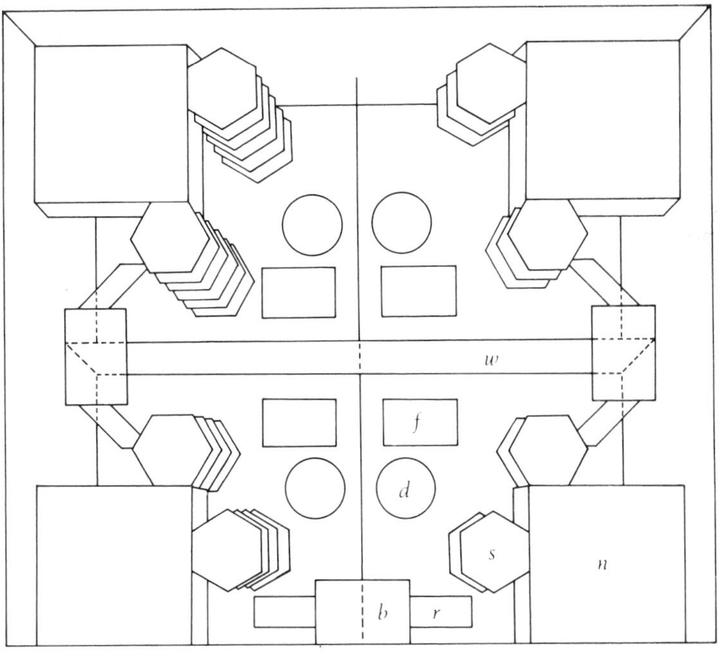

Figure 16.8 CALHOUN'S RAT CROWDING APPARATUS The rat room is viewed from above. w *is a wall across it with a bridge at each end, crossed by another wall (shown by a vertical line) with no bridge at the 'north' end but one at the 'south' end, labelled* b; *a ramp* r *leads up to each bridge. Food* f *and drink* d *are provided in each of the four enclosures, which also have tiered nest-boxes* n *with steps* s *up to each level. Enclosures 1 and 4 (NW and NE) have only one entrance/exit, and remained socially healthy; 2 and 3 (SW and SE) showed the social pathologies (see text)*

taking the form of a dominance system which permits only a certain number of territorial animals to breed, as in grouse[22] (p. 238). Particularly thorough experiments[23] have been performed with rabbits in Australia, and these have shown a great variety of systems contributing to ineffective breeding by crowded or low-caste females, even when high-caste females in the same colony are breeding well. Even in conditions where food is plentiful, therefore, rabbit density does not increase greatly.

In monkey and baboon colonies, however, the situation is much less clear. Supplied with ample food, colonies of rhesus monkeys transported to small islands have engaged in frank warfare, with some animals being wounded or even killed[24]. This violence never happens normally except by default, the outfaced animal usually starving because he is excluded from the territory. There are in fact many social mammals, notably elephants, lions, deer and antelopes, which seem to increase in numbers beyond the carrying capacity of their land. One wonders whether this situation, seen in 'game reserves', is a real part of the African ecology or an artefact caused by crowding these long-lived animals[25]. A social pathology caused by an abnormal situation like this crowding could be part of a feedback loop some tens of years long. If it *is* a 'natural' result of

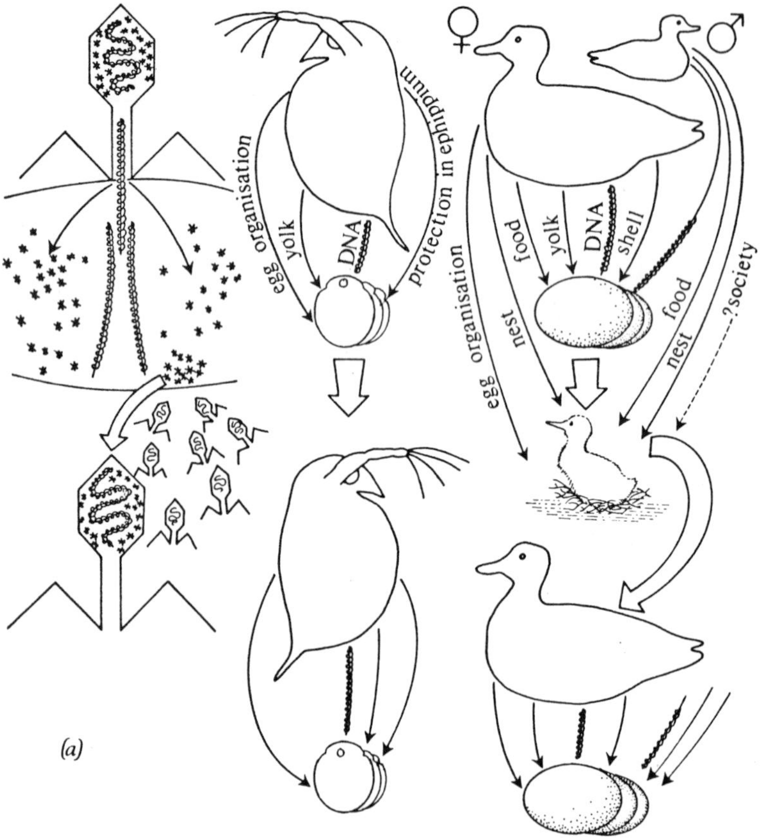

Figure 16.9 AN ATTEMPT TO SHOW PRIMITIVE AND ADVANCED MECHANISMS OF INHERITANCE (a) A virus particle (left) uses its protein to inject DNA into a cell. Much of the protein for the next generation is coded by early DNA, and it associates with later DNA to re-produce infective particles. Daphnia *parthenogenetic females give a DNA genome to their eggs, but they also give an egg organisation to accomplish read-out of this DNA prescription, and yolk to fuel this process. The eggs are protected under mother's carapace while they develop, and may even be shed in an ephippium (protective case) if she dies. Ducks contribute much more to their offspring. Two DNA sources contribute to the zygote nucleus (but much more DNA is included in yolk and supernumery sperms – not informational but very helpful), egg organisation probably determines development up to primitive streak and the beginning of differentiation. Yolk, vitelline membrane, and albumen are all involved from earliest development. Shell, nest, warmth and turning result in a duckling, which requires additional food,* and *an imprinting object before it is properly a duck. (b) Human inheritance is much more complex, and a more abstract diagram is required. Time proceeds downwards, as for the other species, but many more organisms are involved in the inheritance of each. The woman at top right has a husband and produces a baby (centre) who herself produces a baby. Social life encompasses all the individuals, giving stability or at least security, as each mother does for her embryo (arrows). Non-genetic inheritance from both parents, from teachers and perhaps even from grandparents is required before a human being is reproduced*

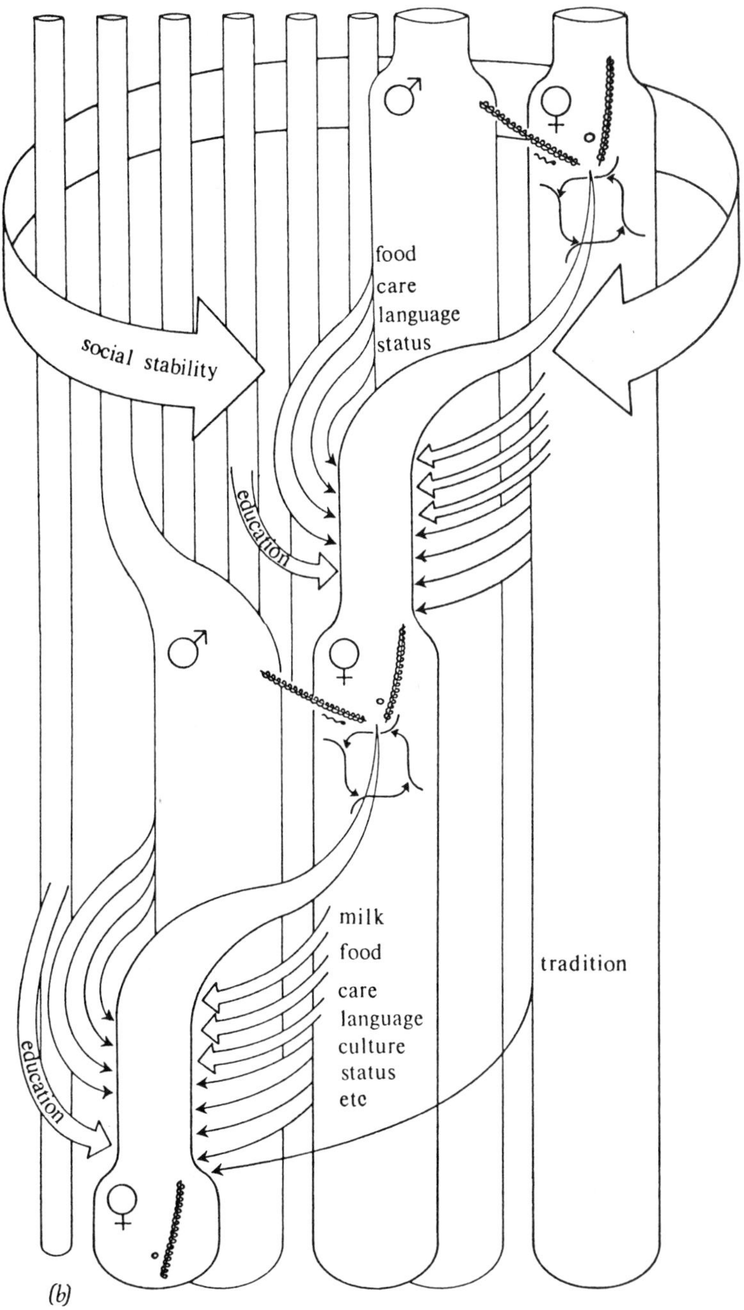
social stability
food
care
language
status
education
milk
food
care
language
culture
status
etc
tradition
education

(b)

breeding without the operation of internal controls, then perhaps it was inevitable that a social mammal would eventually arise who would intemperately take over all other territories, and increase in numbers beyond all reason, up to the starvation level.

The lesson from Calhoun's experiment is not that pathologies are produced in rats which resemble kinds of human antisocial or criminal behaviour. If we did have a built-in control of that kind, cities would be impossible. A much more worrying possibility is that, perhaps like baboons, elephants and lions, we have *no* such regulation and so will be limited only by the failure of our resources.

16.9 The Reproduction of Cultural Patterns

All organisms have some extra-genic heredity. Bonner[26] has distinguished various threads in the tangled cycles of life histories, and emphasised that the several cycles must be out of phase (*Figure 16.9*). Thus the virus must carry its passive nucleic acid in actively infective protein, then replicate its nucleic acid (using a cell's machinery) while the previous generation's protein is lost and before a new batch is made, usually from the earlier DNA coding. In the same way, bacteria separate their replicated DNA into daughter cells, each of which is already a working set of components specified by the mother-'cell' DNA. During the metabolic phase of the new cycle, while the DNA is not separating (although it is often replicating during vegetative functioning), these components are progressively replaced in the daughter 'cells' by the newly-specified metabolic machinery. It is *never* only the nucleic acid which passes between generations, there is always at least a 'cassette' holding the 'instruction tape'; and usually a 'tape recorder', much of which is made on the previous tape's instructions, is included. Many protozoans show this degree of reproductive complexity, but we know the ciliates possess another independent, again always out-of-phase, hereditary mechanism. This is the cortical ciliary field with its specialisations (p. 147); it uses products of the nuclear DNA, and *therefore* cannot replicate at the same time. Micronuclear DNA replicates during the metabolic phase, then duplication of cortical organelles usually precedes the mitotic disarray of the metabolic nucleus (meganucleus).

Clearly the process is more reliable if the various divisions of material and machinery, following replication of each system, proceed in sequence and not all together. Even in those organisms which do seem to 'come to pieces' altogether in the link between generations, there is always a cyst wall or other organisation, remaining from a previous state to form the skeleton of the organisation of the new individual. Examples include gemmule or statoblast walls (*Figure 1.1*, p. 6), and even the spherical zona pellucida of *Perameles* (bandicoot) eggs, whose blastomeres all separate and then re-aggregate on the inner wall of the zona to form the blastocyst[27]. Centrifugation experiments on mollusc eggs demonstrate the 'skeletal' nature of the rigid cortex, laid down in the ovary, from which the more fluid cytoplasmic components apparently derive their organisation. Curtis' experiments with *Xenopus* grey crescent[28], taken with experiments on dorsal lip transplants[29] (*Figure 9.3*, p. 168), demonstrate the survival of maternal organisation to initiate the most important and basic organisation of the new vertebrate individual (Chapter 8).

In metazoa, generally, there are at least three parallel transmissions between generations: nucleic acid in the chromosomes, and the nucleic acid 'informosomes'; the cellular machinery, including 'household' enzymes as well as translation and transcription machinery; and the cytoplasmic and cortical organisation which constrains the interactions of cytoplasm and nucleus. Centrioles, mitochondria and frequently other organelles also pass, and these too have a separate, out-of-phase, replication cycle. More recent symbionts than mitochondria also usually have replication cycles related to, but out of phase with, replication of the major species DNA. Symbionts of *Drosophila* in egg cytoplasm[30], or termite cellulose-digesting symbionts which are 'caught' by licking, are examples.

We have seen that most metazoa pass additional material and energy between the generations as well (Chapters 8 and 12) usually as yolk, and that this permits delay in development of the progeny's own feeding apparatus. Thereby more complex development is fostered; maternal egg organisation, membranes, viviparity, homeothermy of the developing embryo form a series which interacts with other parental care to produce predictably complex organisms by carefully controlled development. Developmental constraints are removed by this parental control extended across the generations, until most of the homeostasis of the developmental environment is determined by parental, usually maternal, abilities. That developing embryonic complexity which is autonomously programmed from the zygote genome, the DNA prescription, itself becomes a smaller part of each reproductive cycle as the organisms become more complex during evolution. The initiation of reading of this DNA prescription, delayed to phyletic stage in most animals, is controlled by maternal egg organisation; while its 'normal' expression is continually permitted only by maternal factors like yolk adequacy, time and place of egg deposition, maternal oxygen supply and so on (*Figure 16.9*).

In the mammals, part of this function is taken over by embryonic membranes which seem at first sight, and on phylogenetic considerations, to be part of the zygote's own provision for its own needs. But in many mammals, especially primates, the membranes *precede* the embryo in development; the mother has produced an egg architecture whose initial effect is to provision the egg, like yolk (and yolk sac). That the zygote genome may be expressed in trophoblast is irrelevant to this point; maternal strategy seems to determine the order of expression of egg functions, and the use of zygote genome may only reflect ancestral embryonic membranes built, like those of birds, around a formed, post-phyletic, embryo.

Previous authors[31,32] have distinguished two modes of inheritance. One, **somatic**, was considered to be entirely genetic but realised through 'interaction with the environment'; the other, **extrasomatic**, ranged from such fancies as the evolution of bicycles to the extravagant comparison of neckties and cuff-buttons with somatic 'vestigial organs' like the human appendix. The evolution of cultures, mores, and languages has also been supposed to correspond in some extrasomatic way with somatic, genetic evolution as if the two were separate but comparable. The view presented here portrays extrasomatic evolution of these latter kinds as a sophisticated reproductive tactic, forming the most recent term in a series whose first term is viral protein and which includes termite symbionts and other animal societies.

Some animal societies, particularly those of social insects, have their character very largely determined by the genotypes of the organisms which compose them. Behaviour in a large colony, however, may differ from that in a small colony of the same organisms because, although the genotypes may be identical, the stimuli eliciting behaviour from each individual may be different.

In many mammals which live at such densities that individual interactions are common, like rats, mice and other small rodents, many density-dependent effects on individual reproduction are found (p. 282) and these are often mediated via pheromones and a complex adrenal stress syndrome[33]. In wolf or wild dog packs and multi-male baboon troupes, however, the restriction of effective reproduction to the high-status individuals maintains a reproductive heterogeneity in the groups all the time; this contrasts with the oscillations between high and low rates of breeding from *all* the individuals, which is often seen in small rodents. This stratification of the society must itself be reproduced, or at least maintained, and in mammals this seems to be achieved by the status interactions of each juvenile individual[24,34] instead of by direct programming as in insects. The spectrum of responses is probably programmed, but those social interactions which elicit one response rather than another from an animal form part of its culture, and cannot be referred back to any individual, certainly not to a genomic pattern.

Many human cultures have ritualised this reproductive heterogeneity, assigning roles which may be hereditary, but often are not, to different individuals who each ensure reproduction of part of the cultural pattern. The transmission of martial skills by a military establishment, or of intellectual skills by medieval clerics, the cobbler with his 'boy', the engineers' apprentice and the artisans' guild all are part of the reproduction of human cultures. One effect of all these tactics is that each juvenile grows up insulated from direct interaction with his whole environment, just as the viral DNA is protected by the infected cell's membrane, or the fetus by its mother. He need not hunt, for hunters (or butchers) of previous generations provide his meat, nor cook for others do this for him; soldiers (or police) protect him from violence, and other specialists maintain his environment in other ways. He is freed from constraints, and may take tortuous but secure paths to a great variety of adult occupations. The culture usually allots resources overtly to warrior or geisha training, universities or village cricket clubs; but some resources are allotted in clandestine fashion, as by the ancient freemasons or by 'anonymous' donation to Church or State establishments.

The soma was the great invention of the metazoa, as the 'cell' wall probably was of the bacteria. Somas competed, died, developed, out-of-phase with the germ cells they contained and sometimes managed to pass on cloaked in somatic protection. This in itself was not a new trick; even virus DNA is usually cloaked in protein coded by the previous DNA generation. Mortality, and senility, became useful tactics, and the way was laid for the use of non-reproductive individuals in societies. The insect society is a by-product of the genotype, and is in many ways comparable to the soma of other metazoans. Its existence, and much of its form, hinges upon the genotype of the reproductives; traditions, rituals or historical necessity can only occur in a restricted sense, if these words can indeed be used at all. In contrast, human society has reproduction of its forms, with only the subtlest involvement of the genotype. This additional protective cloak around the developing human organism does have historical, traditional, ritual as well as genotype-reactive sources: each child's genotype determines that he can learn oral language which conforms to his inbuilt grammar, but the nature of that language is not usefully referred to ancient genotype but to cultural history. This independence of cultural reproduction from biological reproduction is exemplified well by medieval Christianity in Western Europe. This was a major cultural protection for nearly all people, yet its clericy was mostly celibate. The traditions of Christianity were reproduced through a 'germ plasm' which was, sexually, effectively sterile. This happens surprisingly often in human cultures, where post-sexually-reproductive individuals, or sometimes sexually dysfunctional individuals like the shaman, constitute the 'germ plasm' of the cultural reproduction.

As we see somas of two stags fighting, not the germ plasm they are representing, so we see ant colony enslaving ant colony, not a tactic of one genotype in using another's vegetatives. Equally, we see 'cowboys' against 'indians', not one genotype against another but two ways of life, two cultural species, in conflict. Man's great invention has been **cultural reproduction**, which should be added to the more ancient vegetative, asexual, and sexual; for it reproduces way of life *without* organic continuity, yet it combines and assorts properties of ancestors in progeny. Some characters persist in combination, like Latin in biological English or voodoo in Haitian Christianity. Competition, succession and evolution all occur in this reproductive mode as in the others.

Cultural species have another mode of breeding too, however; even without the sexual transfer of genotypes (which *was* usual, however, if only to a limited extent), Christianity, Islam, Spanish and English have been forced upon, or taken over by, many peoples. The process resembles transformation in bacteria, but this should not be forced; cultural transformation does not affect genomes primarily, but does affect nearly all the particularly human aspects of the lives of the people concerned. Usually a ruling class or a clericy is converted by conquest or persuasion, and the cultural species may change as it takes in the new peoples, like the new versions of Christianity in the American Negro slave populations. Speciation and variety formation have many parallels with the sexual mode: Christianity and Islam are successive speciations, but Kabbalism or Methodism are varieties. Higher-level taxa also seem usually to represent earlier separations. Barriers to cultural cross-fertilisation are too obvious to need emphasis here; such reproductive isolation is a very important part of sexual strategies too.

So far, technical/industrial western society has proved a very effective protector of the young, and has enabled more diversification within our society than within any other in human history. Now, like the first sexual organisms, metazoa, social insects, teleosts, or early mammals, our explosion of numbers and our diversification has resulted in competition between species for the newly-available resources. It is to be hoped that cultural reproduction leads to as rich a species mixture as has the sexual reproduction upon which it depends for its substrate.

On this planet, only some mammals have combined the use of parental abilities to protect the juvenile with learning by the juvenile of parental culture. A kind of cultural speciation occurs in birds, where breeding between similar species may be prevented by incompatible behaviour learnt by each juvenile from its parents; this has only the status of an incompatibility tactic[33]. The entirely new reproductive strategy of a few mammals, culture, firmly based on and derived from the parental instruction of these organisms, transcends sexuality as sexuality transcends mitotic reproduction. The specific character of the life of a man, perhaps of a dolphin, possibly but doubtfully of beavers or otters or elephants or wild horses, is the extragenic cultural inheritance. This is not true of most mammals, for the gnu, the mouse and probably the fox only use the extragenic inheritance to protect juvenile genotypes by parental abilities; cultural reproduction is minimal.

This evolutionary experiment in advanced reproductive strategy is still young. Its early success has made the human cultural species and varieties the dominant life form on major areas of our planet. The reproductive success of certain cultures, interacting with the various human genotypes, has produced a great increase in the numbers of human individuals and cultures. Evolutionary success of this innovation can only happen if cultural reproduction on our planet can achieve a stable cultural ecology, which on present evidence seems doubtful.

REFERENCES

1. Dobell, 1911
2. Goodrich, 1931
3. Van Tienhoven, 1968
4. Ball, 1960
5. Bermant and Davidson, 1974
6. Short, 1972
7. Biggers and McFeely, 1966
8. Ebling, Ebling and Skinner, 1969
9. Herbert, 1972
10. Milgrom *et al.*, 1973
11. Filburn and Wyatt, 1974
12. Findlay, 1974
13. Turkington and Kadohama, 1972
14. Wakerley and Lincoln, 1973
15. Allen, 1923
16. Asdell, 1964
17. Ford and Beach, 1952
18. Breed, 1967
19. Bruce, 1960
20. Calhoun, 1962
21. Barnett, 1956
22. Watson and Miller, 1971
23. Mykytowycz and Fullager, 1973
24. Ardrey, 1970
25. Ehrenfeld, 1970
26. Bonner, 1974
27. Nelsen, 1953
28. Curtis, 1965
29. Spemann and Mangold, 1924
30. King, 1970
31. Lotka, 1945
32. Mather, 1964
33. Eibl-Eibesfeldt, 1970
34. Koford, 1963

16.10 Questions

1. In what ways do the homeostatic arrangements of mammals give them peculiar reproductive problems?

2. Neither external genitalia nor sex hormones of mammals diverged from a common stock phylogenetically, yet they are clearly variations on a theme. Why?

3. How can induced ovulation arise, in evolution, from a complexly controlled cyclic ovulator?

4. Contrast the mammals whose young are born 'helpless' (i.e. requiring help) with those which are born able.

5. What passes between generations to reproduce a bacterium, a trout, a cuckoo, a Frenchman or what?

REFERENCES

Note: the page numbers in brackets at the end of each reference indicate where in the text it is mentioned.

ADAMS, C.E. (1969). 'Intraperitoneal insemination in the rabbit.' *J. Reprod. Fert.,* **18**, 333–339 (139)

ALLEN, E. (1923). 'The estrous cycle in the mouse.' *Am. J. Anat.,* **30**, 297–371 (276)

ALLISON, A.C. (1955). 'Aspects of polymorphism in man.' *Cold Spring Harbour Symp. Quant. Biol.,* **20**, 239–55 (22)

ALTMAN, P.L. and DITTMER, D.S. (1962). 'Growth; including reproduction and morphological development.' *Fedn. Am. Socs Exp. Biol., Washington,* 608pp (16, 280)

ALTMAN, P.L. and DITTMER, D.S. (1972). 'Biology Data Book.' (2nd edn) *Fedn. Am. Socs Exp. Biol., Washington,* 606pp (16, 280)

AMOROSO, E.C. (1952). 'Placentation.' In: Marshall's *Physiology of Reproduction* (ed. Parkes) Vol. II, 127–312 (204)

AMOROSO, E.C. (1960). 'Viviparity in fishes.' *Symp. Zool. Soc. Lond.,* **1**, *Hormones in fish,* 153–181 (200)

ARDREY, R. (1967). *The Territorial Imperative; a personal enquiry into the animal origins of property and nations.* 390pp. Collins; London (237, 239)

ARDREY, R. (1970). *The Social Contract; a personal enquiry into the evolutionary sources of order and disorder.* 405pp. Collins; London (283, 288)

ASDELL, S.A. (1964). *Patterns of mammalian reproduction.* 670pp. Cornell University Press; Ithaca, NY (35, 59, 62, 202, 278, 280)

ASIMOV, I. (1967). A conservational bon-mot (2)

AUSTIN, C.R. (1965). *Fertilisation.* 145pp. Prentice-Hall; New Jersey (27, 122, 128, 133, 140)

AUSTIN, C.R. (1972). 'Pregnancy losses and birth defects.' In: *Reproduction in Mammals* (ed. Austin and Short) 2, 134–153 (34)

AUSTIN, C.R. (1975). 'Fertilisation.' In: *Concepts of Development.* 48–75. (ed. Lash and Whittaker). Sinauer Associates; Stamford, Conn. (102, 142)

AUSTIN, C.R. and BAVISTER, B.D. (1975). 'Preliminaries to the acrosome reaction in mammalian spermatozoa.' In: *Functional Morphology of the Spermatozoon.* 83–87. (ed. Afzelius). Pergamon; Oxford (140)

AUSTIN, C.R. and SHORT, R.V. (1972). *Reproduction in mammals.* Cambridge University Press; London. (Book 1. Germ Cells and Fertilisation. Book 2. Embryonic and Fetal Development. Book 3. Hormones in Reproduction. Book 4. Reproductive Patterns. Book 5. Artificial Control of Reproduction) (vi, 266)

BACCETTI, B. *et al.*, (1975). 'Motility patterns in sperms with different tail structure.' In: *Functional Morphology of the Spermatozoon.* 141–150. (ed. Afzelius). Pergamon; Oxford (95)

BAKER, T.G. (1972a). 'Primordial germ cells.' In: *Reproduction in Mammals.* (ed. Austin and Short) **1**, 1–13 (27, 74)

BAKER, T.G. (1972b). 'Oogenesis and ovulation.' In: *Reproduction in Mammals* (ed. Austin and Short). **1**, 14–45 (27)

BALL, J.N. (1960). 'Reproduction in female bony fishes.' In: *Hormones in Fish* (ed. I. Chester Jones). *Symp. Zool. Soc. London.* **I**, 105–135 (263)

BALLARD, W.W. (1964). *Comparative Anatomy and Embryology.* 619pp. Ronald; New York (210)

• BANNISTER, R.C.A., HARDING, D. and LOCKWOOD, S.J. (1974). 'Larval Mortality and subsequent year-class strength in the plaice (*Pleuronectes platessa*).' In: *The Early Life History of Fish.* 21–37. (ed. Blaxter). Springer-Verlag; Berlin (18, 237)

BARNETT, S.A. (1956). 'Endothermy and ectothermy in mice at −3 °C.' *J. Exp. Biol.*, **33**, 124–33 (282)

BARRINGTON, E.J.W. (1967). *Invertebrate Structure and Function.* 547pp. Nelson; London (192)

BEARDMORE, J.A. (1976). Personal communication: in guppy and plaice, heterozygosity is higher in progressively older samples (238)

BEATTY, R.A. (1967). 'Parthenogenesis in Vertebrates.' In: *Fertilisation; comparative morphology, biochemistry and immunology* **1**, 413–440. (ed. Metz and Monroy). Academic Press; New York (143)

BEDFORD, J.M. and CALVIN, H.I. (1974). 'The occurrence and possible functional significance of –S–S– crosslinks in sperm heads, with particular reference to eutherian mammals.' *J. Exp. Zool.*, **188**, 137–56 (98, 133)

BEDFORD, J.M. (1971). 'The rate of sperm passage into the cervix after coitus in the rabbit.' *J. Reprod. Fert.*, **25**, 211–8 (85, 112)

BEERMAN, W. (1956). 'Nuclear differentiation and functional morphology of chromosomes.' *Cold Spring Harbour Symp. on Quant. Biol.*, **21**, 217–32 (169)

BENNETT, D. (1975). 'The T-locus of the mouse.' *Cell*, **6**, 441–54 (104, 106)

BERMANT, G. and DAVIDSON, J.M. (1974). 'Biological Bases of Sexual Behaviour.' 306pp. Harper and Row; London (60, 298, 263)

BERNSTEIN, G.S. (1952). 'Sperm agglutinins in the egg jellies of the frogs *Rana pipiens* and *R. clamitans*.' *Biol. Bull.*, **103**, 285 (abstr.) (111)

BERRILL, N.J. (1950). *The Tunicata; with an account of the British species.* 354pp. Ray Society; London (207, 88, 195)

BERRILL, N.J. (1961). *'Salpa.' Sci. Am.*, **204**, 150–163 (207)

BIGGERS, J.D. and MCFEELY, R.A. (1966). 'Intersexuality in domestic animals.' In: *Advances in Reproductive Physiology.* **1**, 29–60. (ed. McLaren). Logos Press; London (269)

BLACKLER, A.W. (1966). 'Embryonic sex cells of amphibia.' In: *Advances in Reproductive Physiology.* **1**, 9–28. (ed. McLaren). Logos Press; London (74, 295)

BONNER, J.T. (1974). *On Development; the biology of form.* 282pp. Harvard University Press; Cambridge, Mass. (170, 286)

BOUNOURE, L. (1939). *L'Origine des Cellules Reproductrices et la Problème de la Lignée Germinale.* Gauthier-Villars; Paris. See also Blackler, 1966 (10)

BOYCOTT, A.E., DIVER, C., GARSTANG, S.L. and TURNER, F.M. (1930). 'The inheritance of sinistrality in *Limnaea peregra.*' *Phil. Trans. Roy. Soc. B.,* **219,** 51–130 (155)

BRADEN, A.W.H. and AUSTIN, C.R. (1954). 'Number of sperms about the eggs in mammals and its significance for normal fertilisation.' *Austr. J. Biol. Sci.,* **7,** 543–551 (112, 140)

BREED, W.G. (1967). 'Ovulation in the genus *Microtus.*' *Nature, Lond.,* **214**, 826 (279)

BROWN, G.G. and KNOUSE, J.R. (1973). 'Effects of sperm concentration, sperm aging and other variables of fertilisation in the horse-shoe crab *Limulus polyphemus* L.' *Biol. Bull.,* **144,** 462–470 (32, 110)

BROWN, J.L. (1975). *The Evolution of Behaviour.* 761pp. Norton; New York (63, 176, 224, 241, 243, 244)

BRUCE, H.M. (1960). 'A block to pregnancy in the mouse caused by proximity of strange males.' *J. Reprod. Fert.* **1,** 96–103 (282)

BURNET, Sir Macfarlane (1974). *Intrinsic mutagenesis; a genetic approach to ageing.* 244pp. MTP; Lancaster (231)

CALHOUN, J. (1962). 'Population density and social pathology.' *Sci. Am.,* **206,** 139–142 (282)

CAMPBELL, J.W. Jr. (1961). 'On the selective breeding of human beings.' *Analog, Science Fact and Fiction.* February 1961 (British Edn). Editorial, pp.2–4 (227)

CAVE, A.J.E. (1964). 'The processus glandis in the *Rhinoceratidae.*' *Proc. Zool. Soc. Lond.,* **143,** 569–586 (59)

CHANG, M.C. and SCHAEFFER, D. (1957). 'Number of spermatozoa ejaculated at copulation, transported into the female tract, and present in the male tract of the golden hamster.' *J. Hered.,* **48,** 107–9 (60)

CHAPMAN, R.F. (1969). *The Insects: Structure and Function.* 819pp. English University Press; London (77, 180, 219)

CLERMONT, Y. (1972). 'Kinetics of spermatogenesis in mammals; seminiferous epithelium cycle and spermatogonial renewal.' *Physiol. Rev.,* **52,** 198–236 (81)

CLOUDESLEY-THOMPSON, J.L. (1968). *Spiders, Scorpions, Centipedes and Mites.* 278pp. Pergamon Press; Oxford (59)

COHEN, J. (1963). *Living Embryos. An introduction to the study of animal development.* 116pp. Pergamon Press; Oxford (161)

COHEN, J. (1967). 'Correlation between sperm 'redundancy' and chiasma frequency.' *Nature, Lond.,* **215,** 862–3 (27)

COHEN, J. (1971). 'Comparative physiology of gamete populations.' *Adv. Comp. Physiol. Biochem.,* **4,** 267–380 (9, 40, 78, 84)

COHEN, J. (1973). 'Crossovers, sperm redundancy, and their close association.' *Heredity,* **31,** 408–413 (27, 29, 108)

COHEN, J. (1975a). 'Gametic diversity within an ejaculate.' In: *The Functional Anatomy of the Spermatozoon.* 329–339. (ed. Afzelius). Pergamon; Oxford (141)

COHEN, J. (1975b). 'Gamete redundancy – wastage or selection?' In: *Gametic Competition in Plants and Animals.* 99–112. (ed. Mulcahy). Elsevier; New York (107, 108, 110, 118, 127)

COHEN, J. and McNAUGHTON, D.C. (1974). 'Spermatozoa; the probable selection of a small population by the genital tract of the female rabbit.' *J. Reprod. Fert.,* **39**, 297–310 (27, 102)

COHEN, J. and WERRETT, D.J. (1975). 'Antibodies and sperm survival in the female tract of the mouse and rabbit.' *J. Reprod. Fert.,* **42**, 301–310 (111, 112)

COLWIN, L.H. and COLWIN, A.L. (1967). 'Membrane fusion in relation to sperm–egg association.' In: *Fertilisation; comparative morphology, biochemistry and immunology.* **1**, 295–367. (ed. Metz and Monroy). Academic Press; New York (99, 128)

COMFORT, A. (1964). *Ageing: the biology of senescence.* 365pp. Routledge and Kegan Paul; London (228, 229)

COOKE, J.V. (1954). 'The rate of growth in progeria; with a report of two cases.' *J. Pediat.,* **42,** 26–37 (230)

CURTIS, A.S.G. (1965). 'Cortical inheritance in the amphibian *Xenopus laevis*: preliminary results. *Arch. Biol.,* **76,** 523–46 (147, 287)

CUSHING, D.H. (1974). 'The possible density-dependence of larval mortality and adult mortality in fishes.' In: *The Early Life History of Fish.* 103–111. (ed. Blaxter). Springer-Verlag; Berlin (237)

DALLAI, R., BACCETTI, B., *et al.,* (1975). 'New models of aflagellate arthropod spermatozoa.' In: *The Functional Morphology of the Spermatozoon.* 279–287. (ed. Afzelius). Pergamon; Oxford (95)

DAN, J.C. (1967). 'Acrosome reaction and lysins.' In: *Fertilisation; comparative morphology, biochemistry and immunology.* 237–294. (ed. Metz and Monroy). Academic Press; New York (98, 99, 103)

D'ANCONA, V. (1960). 'The life-cycle of the atlantic eel.' In: *Hormones in Fish, Symp. Zool. Soc. Lond.,* **1**, 61–76 (47, 48)

DARLINGTON, C.D. (1969). *The Evolution of Man and Society.* 753pp. George Allen and Unwin; London (36)

DARWIN, C. (1859). *Origin of Species.* John Murray; London. Several modern reprints, e.g. Mentor (1958) by New American Library of World Literature (14)

DAVEY, K.G. (1965). *Reproduction in the Insects.* 96pp. Oliver and Boyd; London (103)

DAVIDSON, E.H. and HOUGH, B.R. (1971). 'Genetic information in oocyte RNA.' *J. Mol. Biol.,* **56,** 491–506 (115)

DE BEER, G. (1958). *Embryos and Ancestors.* 197pp. Clarendon; Oxford (178, 186, 248, 253)

DEHAAN, R.L. (1964). 'Cell interactions and oriented movements during development.' *J. Exp. Zool.,* **157**, 127–138 (161)

DEEVEY, E.S. (1947). 'Life tables for natural populations of animals.' *Quart. Rev. Biol.,* **22**, 283–314 (228)

DEEVEY, E.S. (1960). 'The human population.' *Sci. Am.,* **203**, 195–204 (37)

DENENBERG, V.H. (1963). 'Early experience and emotional development.' *Sci. Am.,* **208**, 138–146 (216)

DE ROPP, R.S. (1969). *Sex Energy; the sexual force in man and animals.* 236pp. Jonathan Cape; London (40, 67)

DEWSBURY, D.A. and JANSEN, P.E. (1972). 'Copulatory behaviour of Southern Grasshopper Mice (*Onychomys torridus*)' *J. Mammal.,* **53,** 267–275, and other papers by Dewsbury (60)

DOBELL, C. (1911). 'The principles of protistology.' *Arch. Protistenk.,* **21,** 269–310. (Also in 'Collected Works' as 'Status of the Protista') (254, 263)

DONOVAN, B.T. (1970). *Mammalian Neuroendocrinology.* p.108. McGraw-Hill; London (267)

DUNN, L.C. and BENNETT, D. (1969). 'Studies of effects of t-alleles in the house mouse on spermatozoa. II Quasi-sterility caused by different combinations of alleles.' *J. Reprod. Fert.,* **20,** 239–46 (106)

DUPOUY, J. (1964). 'La teratogénèse germinale male des gasteropodes et ses rapports avec l'oogenese atypique et la formation des oeufs nourriciers.' *Arch. Zool. Exp. Gen.,* **103,** 217–368 (93)

EBLING, F.J., EBLING, E. and SKINNER, J. (1969). 'The influence of the pituitary on the response of the sebaceous and preputial gland of the rat to progesterone.' *J. Endocr.,* **45,** 257–63 (269)

EHRENFELD, D.W. (1970). *Biological Conservation.* 226pp. Holt, Rhinehart and Winston; London (283)

EIBL-EIBESFELDT, I. (1970). *Ethology, the biology of behaviour.* 530pp. Holt, Rhinehart and Winston; London (59, 216, 226, 288, 290)

ENGELMANN, F. (1970). *The Physiology of Insect Reproduction.* 307pp. Pergamon Press; Oxford (16, 40, 74, 182, 219, 221, 241, 242)

EPEL, D. (1975). 'The program of, and mechanisms of, fertilisation in the echinoderm egg.' *Am. Zool.,* **15,** 507–22 (135, 139)

ETKIN, W. and GILBERT, C.I. (eds) (1968). *Metamorphosis.* 459pp. North Holland Publishing Company; Amsterdam (194)

FANKHAUSER, A. and HUMPHREY, R.R. (1954). 'Chromosome number and development of progeny of triploid axolotl males crossed with diploid females.' *J. exp. Zool.,* **126,** 33–58 (110)

FELLOUS, M. and DAUSSET, J. (1970). 'Probable haploid expression of HL-A antigens on human spermatozoa.' *Nature, Lond.,* **225,** 191–193 (107)

FILBURN, C.R. and WYATT, G.R. (1974). 'Developmental endocrinology.' In: *Concepts of Development.* 321–348. (eds. Lash and Whittaker). Sinauer Associates; Stamford, Conn. (272)

FINDLAY, A.L.R. (1974). 'Lactation.' *Research in Reproduction,* **6,** No.6. I.P.P.F.; London (274, 275)

FISHER, R.A. (1930a). 'Mortality amongst plants and its bearing on Natural Selection.' *Nature, Lond.,* **125,** 972–3 (21)

FISHER, R.A. (1930b). *The Genetical Theory of Natural Selection.* (Especially p.134). Oxford University Press; London (21)

FISHER, R.A. (1936). 'The measurement of selective intensity.' *Proc. Roy. Soc. B,* **121,** 50–62 (26)

FLETCHER, L.D. Jnr and JOHNSON, J. (1974). 'Re-examination of "normal" sperm count.' *Urology,* **3,** 99–100 (112)

FORD, C.S. and BEACH, F.A. (1952). *Patterns of Sexual Behaviour.* 330pp. Eyre and Spottiswoode; London. (1965) Methuen; London (62, 65, 67, 244, 279)

FOX, C.A. and FOX, B. (1971). 'A comparative study of coital physiology, with special reference to the sexual climax.' *J. Reprod. Fert.,* **24,** 319–36 (71, 101)

FRANZÉN, A. (1970). 'Phylogenetic aspects of the morphology of spermatozoa and spermiogenesis.' In: *Comparative Spermatology.* 29–46. (ed. Baccetti). Academic Press; New York (92)

FREMLIN, J. (1972). *Be fruitful and multiply.* 241pp. Granada; St. Albans (36)

FRETTER, V. and GRAHAM, A., (1964). 'Reproduction.' In: *Physiology of Mollusca,* **I,** 127–164. (Wilbur and Yonge, eds). Academic Press; London (9

FREUD, S. (1905). 'Three essays on the theory of sexuality. Quoted and criticised in Bermant and Davidson (1974) (67)

FRYER, G. and ILES, T.D. (1972). *The Cichlid fishes of the Great Lakes of Africa; their biology and evolution.* 641pp. Oliver and Boyd; Edinburgh (52, 256)

GALTSOFF, P.S. (1964). 'The American oyster *Crassostrea virginica* Gmelin.' *Fish. Bull. of Fish and Wildlife Service,* **64,** 1–102 (19)

GATENBY, J.B. (1920). 'The germ-cells, fertilisation and early development of *Grantia (Sycon) compressa.*' *J. Limn. Soc. Zool.,* **34,** 26–58 (103)

GIBBONS, I.R. (1975). 'Mechanisms of flagellar motility.' In: *The Functional Anatomy of the Spermatozoon.* 127–140. (ed. Afzelius). Wenner-Gren. Pergamon Press; Oxford (104, 134)

GINZBERG, A.S. (1968), (transl. 1972). *Fertilisation in fishes and the problem of polyspermy.* Israel Program for Scientific Translations; Jerusalem (142)

GLOVER, T.D., SUZUKI, F. and RACEY, P.R. (1975). 'The role of the epididymal cells in sperm survival.' In: *Functional Anatomy of the Spermatozoon.* 359–372. Pergamon Press; Oxford (81)

GOFFMAN, E. (1971). *Interaction ritual.* 192pp. Penguin; London (312)

GOLDBERG, E.H., AOKI, T., BOYSE, E.A. and BENNETT, D. (1970). 'Detection of H2 antigens on mouse spermatozoa by the cytotoxic antibody test.' *Nature, Lond.,* **228,** 570–572 (107)

GOODRICH, E.S. (1931). *Studies on the structure and development of vertebrates.* 937pp. Dover; New York; (Reprint, 1958) (263)

GOODWIN, B.C. (1963). *Temporal Organisation in Cells; a dynamic theory of cellular control processes.* 163pp. Academic Press; London (172)

GOODWIN, B.C. and COHEN, M.H. (1969). 'A phase-shift model for the spatial and temporal organisation of developing systems.' *J. Theoret. Biol.,* **25,** 49–107 (172)

GORDON, H. and GORDON, M. (1957). 'Maintenance of polymorphism by potentially injurious genes in eight natural populations of the platyfish *Xiphophorus maculatus.*' *J. Genet.,* **55,** 1–44 (22)

GREENWOOD, P.H. (1974). 'Cichlid fishes of Lake Victoria, E. Africa: the biology and evolution of a species flock.' *Bull. Br. Mus. Nat. Hist. Zoology series,* Suppl. 6. 134pp (256)

GURDON, J.B. (1974). *The Control of Gene Expression in Animal Development.* 160pp. Clarendon; Oxford (74, 163, 171)

GUSTAVSON, T. and WOLPERT, L. (1961). 'The forces that shape the embryo.' *Discovery – New Series,* **22**, 470–477 (158)

HADZI, J. (1963). *The Evolution of the Metazoa.* 224pp. Macmillan; London (9, 52, 91)

HAECKEL, E. (1874). 'The Gastraea-Theory, the phylogenetic classification of the animal kingdom and the homology of the germ-lamellae.' (Trans. E.P. Wright). *Quart. J. Microsc. Sci.,* **14**, 142–165 and 223–247 (248)

HAFEZ, E.S.E. (1973). 'Gamete transport.' In: *Human Reproduction: Conception and Contraception.* 82–98. (eds Hafez and Evans). Harper and Row; New York (84)

HALDANE, J.B.S. (1954). 'The measurement of natural selection.' *Proc. 9th Intnl. Congr. Genet.,* 480–503 (18)

HALDANE, J.B.S. (1957). 'The cost of natural selection.' *J. Genet.,* **55**, 511–24 (26, 35)

HAMILTON, W.D. (1964). 'The genetical evolution of social behaviour, I and II.' *J. Theoret. Biol.,* **7**, 1–52 (63)

HAMILTON, W.D. (1966). 'The moulding of senescence by natural selection.' *J. Theoret. Biol.,* **12**, 12–45 (228, 230)

HARDY, Sir A. (1956). *The Open Sea; its Natural History. 1. The World of Plankton.* Collins; London (188, 192)

HARDY, Sir A. (1960). 'Was man more aquatic in the past?' *New Scientist,* **7**, 642–5 (66)

HARLOW, H.E. and HARLOW, M.K. (1962). 'Social deprivation in monkeys.' *Sci. Am.,* **207**, 137–146 (216)

HARRIS, H. (1970). *Nucleus and Cytoplasm* (2nd ed.) 181pp. Clarendon; Oxford (170, 171)

HARRIS, H. (1971). 'Protein polymorphism in man.' *Can. J. Genet. Cytol.,* **13**, 351–396 (26)

HARTL, D.L. (1975). 'Segregation distortion in natural and artificial populations of *Drosophila melanogaster.*' In: *Gamete Competition in Plants and Animals.* 83–92. (ed. Mulcahy). Elsevier; Amsterdam (104)

HAYFLICK, L. (1965). 'The limited *in vitro* lifetime of human diploid cell strains.' *Exp. Cell. Res.,* **37**, 614–636 (173, 229)

HEINLEIN, R. (1955). *Tunnel in the Sky.* Charles Scribners; New York (246)

HERBERT, J. (1972). 'Behavioural patterns.' In: *Reproduction in mammals* (ed. Austin and Short), **4**, 34–68 (271)

HERSKOWITZ, I.H. (1967). *Basic Principles of Molecular Genetics.* p.264. Nelson; London (14)

HICKMAN, C.P. (1973). *Biology of the Invertebrates.* C.V. Mosby Co.; St Louis (185)

HINTON, H.E. (1964). 'Sperm transfer in insects and the evolution of haemocoelic insemination.' *Symp. R. Ent. Soc. Lond.,* **2**, 95–107 (103)

HUBBS, C.L. (1955). 'Hybridization between fish species in nature.' *Syst. Zool.,* **4**, 1–20 (40, 138)

HUXLEY, J.S. (1972). *Problems of Relative Growth.* 312pp. Dover; New York. (original Methuen, 1932) (222, 312)

HYMAN, L.H. (1940). *The Invertebrates I. Protozoa through Ctenophora.* McGraw-Hill; New York (9, 46, 260)

HYMAN, L.H. (1951). *The Invertebrates. II. Platyhelminthes and Rhynchocoela. The acoelomate Bilateria.* 548pp. McGraw-Hill; New York (201)

INOUE, S. and SATO, H. (1962). 'Arrangement of DNA in living sperm; a biophysical analysis.' *Science,* **136,** 1122–1124 (98)

IVANOVA-KASAS, O.M. (1972). 'Polyembryony in insects.' In: *Developmental Systems: Insects. I.* 243–71. (ed. Counce and Waddington). Academic Press; London (90, 200)

JACOB, F. and MONOD, J. (1961). 'Genetic regulatory mechanisms in the synthesis of proteins.' *J. molec. Biol.,* **3,** 318–356

JENNESS, R. (1974). 'The composition of milk.' In: *Lactation.* 3–107. (ed. Larson and Smith). Academic Press; New York

JINKS, J.L. (1964). 'Extrachromosomal inheritance. 177p. Prentice-Hall; New Jersey (1)

JOHNSON, D.R. (1974). 'Hairpin-tail: a case of post-reductional gene action in the mouse egg?' *Genetics,* **76,** 795–805 (107)

JOHNSON, W.C. and HUNTER, A.G. (1972). 'Seminal antigens: their alteration in the genital tract of female rabbits and during partial *in vitro* capacitation with β-amylase and β-glucuronidase.' *Biol. Reprod.,* 7, 332–340 (140)

JONES, D.A. and WILKINS, D.A. (1971). *Variation and adaptation in plant species.* 184pp. Heinemann; London (19)

JONES, A., ALDERSON, R. and HOWELL, B.R. (1974). 'Progress towards the development of a successful rearing technique for larvae of the turbot *Scophthalmus maximus* L.' In: *The Early Life History of Fish.* 731–737. (ed. Blaxter). Springer-Verlag; Berlin (236)

JONES, J.C. (1968). 'Sexual life of a mosquito.' *Sci. Am.,* **218,** 108–118 (32, 108)

JONES, J.W. and ORTON, J.H. (1940). 'The paedogenetic male cycle in *Salmo salar* L.' *Proc. Roy. Soc. B,* **128,** 485–499 (48)

KATCHADOURIAN, H.A. and LUNDE, D.T. (1975). *Fundamentals of human sexuality.* (2nd edn). 595pp. Holt, Rhinehart and Winston; New York (71)

KEARN, G.C. (1970). 'The production, transfer and assimilation of spermatophores by *Entobdella soleae*, a monogenean skin parasite of the common sole.' *Parasitology,* **60,** 301–311 (58)

KIHLSTROM, J.G. (1966). 'Diurnal variation and the spontaneous ejaculations of the male albino rat.' *Nature, Lond.,* **209,** 513–4 (100)

KILLE, R.A. (1960). 'Fertilisation of the lamprey egg.' *Exptl. Cell Res.,* **20,** 12–27 (122)

KIMURA, M. and CROW, J.F. (1964). 'The number of alleles that can be maintained in a finite population.' *Genetics,* **49,** 725–738 (35)

KING, R.C. (1970). *Ovarian Development in* Drosophila melanogaster, 227pp. Academic Press; New York (287)

KIERSZENBAUM, A.L. and TRES, L.L. (1975). 'Structural and transcriptional features of the mouse spermatid genome.' *J. Cell. Biol.,* **65,** 258–270 (100)

KNIGHT-JONES, E.W. and MOYSE, J. (1961). 'Intraspecific competition in sedentary marine animals.' *Symp. Soc. Exp. Biol.,* **15,** 72–95 (194)

KOFORD, C.B. (1963). 'Rank of mothers and sons in bands of rhesus monkeys.' *Science,* **141**, 356–7 (243, 288)

KORSCHELDT, E. and HEIDER, K. (1892). *Lehrbuch der Vergleichenden Entwicklungsgeschichte.* Gustav Fischer; Jena (184)

KUHN, T.S. (1970). *The Structure of Scientific Revolutions.* 210pp. University of Chicago Press; London (315)

LACK, D. (1954). 'The evolution of reproductive rates.' In: *Evolution as a Process.* 143–156. (ed. Huxley, Hardy and Ford.) Allen and Unwin; London (176, 233)

LACY, D. and PETTITT, A.J. (1970). 'Sites of hormone production in the mammalian testis, and their significance in the control of male fertility.' *Br. med. Bull.,* **26,** 87–94 (81)

LENNENBERG, E.H. (1967). *Biological foundations of language.* Wiley; New York (216)

LEWONTIN, R.C. (1974). *The Genetic Basis of Evolutionary Change.* 346pp. Columbia University Press; New York (35, 238, 240, 258)

LEWONTIN, R.C. and HUBBY, J.L. (1966). 'A molecular approach to the study of genic heterozygosity in natural populations of *Drosophila pseudoobscura*.' *Genetics,* **54,** 595–609 (26, 35)

LIGGINS, G.C. (1972). 'The fetus and birth.' In: *Reproduction in Mammals* 2, 72–109. (eds. Austin and Short) (206)

LILLIE, F.R. (1902). 'Differentiation without cleavage in the egg of the annelid *Chaetopterus pergamentaceus.* *Roux Arch. Entwickl.,* **14,** 477–499 (189)

LILLIE, F.R. (1919). *Problems of Fertilisation.* 278pp. University of Chicago Press; Chicago, Illinois (134)

LLEWELLYN, J. (1965). 'The evolution of parasitic platyhelminths.' In: *Evolution of Parasites.* 47–78. (ed A.E.R. Taylor). Blackwell Scientific Publications; Oxford (121)

LONGO, F.J. and ANDERSON, E. (1974). 'Gametogenesis.' In: *Concepts of Development.* 3–47. (ed. Lash and Whittaker). Sinauer Associates; Stamford, Conn. (97)

LORENZ, K. (1965). *Evolution and Modification of Behaviour.* University of Chicago Press; Chicago, Illinois (216)

LOTKA, A.J. (1945). 'The law of evolution as a maximal principle.' *Human Biol.,* **17**, 164–194 (287)

LOVELACE, L. (1974). *Inside Linda Lovelace.* 191pp. Pinnacle Books; New York (67)

LOWE-McCONNELL, R.H. (1975). *Fish Communities in Tropical Freshwaters; their distribution, ecology and evolution.* 337pp. Longmans; London (218, 219)

McLAREN, H.C. (1975). 'Liverpool flats' story in undergraduate lectures. (245)

MACRIDES, F., BARTKE, A. and DALTERIO, S. (1975). 'Strange females increase plasma testosterone levels in male mice.' *Science,* **189,** 1104–1106 (225)

MAHOWALD, A.P. (1972). 'Oogenesis.' In: *Developmental Systems: Insects.* **1**, 1–47. (eds. Counce and Waddington) Academic Press; London (75, 147, 155)

MALTHUS, T.R. (1798). *An essay on the principle of population as it affects the future improvement of society.* Published anon. 2nd edition

1803. Modern reprints, e.g. J.M. Dent and Sons, London (1973), are of 7th edition (14)

MANN, T. (1964). *The Biochemistry of Semen and of the Male Reproductive Tract.* Methuen; London (101, 135)

MANN, T., MARTIN, A.W. Jr. and THIERSCH, J.B. (1970). 'Male reproductive tract, spermatophores and spermatophoric reaction in the giant octopus of the North Pacific, *Octopus dofleini martini.*' *Proc. Roy. Soc. Ser. B.*, **175**, 31–61 (53, 95)

MANWELL, C. and BAKER, C.M.A. (1970). *Molecular Biology and the Origin of Species; heterosis, protein polymorphism and animal breeding.* University of Washington Press; Seattle (22)

MARCUS, S. (1969). *The Other Victorians.* Corgi Paperbacks; London (318)

MARINKOVIC, D. (1967). 'Genetic loads affecting fecundity in natural populations of *D. pseudo-obsura.*' *Genetics,* **56**, 61–71 (22)

MARSHALL, F.H.A. (1910). *Physiology of Reproduction.* 1st edition, Longmans Green; New York. 2nd edition, 1922. 3rd edition (ed. A.S. Parkes), 1956 (4)

MARTINET, L. and RAYNAUD, R. (1974). 'Survie prolongee des spermatozoides dans l'uterus de la Hase: explication de la superfoetation.' In: *Transport, Survie et Pouvoir Fécondant des Spermatozoides.* INSERM, 295–308 (202)

MASTERS, W.H. and JOHNSON, V.E. (1966). *Human Sexual Response.* 366pp. Churchill; London (70)

MATHER, K. (1964). *Human Diversity; the nature and significance of differences among men.* 126pp. Oliver and Boyd; Edinburgh (20, 287)

MATHEWS, C.P. (1971). 'Contribution of young fish to total production of fish in the River Thames near Reading.' *J. Fish. Biol.*, **3**, 157–180 (237)

MAY, R.C. (1974). 'Larval mortality in marine fishes and the critical period concept.' In: *The Early Life History of Fish.* 3–20. (ed. Blaxter). Springer-Verlag; Berlin (235)

McVAY, S. (1964). Personal communication (60)

MEDAWAR, P.B. (1957). *The Uniqueness of the Individual.* Constable; Edinburgh (227, 229, 230, 232)

MEDAWAR, P.B. (1963). 'Theories destroy facts.' *Mensa Annual Lecture,* 1963 (2)

MEDAWAR, P.B. (1972). *The Hope of Progress.* Methuen; London (67)

MESKE, C. (1973). 'Experimentally induced sexual maturity in artificially reared male eels (*Anguilla anguilla*).' In: *Genetics and Mutagenesis of Fish.* 161–170. (ed. Schroder). Springer-Verlag; Berlin (47)

METZ, C.B. (1967). 'Gamete surface components and their role in fertilisation.' In: *Fertilisation; comparative morphology, biochemistry and immunology,* **I**, 163–236. (eds Metz and Monroy). Academic Press; New York (135)

MILES, Bernard (c.1950). *'Nice goings on: the rudiments of Greek mythology.'* HMV record no C4138 (194)

MILGROM, E., THI, M.L. and BAULIEU, E.E. (1973). 'Control mechanisms of steroid hormone receptors in the reproductive tract.' In: *Protein Synthesis in Reproductive Tissue, Karolinska Symposia on Research Methods in Reproductive Endocrinology,* **6**, 380–403 (271)

MILLER, R.L. and O'RAND, M.G. (1975). 'Utilisation of chemical specificity during fertilisation in the Hydrozoa.' In: *The Functional Anatomy of the Spermatozoon.* 15–26. (ed. Afzelius). Pergamon Press; Oxford (111, 122)

MINTZ, B. (1960). 'Embryological phases of mammalian gametogenesis.' *J. Cell. Comp. Physiol.,* **56,** Suppl. I. 31–47 (74)

MIYAMOTO, H. and CHANG, M.C. (1972). 'Development of mouse eggs fertilised *in vitro* by epididymal spermatozoa.' *J. Reprod. Fert.,* **30,** 135–7 (140)

MONROY, A. (1973). 'Fertilisation and its biochemical consequences.' *An Addison-Wesley Module in Biology,* **7,** 37pp (131, 132)

MOORE, J.A. (1955). 'Abnormal combinations of nuclear and cytoplasmic systems in frogs and toads.' *Adv. Genet.,* **7,** 139–82 (155)

MOREHEAD, D.M. and MOREHEAD, A. (1974). 'From signal to sign: a Piagetian view of thought and language during the first two years.' In: *Language Perspectives – Acquisition Retardation and Intervention.* 154–190. (eds Schiefelbusch and Lloyd). Macmillan; London (216)

MORGAN, E. (1974). *Descent of Woman.* 288pp. Corgi Paperback; London (66)

MORRIS, D. (1967). *Naked Ape.* 250pp. Jonathan Cape; London (66, 67)

MYKYTOWYCZ, R. and FULLAGER, P.J. (1973). 'Effect of social environment on reproduction in the rabbit, *Oryctolagus cuniculus.*' *J. Reprod. Fert. Suppl.,* **19,** 503–22 (283)

NATH, V. (1965). *Animal Gametes (Male); a morphological and cytochemical account of spermatogenesis.* 162pp + 185 fig. Asia Publishing House; London (95, 103)

NEFZAWI, (1967). *The Perfumed Garden.* 192pp. (trans. Sir Richard Burton). Panther Books; London (66)

NELSEN, O.E. (1953). *Comparative Embryology of the Vertebrates.* 982pp. Blakiston Co.; New York (32, 101, 119, 286)

NEYFAKH, A.A. (1971). 'Steps of realisation of genetic information in early development.' *Curr. Topics in Devel. Biol.,* **6,** 45–78 (163)

ODELL, W.D. and MOYER, D.L. (1971). *Physiology of Reproduction.* 152pp. C.V. Mosby Co.; St Louis (268)

OHNO, S. (1970). *Evolution by Gene Duplication.* 150pp. Allen and Unwin; London (22, 257, 258, 259)

OPIE, I. and OPIE, P. (1959). *The lore and language of schoolchildren.* Oxford University Press; London (199)

ORWIN, I. (1975). Personal communication (235)

PAUL, J. and GILMOUR, R.S. (1968). 'Organ-specific restriction of transcription in mammalian chromatin.' *J. Mol. Biol.,* **34,** 305–316 (163)

PAUL, J., CARROLL, D., GILMOUR, R.S., MORE, J.A.R., THRELFALL, A., WILKIE, M. and WILSON, S. (1972). 'Functional studies on chromatin.' In: *Karolinska Symposia on Research Methods in Reproductive Endocrinology. 5: Gene Transcription in Reproductive Tissue.* 277–297. (ed. Diczfalusy) (163)

PHILLIPP, E.E. (1973). 'Discussion on moral, social and ethical issues.'

In: *Law and Ethics of A.I.D. and Embryo Transfer; Ciba Foundation Symposium,* **17,** (New Series), pp63 and 66 (245)

PIVETEAU, J. (1958). 'Traité de Palaeontologie.' T.6 Vol.2. *Mammifères; evolution.* p.399. Masson et Cie; Paris (254)

POPPER, K.R. (1963). *Conjectures and Refutations.* 426pp. Routledge; London (67)

PORTER, K.R. (1972). *Herpetology.* 524pp. W.B. Saunders Co.; Pennsylvania (16, 228)

POTTS, D.M. (1972). 'Limiting human reproductive potential.' In: *Reproduction in Mammals,* **5**, 32–66. (ed. Austin and Short) (245)

RACEY, P.A. (1975). 'The prolonged survival of spermatozoa in bats.' In: *Biology of the Male Gamete,* (eds Duckett and Racey). *J. Linn. Soc.,* **7**, Suppl. 1, 385–416 (62, 142, 202)

RAVEN, C.P. (1961). *Oogenesis: the storage of developmental information.* 274pp. Pergamon; Oxford (121, 125)

RAVEN, C.P. (1966). *Morphogenesis; the analysis of molluscan development.* 365pp. 2nd edition. Pergamon; Oxford (147, 155)

REES H. and JONES, R.N. (1972). 'The origin of the wide species variation in nuclear DNA content.' *J. intern. Rev. Cytol.,* **32**, 53–92 (258)

REICH, W. (1968). *The Function of the Orgasm.* 218pp. Panther Books; London (67)

ROMER, A.S. (1959). *The Vertebrate Story.* 437pp. University of Chicago Press; Chicago, Ill. (254, 255)

ROOSEN-RUNGE, E.C. (1962). 'The process of spermatogenesis in mammals.' *Biol. Revs. (Camb.),* 343–377 (29)

ROOSEN-RUNGE, E.C. (1969). 'Comparative aspects of spermatogenesis.' *Biol. Reprod.,* Supp. I, 24–39 (81)

ROTH, P. (1967). *Portnoy's Complaint.* 309pp. Corgi Paperbacks; London (67)

ROTHSCHILD, Lord (1956). *Fertilisation.* 170pp. Methuen; London (27, 130)

ROTHSCHILD, Lord (1961). 'Structure and movements of tick spermatozoa (*Arachnida, Acari*)', *Quart. J. micr. Sci.,* **102**, 239–247 (93, 95)

ROTHSCHILD, Lord and SWANN, M.M. (1949). 'The fertilisation reaction in the sea urchin egg.' *J. Exp. Biol.,* **26,** 164–176 (139)

ROTHSCHILD, Miriam (1965). 'Fleas.' *Sci. Am.,* **213,** 44–53 (58)

RUNNSTROM, J. (1952). 'The cell surface in relation to fertilisation.' *Symp. Soc. Exp. Biol.,* **6,** 39–88 (135)

SADLEIR, R.M.F.S. (1973). *The Reproduction of Vertebrates.* 227pp. Academic Press; New York (35)

SADOGLU, P. (1967). 'The selective value of eye and pigment loss in Mexican cave fish.' *Evolution,* **21,** 541–9 (19, 251)

SALISBURY, E.J. (1930). 'Mortality amongst plants and its bearing on Natural Selection.' *Nature, Lond.,* **125,** 817 (20)

SALTHE, S.M. (1972). *Evolutionary Biology.* 437pp. Holt, Rhinehart and Winston Inc; New York (240)

SAUNDERS, J.W. (1970). *Patterns and Principles of Animal Development.* 282pp. Collier-Macmillan; London (35)

SELANDER, R.K., HUNT, W.G. and YANG, S.Y. (1969). 'Protein polymorphism and genetic heterozygosity in two European species of the house mouse.' *Evolution,* **23**, 379–390 (26, 35)

SETCHELL, B.P., SCOTT, T.W., VOGLMAYR, J.R. and WAITES, G.M.H. (1969). 'Characteristics of testicular spermatozoa and the fluid which transports them into the epididymis.' *Biol. Reprod.,* Supp. **I**, 40–66 (100)

SHEPHERD, P.M. (1975). *Natural Selection and Heredity.* 239pp. (4th edition). Hutchinson; London (238, 240, 256, 257)

SHORT, R.V. (1972). 'Sex determination and differentiation.' In: *Reproduction in Mammals,* 2, 43–71. (eds Austin and Short) (265, 266, 269, 272)

SHULMAN, S. (1971). 'Immunity and infertility: a review.' *Contraception,* **4**, 135–154 (245)

SIMPSON, G.G. (1950). *The Meaning of Evolution.* 364pp. Oxford University Press; London. Especially pp.130–136 (255)

SINGER, I. (1973). *The Goals of Human Sexuality.* 224pp. Wildwood House; London (70, 72)

SLYPER, E.J. (1962). *Whales.* 475pp. Hutchinson; London (60)

SMITH, J.M. (1958). *The Theory of Evolution,* 320pp. (1975) 3rd edition 344pp. Penguin Books; London (22, 25, 35)

SMITH, J.M. (1964). 'Group selection and kin selection.' *Nature, Lond.,* **201**, 1145–7 (63)

SPECTOR, W.S. ed. (1956). *Handbook of Biological Data.* Saunders; Pennsylvania (16, 32)

SPEMANN, H. and MANGOLD, H. (1924). 'Ueber Induktion von Embryonenanlagen durch Implantation artfremder Organisatoren.' *Roux Arch. f. Ent. – mech.,* **100**, 599–638. (168, 287) See also Ballard (1964)

SPIRIN, A.S. (1966). 'On "masked" forms of messenger RNA in early embryogenesis and in other differentiating systems.' *Curr. Top. in Devel. Biol.,* **1**, 1–38 (132)

SPURWAY, H. (1948). 'Genetics and cytology of *Drosophila sub-obscura.* IV. An extreme example of delay in gene action, causing sterility.' *J. Genet.,* **49**, 126–140 (107)

SPURWAY, H. (1953). 'Spontaneous parthenogenesis in a fish.' *Nature, Lond.,* **171**, 437 (44)

STERBA, G. (1962). *Freshwater Fishes of the World.* 878pp. Transl. and rev. by D.W. Tucker. Vista Books; London (249, 251)

TARTAR, V. (1960). *The Biology of Stentor.* Pergamon Press; Oxford (91, 145)

THOM, R. (1972). *Stabilite Structurelle et Morphogénèse.* Benjamin. (1975) *Structural Stability and Morphogenesis* (trans. D.H. Fowler). Addison-Wesley; Reading, Mass. (175)

THOMPSON, Sir D'Arcy W. (1961). *On Growth and Form.* 345pp. (abridged, ed. Bonner). Cambridge University Press; London (231)

TIMOURIAN, H., HUBERT, C.E. and STUART, R.N. (1972). 'Fertilisation in the sea urchin as a function of sperm-to-egg ratio.' *J. Reprod. Fert.,* **29**, 381–5 (32)

TUCKER, D.W. (1959). 'A new solution to the Atlantic eel problem.' *Nature, Lond.,* **183**, 495–501, and **184**, 1281–3 (47, 48)

TURING, A.M. (1952). 'The chemical basis of morphogenesis.' *Phil. Trans. Roy. Soc. B,* **237**, 37–72 (169)

TURKINGTON, R.W. and KADOHAMA, N. (1972). 'Gene activation in mammary cells.' In: *Gene Transcription in Reproductive Tissue, Karolinska Symposium on Research Methods in Reproductive Endocrinology,* **5**, 346–368 (274)

TYLER, A. (1961). 'Immunological phenomena and fertility control.' *J. Reprod. Fert.,* **2**, 473–506 (111, 135, 139)

ULLMAN, S. (1976). 'Anomalous litters in hybrid mice and the retention of spermatozoa in the female tract.' *J. Reprod. Fert.,* **47**, 13–18 (202)

VAN DER VELDE, T. (1926). *Ideal Marriage, Its Physiology and Technique.* (Trans. Stella Brown, 1930). Random House; New York (70, 71)

VAN LAWICK-GOODALL, H. and VAN LAWICK-GOODALL, J. (1971). *Innocent Killers.* Houghton Mifflin; Boston, and the BBC/Time–Life Films of their Wild Dogs (239, 242, 243)

VAN TIENHOVEN, A. (1968). *Reproductive Physiology of Vertebrates.* 498pp. Saunders; Pennsylvania (212, 263)

VATSAYANA (called Malla-naga) (1963). *The Kama Sutra.* 200pp. (trans. Sir R. Burton and F.F. Arbuthnot, ed. Archer, 1963). Panther Books; London (66)

WADDINGTON, C.H. (1957). *The Strategy of the Genes; a discussion of some aspects of theoretical biology.* 262pp. George Allen and Unwin; London (169)

WADDINGTON, C.H. (1966). *Principles of Differentiation and Development.* 115pp. Macmillan; New York (166, 167, 168)

WAKERLEY, J.B. and LINCOLN, D.W. (1973). 'The milk-ejection reflex of the rat: a twenty-to-forty-fold acceleration in the firing of paraventricular neurones during oxytocin release.' *J. Endocr.,* **57**, 477–93 (276)

WALKER, K. (1949). *The Physiology of Sex.* 161pp. Penguin Books; London (70, 71)

WALLACE, B. (1970). *Genetic Load, its biological and conceptual aspects.* 116pp. Prentice-Hall; New Jersey (19, 21, 35, 239, 240)

WALLACE, H. (1974a). Unpublished (110)

WALLACE, H. (1974b). 'Chiasmata have no effect on fertility.' *Heredity,* **33**, 423–429 (107, 108)

WALLACE, H. (1974c). 'The chiasmata of axolotls.' *Archiv. für Genetik,* **47**, 86–95 (110)

WALNE, P.R. (1961). 'Observations on the mortality of *Ostrea edulis.*' *J. mar. biol. Ass., U.K.,* **41**, 113–22 (233)

WATSON, A. and MOSS, R. (1970). 'Dominance, spacing behaviour and aggression in relation to population limitation in vertebrates.' In: *Animal Populations in Relation to their Food Sources.* 167–218. (ed. Watson). Blackwell; Oxford (225)

WATSON, A. and MILLER, G.R. (1971). 'Territory size and aggression in a fluctuating red grouse population.' *J. Anim. Ecol.,* **40**, 367–383 (238, 240, 283)

WEIR, B.J. (1971). 'The reproductive organs of the female Plains Viscacha (*Lagostomus maximus*).' *J. Reprod. Fert.,* **25**, 365–373 (125, 126)

WEISSMANN, A. (1904). *The Evolution Theory* (trans. J. Arthur Thomson). 416 + 405pp. Edward Arnold; London (10)

WENDT, H. (1965). *Sex Life of the Animals.* 362pp. Simon and Schuster; New York (1, 40)

WHITE, M.J.D. (1954). *Animal Cytology and Evolution.* 454pp. 2nd edition. Cambridge University Press; London (259)

WHITEHOUSE, H.L.K. (1973). *Towards an Understanding of the Mechanism of Heredity.* 528pp. 3rd edn. Arnold; London (172)

WHITTAKER, J.R. (1974). 'Aspects of differentiation and determination in pigment cells.' In: *Concepts of Development.* 163–178. (ed. Lash and Whittaker). Sinauer Associates; New Jersey (171)

WHYTE, L. (1965). *Internal Factors in Evolution.* 81pp. Tavistock Publications Ltd.; London (19)

WICKLER, W. (1962). ' "Egg dummies" as natural releasers in mouth-breeding cichlids.' *Nature, Lond.,* **194,** 1092–1093 (55)

WICKLER, W. (1973). *Breeding Behaviour of Aquarium Fishes.* T.F.H.; New Jersey (55)

WILKES, A. (1965). 'Sperm transfer and utilisation by the arrhenotokous Wasp *Dahlbominus fuscipennis* (Zelt) (*Hymenoptera Eulophidae*).' *Can. Entomol.,* **97**, 647–657 (32, 108)

WILLIAMS, G.C., KOEHN, R.K. and MITTON, J.B. (1973). 'Genetic differentiation without isolation in the American eel *Anguilla rostrata*.' *Evolution,* **27**, 192–204 (26, 48, 238)

WILLIAMS, J. (1967). 'Chemical constitution and metabolic activities of animal eggs.' In: *The Biochemistry of Animal Development.* 13–71. (ed. Weber). Academic Press; New York (119)

WILLIAMS, W.L., ROBERTSON, R.T. and DUKELOW, W.R. (1969). 'Decapacitation factor and capacitation.' *Schering Symp. on Mechanisms involved in Conception, Berlin 1969. Adv. Biosciences,* **4,** 61–72 (140)

WITSCHI, E. (1948). 'Migration of the germ cells of human embryos from the yolk sac to the primitive gonadal folds.' *Contr. Embryol. Carnegie Instn.,* **32**, 67–80 (265)

WOOLLEY, D.M. (1975). 'The rat sperm tail after fixation by freeze-substitution.' In: *Functional Morphology of the Spermatozoon.* 177–178. (ed. Afzelius). Pergamon; Oxford (103)

WYNNE-EDWARDS, V.C. (1962). *Animal dispersion in relation to social behaviour.* Hafner; New York (225, 237)

WYNNE-EDWARDS, V.C. (1964). 'Population control in animals.' *Sci. Am.,* August 1964 (225)

YANAGISAWA, K., POLLARD, D.R., BENNETT, D., DUNN, L.C. and BOYSE, E.A. (1974). 'Transmission ratio distortion at the t-locus: serological identification of two sperm populations in t-heterozygotes.' *Immunogenetics,* **1**, 91–96 (106)

YOUNG, J.Z. (1957). *The Life of Mammals.* 820pp. Clarendon Press; Oxford (55)

ZEEMAN, E.C. (1976). 'Catastrophe theory.' *Sci. Am.,* **234,** 65–83 (175)

GLOSSARY

Adolescent: used of human beings who have attained function of the primary and secondary sex organs (usually menarche or ejaculation) but have not yet been accepted as responsible adults in society. Widened to include apes but not, generally, other animals.

Alleles: alternative forms of a gene at a particular genetic locus e.g. coloured/albino, or normal/sickle haemoglobin.

Anthropology: the study of man.

Antibody: a protein, produced by lymphocytes of mammals, usually a globulin, which attaches specifically to another molecule, usually foreign, called an antigen. Antibody molecules come in several forms (e.g. IgG in blood plasma, secretory IgA on epithelial surfaces of gut and genitalia, IgE producing allergies), and there are similar substances thought to be associated with cell attack on foreign invaders.

Artiodactyls: cloven-hoofed mammals, with two effective toes on each foot, e.g. deer, pigs, cattle.

Attrition: progressive reduction in quantity, especially by wearing away. Used here by analogy, to describe the progressive fall in numbers as offspring grow up. 'Ten green bottles' is the classic case.

Axiom: a proposed rule for erecting a generalisation; frequently, as in Euclidean geometry, a *tautology* dependent upon definitions.

Bacteria: microscopic organisms without a nucleus and with a simple structure, e.g. *Escherichia coli* on which most biochemical work on genetic mechanisms has been performed. A few bacteria are pathogenic (cause illness), e.g. *Gonococcus* (gonorrhoea) and *Salmonella* (food poisoning) but most live on the products of the decay of other organisms.

Benthic: living on the bottom of the sea.

Berdache: a man who adopts female gender role, including female dress, and is involved in approved homosexual relations with another male.

Bestiality: sexual practices between human beings and other animals; coitus is usually proscribed, but the range could be seen as including the stroking of cats.

Blastodisc (-derm): the cytoplasmic cap on very yolky (telolecithal) eggs; called blastoderm after cleavage into cells (blastomeres).

Breeding: (a) the production of young animals by people, e.g. in 'tropical fish breeders'. (b) Mating (usually copulating) as in 'cattle breeding station'.

Budding: the production, by an organised individual, of tissue which undergoes morphogenesis to produce another individual, which usually remains attached to the parent stock at least until fully organised.

Centriole: organelle of eukaryote cells which organises the mitotic or meiotic spindle and its fibres, and is usually found associated with the flagellar or ciliary 9 + 2 fibre array. Multiplies in the cell by a strange replication involving the appearance of 9 new fibres at 90° to the 9 old ones; each functional centriole seems to be such a pair, at right angles.

Centromere: special part of a chromosome which serves for attachment to the spindle fibres at cell division. Divides before mitotic divisions as the chromosome becomes two chromatids, but does not divide in the same plane before meiosis. The centromere probably always consists of thousands of repeats of a special short DNA sequence. In the mouse the same sequence occurs on all chromosomes, but the chromosomes of man have several different centromeric sequences.

Chelae: the large claws of some arthropods, notably decapod crustaceans (e.g. lobster) and scorpions.

Chiasma: the pattern formed where two (non-sister) chromatids have been involved in an exchange of segments early in meiotic prophase, observed at late prophase or metaphase. Such chiasmata can be counted in chromosome spreads of meiotic cells, and their number agrees with the extent of genetic crossing-over found between the homologous chromosomes.

Chinese boxes: toys, like the wooden Russian dolls, whose fascination is the box-within-box-within-box paradox.

Chitin: the polysaccharide material which gives strength to the exoskeleton of arthropods, especially insects and spiders. Mass (e.g. in crustaceans like crabs) is achieved by deposition of calcium salts, and waterproofing (as in most insects) by a waxy layer on the outside.

Chloroplasts: organelles in plant cells which contain chlorophyll and other photosynthetic machinery. Resemble bacteria in their DNA ring and multiplication, and are probably descended from symbiotic bacteria.

Choanocyte: 'collar-cell' of sponges, lining the flagellated chambers and responsible for water passage through the sponge.

Chromosome: the rod- or string-like structures which contain the DNA molecules in the nucleus of eukaryotes; they probably each contain one molecule of DNA, surrounded by protein except where it is replicating or being copied (transcribed) onto RNA. The chromosomes form a set characteristic for each species of eukaryote; the organism's cells normally have two such sets (diploid) but the sperms (rarely the egg) have one set (haploid).

Circumcision: the removal of most of the foreskin from the penis, usually of infants (Judaism and Islam) but sometimes as a puberty ritual.

Clitoridectomy: removal of the glans of the clitoris in female infants. Practised by the Kikuyu.

Clone: all those organisms produced from one ancestor, by budding or other asexual process, by parthenogenesis or even occasionally used for self-fertilising bisexuals. All presumed genetically similar or identical.

Coenocytic: a group of 'cells' without intervening cell walls; nuclei in common cytoplasm; a syncytium.

Coenosarc: the common part of the colony of branching polyps like *Obelia* or ectoprocts; properly restricted to the skeleton of the attached part of hydrozoan coelenterate colonies, but used much more widely for the creeping attachment stems of a variety of animals.

Coitus: sexual intercourse with intromission, especially in mammals.

Copulation: the transfer of male gametes into another organism, where they may later be used to fertilise the eggs of the second organism. Contrast amplexus.

Crepuscular: active at dawn and dusk; contrast diurnal, nocturnal.

Deoxyribonucleic acid (DNA): the long paired molecule in the chromosomes of higher organisms (eukaryotes), and the genome of bacteria, which carries the genetic message between generations of cells or organisms. In the pairing of the double molecule, adenine and thymine bases complement each other on the two chains, as do guanine and cytosine. RNA is the molecule used for transcribing the DNA genetic message.

Development: the processes by which an egg or spore becomes, or forms, an organism.

Dogma: a central and necessary belief within a system of beliefs. Compare paradigm.

Droit de seigneur: the right of the lord of the manor to deflorate virgins marrying in his domain. Practised overtly until nineteenth century; see the novel by Beaumarchais 'Marriage de Figaro', or the Mozart opera libretto.

Ecology: the study of natural populations of organisms, and of simpler laboratory models of them.

Embryology: the study of the development of organisms; how the interactions of heredity and environment produce the succession of forms which organisms pass through.

Endemic: of a disease normally present in the population. Contrast epidemic.

Endocrinology: the study of animal hormones, including the stimuli which evoke them, their interplay, and tissue responsiveness to them.

Enzyme: a biological catalyst, almost always a protein.

Epi-oophoron: vestigial, often cyst-like remains of the Wolffian ducts in female mammals, usually close to the ovary or its suspensory 'ligament'. In the male they form epididymes and vas deferens.

Erythrocytes: red blood corpuscles.

Essentialism: the belief in 'one right way' or one essential truth, in general or on one subject like human sexual climax. Contrast pluralism.

Eukaryotes: organisms with an organised nucleus (in most cells); organisms of more complexity than bacteria, and probably originating from (and formally equivalent to) symbiotic associations of prokaryotes like modern bacteria. The nucleus, with the chromosomes, is usually separated from the cytoplasm by a well-defined nuclear membrane. Man, amoeba, dandelion and mushroom are all eukaryotes.

Eversible: capable of extension, by being turned inside out, like a stocking being taken off.

Evolutionary symbionts: those prokaryote organisms which grouped together to form the eukaryote cell, or the fungus and alga which cooperate to become a lichen. Other symbionts, while they may be important to each other, are not commonly dignified by this title.

Exhibitionism: display of the genitalia, usually by men to young girls, as a sexual episode not an overture.

FSH: follicle-stimulating hormone of mammals, produced in pituitary after stimulation with FSH-releasing hormone from hypothalamus.

Fatality: death ascribable to a cause. Used here as percentage fatality during various phases of life history and ascribed to external accident and/or internal differences between organisms for some of which the accident is fatal. Contrast mortality.

Fecundity: the number of offspring actually produced. Contrast fertility.

Fertility: the ability to produce offspring. Contrast fecundity, 'fitness'.

Flirting: use of sexual preliminaries for other social functions, either to reinforce (test) roles (Goffman) or to test the climate for more intimate relationship which may not be overtly sexual.

Gamete: a specialised product of an organism, adapted to take part in fertilisation; e.g. sperms or eggs. Contrast spore.

Gamones: substances released with gametes, and with physiological effects upon gametes.

Gene: that chromosome length which controls the development of a variant character, like the gene for albinism in many organisms including mice and men. Alternatively, that DNA nucleotide sequence which, when transcribed by DNA and translated by ribosomes into protein, produces one functional protein molecule. The first usage, as in neo-Mendelian theory, may be the breeding consequence of variants of the second usage.

Genetics: the study of heredity, particularly chromosomal (DNA) heredity.

Genotype: the array of genes (hereditary factors carried on chromosomes, and perhaps other hereditary factors) possessed by an organism. Contrast phenotype.

Heredity: the control of an organism's life by that structure and environment determined by its parent organisms. Contrast genotype.

Heterogony: variation of the breeding form, usually in different 'generations', e.g. polyp and medusa. (Was also used by Huxley to mean non-isometric growth.)

Heterozygote: an organism which receives dissimilar alleles of a particular gene from its parents, e.g. a brown-eyed child one of whose parents is blue-eyed. Contrast homozygote.

Histone: protein, rich in lysine, which are associated with the DNA in chromosomes. There are five kinds in most eukaryotes and they are probably used non-specifically to regulate gene function; specificity is conferred by other, acidic, proteins.

Holometabolous: those insects whose metamorphosis to the adult is drastic, involving the development of adult organs from 'imaginal discs', like *Diptera* and *Lepidoptera*. Contrast hemimetabolous.

Homozygote: an organism which receives identical alleles of a particular gene from both parents, for example blue-eyed humans, albino rabbits or mice. Contrast heterozygote.

Hydrorhiza: the branching stem of hydroid colonies which anchors them to the substrate.

Immunoglobulin: see Antibody.

Implantation: the attachment of the mammalian blastocyst to the uterine wall. See Trophoblast.

Incest: prohibited sexual intercourse between close (human) relations, e.g. father–daughter or between siblings. Cousin marriages are usually tolerated; some societies have allowed brother–sister marriage in special circumstances.

Infraciliature: the complex system of interconnecting fibres and granules underlying and coordinating the cilia, particularly those of ciliophoran protozoans.

Intromission: the insertion of a male organ into the female genital tract.

Irritability: also called excitability; the sensitivity of organisms to outside stimuli.

'K-selection': the selection operating on a specialised species in a narrow habitat, pruning deviance (i.e. stabilising selection). Selection for constancy; contrast *r*-selection.

Keratin: the characteristic protein of the horny parts of the skin of terrestrial vertebrates, e.g. hair, feathers, nails, hoofs. Also used as a general term for the whole complex of cellular debris found in these dead tissues.

LH: luteinising hormone, produced by pituitary, converting ovulated follicle remains into corpus luteum.

Lactation: the production of a nutritious secretion by a female mammal (or bird) which feeds the young after birth.

Lamarckian: evolutionary theories which depend upon the 'transmission of acquired characters' e.g. which suppose that the children of a blacksmith will inherit his acquired muscularity. Contrast neo-Mendelian.

Lordosis: reverse flexure of the spine. Used by students of animal behaviour as a 'jargon' for sexual presentation by female mammals, who do commonly elevate the genitalia by hyperextending the body.

Masturbation: stimulation of the genitalia by other than coital means.

Mean: average; that is to say, the sum of the measurements carried out for a population of animals divided by the number of animals measured.

Medusa: jelly fish are examples of large complex medusae. Larger jelly fish are budded usually from a 'degenerate' polyp called a scyphistoma The simple medusae are usually the distributive and sexual phase of hydroid polyps like *Obelia.*

Meiosis: the terminal two cell divisions which produce gametes. There is a long prophase to the first division during which chromosomes are seen to be paired as *n* bivalents (but see Sex chromosomes) and when chiasmata can be observed to have occurred; its stages are leptotene, zygotene, pachytene, diplotene, diakinesis which leads to the first metaphase (M1); during the succeeding anaphase the chromosomes (with their exchanges) part to opposite poles, *n* at each pole (telophase 1). Then a brief P2 leads through M2 (at which many eggs remain till fertilised) and A2, T2, giving four sperms, or an egg and three polar bodies.

Mendelian segregation: the appearance of parental (or in general ancestral) characters in descendants without 'mixing', in specific statistical ratios suggesting that the factors (genes) controlling them are assorted randomly into the parental gametes, which then associate (fertilise) without bias. Contrast segregation distortion, Lamarckian.

Metabolism: those processes which maintain the life of the organism; divided into anabolic (building-up) and catabolic (breaking-down) for some purposes.

Metazoa: those animals which are composed of many cells with different functions, e.g. man, *Hydra*, insects. Contrast Protozoa, Parazoa (sponges).

Mitochondria: organelles in the cytoplasm of eukaryotes, in which much of the oxidation for energy production occurs. Resemble bacteria in their DNA ring and multiplication, and are probably descended from symbiotic bacteria.

Mitosis: the ordinary process of cell division in eukaryotes. The $2n$ chromosomes appear (become stainable) in prophase, each made of two chromatids, align on the spindle equator at metaphase; the chromatids are drawn apart by their centromeres during anaphase, and each group constitutes a new nucleus equivalent to the original in chromosome number, during telophase. Contrast with meiosis.

Molecular biology: the study of biology from the viewpoint of its chemistry; the techniques of molecular biology have elucidated much of the mechanism of gene action, especially in bacteria.

Morphogenesis: the development of structure or shape.

Morphology: 'shape', including the structure of internal organs (anatomy) as well as external appearance.

Mortality: the condition of having a finite lifespan. Percentage mortality is only useful if age-related, e.g. perinatal mortality. Contrast fatality.

Multiplication: in biology, increase of number by mitosis, budding or other asexual means such that the many products are of the same nature as the few starters, e.g. amoebae 'dividing', lymphocytes forming a clone or *Hydras* budding.

Murine: like mice (*Mus*); usually includes myomorph rodents like rats, lemmings, etc.

Mutation: a change in the hereditary material between ancestors and descendants, usually now restricted in usage to changes in DNA nucleotide sequence.

Myosin: one of the two characteristic proteins which mediate the contraction of muscle cells; the other is actin. Contrast spermosin, flagellin, tubulin. The myosin of striped muscle is different from that of smooth muscle, the latter probably occurring in all motile cells as part of the cell surface.

Neo-Darwinism: the belief that Natural Selection acts upon organisms which differ randomly, in ways determined by genetic and environmental factors, the genetic differences being determined by chromosomal events and 'breeding true'. Darwin himself believed in 'blending inheritance' and thought selected differences might be diluted out.

Neonatal: just born.

Non-dysjunction: failure of homologous chromosomes or sister chromatids to part into different daughter cells at meiosis or mitosis, resulting in one daughter cell with $n + 1$, and one with $n - 1$ chromosomes, i.e. an aneuploid pair of cells.

'Normal' distribution: the pattern of distribution of a character in a population such that one measurement is most common and measurements higher or lower are progressively rarer. The commonest value is the mode, if only integers are possible and not fractions, or the mean if the character is continuous, like weight. Such a clustering can be described by two parameters, the mean and the standard deviation: one SD above and below the mean include about 60% of the population, and 2 SDs about 98%.

Nulliparous: of women, having had no children. Contrast multiparous.

Oocyte: an immature egg; the word is used as soon as the cell resulting from germ cell division has finished its last pre-meiotic mitosis.

Orchitis: inflammation of the testes.

Paradigm: normal usage: excellent example, as 'the rat is the paradigm of rodents'. New usage, employed by Kuhn in his theories of revolution in science: the theoretical framework within which hypothesis, experiment and disproof are constrained ('normal science'), before the replacement of the paradigm by a new one ('revolutionary science'), as Newtonian physics was replaced by relativistic physics.

Pelagic: free-swimming. Contrast benthic, planktonic.

'Personal space': the distance outside which courteous people stay. Intrusion by accidental touch or other violation is apologised for; varies greatly among human cultures, Englishmen having a proverbially voluminous one.

Phenotype: the characteristic appearance, morphology or physiology of an organism in relation to its siblings or, more generally, the norm for its species. Contrast genotype.

Pheromones: volatile substances, secreted by organisms into their environment, which modify the behaviour of members of the same species.

Phylogenetic: relationship by descent; 'blood' relationship.

Physiology: functioning of organisms; for example respiration, excretion and nervous activity are physiological processes. 'Physiological' is also used to mean within the normal range met by an organism, e.g. physiological temperatures.

Phytoplankton: see Plankton.

Plankton: the organisms which drift, float or swim in the surface waters of the oceans. The primary producers, phytoplankton (simple plants) eaten by zooplankton (mostly crustaceans); very many marine organisms have a part of their life history in the plankton, usually the larvae (e.g. crabs, oysters) but sometimes the adults (e.g. jelly fish).

Plasmodium: a mass of cytoplasm with several-to-many nuclei not partitioned from one another by cell membranes. (Also the generic name of the malaria parasite.) See Coenocytic.

Pleiotropic: of a gene with several apparent functions. Probably true of nearly *all* genes, because of the prescriptive nature of development.

Ploidy: the number of chromosome sets in a nucleus, e.g. haploid (one set) diploid (two sets) octaploid (eight) etc. Also used are euploid (exact number of sets) and its opposite, aneuploid, for nuclei with peculiar chromosome complements.

Poikilotherms: animals which do not regulate their internal temperature except by behavioural changes; 'cold-blooded' animals, e.g. modern reptiles, most fishes, most insects. Contrast homoiotherms, e.g. birds and mammals.

Poisson series: (or distribution): variation in a quantitative character in a population, in such a way that the numbers of individuals within successive ranges is not related to the absolute magnitude of the range; contrast 'normal distribution' where most individuals are to be found around a mean value. Nor do the numbers of individuals fall or rise evenly or exponentially with magnitude; individuals with higher measurements are rarer by a factor related to *e*.

Polymer: a molecule formed by the association of sub-units, usually identical (collagen) but frequently dissimilar (haemoglobin α and β, LDH A and B chains).

Polymorphonuclear leucocytes: (polymorphs): white blood cells, of mammals, which have strangely lobed nuclei and are often found at sites of tissue destruction or disruption; as in bacterial invasion, oocyte atresia, or the destruction of excess sperms in the uterus and vagina.

Pornography: that category of various artistic media which depends for its appeal upon sexual excitation rather than aesthetic or other cultural enjoyment.

Prokaryotes: organisms without a nucleus, including viruses, bacteria and blue-green algae. Some cell organelles, notably mitochondria and chloroplasts, have a similar structure.

Pronotum: the anterior part (head/thorax) of spiders and mites.

Prostitute: woman (or less frequently man) who receives payment for cooperating in sexual practices. Contrast volunteer, enthusiastic amateur.

Protozoa: animals which are not composed of separate cells with different functions, sometimes called 'unicells', often called unicellular or one-celled; e.g. *Amoeba, Paramecium.* Contrast Metazoa, Protista.

'*r*-selection': that selection operating on a prolific, variable species over a broad or variable habitat; stringent selection of a few breeders in each generation. Selection for versatility; contrast K-selection.

Reductionism: the belief that biology should be explained in chemical and physical terms, chemistry by physics and maths, and so on.

Regeneration: the re-organisation, often involving growth of new tissue, of damaged or lost parts; e.g. the limbs of newts and the tail ends of worms.

Reproduction: the process by which a group of organisms replaces itself; often used in a more restricted sense to mean breeding or mating or producing offspring.

Scyphistoma: the reduced polyp phase of large jelly fish like *Aurelia*, from which the juvenile jelly fishes (ephyrae) are budded by strobilisation. The polyp resembles a pile of saucers, of which the uppermost swim off.

Sex chromosomes: those chromosomes whose presence or absence is associated with one or other sex (e.g. the Y chromosome of man), and their homologues usually present in both sexes (e.g. the X chromosome). X–Y and W–Z (male, homogametic) are most common, but many variants occur.

Sexual congress: the association of males and females for sexually reproductive purposes.

Siblings: members of the same litter, or more generally from the same parents; occasionally used more widely still to mean sharing at least one parent (half-sib is preferable here); brothers and sisters.

Silk: the protein secretion of glands of certain caterpillars, used to spin a fibrous cocoon. The cocoons are unwound and the fibres used for fine cloth.

Sociology: the study of societies; virtually restricted to human societies.

Stochastic: a random process whose overall incidence may be predicted accurately, but whose individual instances may be apparently indeterminable, e.g. deaths from traffic accidents, or the decay of radioactive atoms. Contrast deterministic.

Stolon: that special, usually elongated, part of a budding organism which bears the buds, especially if a series of buds is formed.

Strategy: a long-term or broad plan, usually military. Used here by analogy to mean a reproductive programme whose elements are recognisable broadly in a great variety of organisms (e.g. budding, or senility) or which extend through the life histories of many species (e.g. alternating parthenogenesis and sexual resistant eggs). Contrast tactics.

Tachycardia: fast heart rate.

Tachypnoea: fast breathing rate.

Tactics: planned operations for the solution of problems, usually in warfare. Used here for adaptations by organisms to meet circumscribed or special reproduction needs. Contrast strategy.

Tonus: the normal state of slight contraction of most voluntary muscles.

Torsion: a process occurring in the larval life (veliger or equivalent) of all gastropod molluscs ('snails') during which the visceral mass is rotated over the head-foot, bringing the anus over the head but allowing the velar lobes to be drawn into the shell. This larval trick may save lives in the plankton (the larva can drop suddenly), but all adults change their anatomy to avoid its results, often 'detorting' or relegating the torsion to a small part of the adult anatomy.

Trophoblast: the extra-embryonic part of a mammal blastocyst, which erodes into mother's uterine lining and provides anchorage and perhaps nutrient.

Trophozoite: the growth phase of an organism, especially a protozoan parasite. Contrast sporozoite, gamont.

Tubal pregnancy: development of (human) embryos in the oviducts (Fallopian tubes) instead of the uterus. Very dangerous to mother.

Uterus masculinus: the undeveloped vestige of fusion of the Müllerian ducts in male mammals, usually found embedded in the prostate. In the female this forms the uterus.

Variance: the extent to which members of a population depart from the average, or mean, value of a given character; variability.

Vegetative: pertaining to those activities of an organism relating to its ordinary, non-reproductive, activity; e.g. respiration, feeding, excretion, avoiding its predators or parasites. Contrast reproductive activity.

Victorian morality: the mores of English society in the late nineteenth century, under Queen Victoria. Believed to be a time of unparalleled prudery and repression, at least for the literate. But see *The Other Victorians* by Marcus.

Virus: an organism of very simple structure, containing only protein and nucleic acid (DNA or RNA); viruses require the transcription machinery of other organisms for their replication. Some infect bacteria (bacteriophages or 'phages') and some infect animal and plant cells, e.g. influenza and tobacco mosaic virus.

Zooplankton: see plankton.

Zygote: the product of fusion of (usually) two gametes; in Metazoa, the fertilised egg. But in Protozoa the diploid zygotes are usually transient, undergoing meiosis to produce spores which give normal haploid individuals.

List of Organisms

(only organisms mentioned in the text appear in this list)

PROKARYOTES		
Virus		
Bacteria		*Escherichia coli* *Streptococcus*
EUKARYOTES		
Fungi		*Aspergillus*
Metaphyta (higher plants)		
BRYOPHYTA (mosses and liverworts)		
PTERIDOPHYTA (ferns, cycads)		*Cycas*
ANGIOSPERMS (flowering plants)		*Cucurbita* (gourds) *Helianthus* (sunflower) *Tradescantia*
Protozoa		
MASTIGOPHORA (flagellates)		*Chlamydomonas* *Volvox* *Naegleria* (=*Dimastigamoeba*)
SPOROZOA		
	GREGARINOMORPHA	*Monocystis* *Gregarina*
	COCCIDIOMORPHA	*Adelia* *Plasmodium* (malaria)
CILIOPHORA		
Euciliata		*Paramecium* *Spirochona* *Stentor*
Suctoria		*Dendrocommetes*
Parazoa (sponges)		
Calcarea		
Metazoa		
COELENTERATA (=CNIDARIA)		
Hydrozoa		

	ATHECATA	*Hydra*
		Hydractinia
	THECATA	*Obelia*
	LIMNOMEDUSAE	*Craspedocusta*
	TRACHYMEDUSAE (=trachyline medusae)	
	SIPHONOPHORA	*Halistemma*
		Physalia (Portuguese man of war)
Scyphozoa (jelly fish)		*Aurelia* (common British jelly fish)
Anthozoa		sea anemones
		corals
PLATYHELMINTHES (flatworms)		
Turbellaria		*Polycelis*
		other planarians
Monogenea (fish flukes)		*Entobdella*
		Gyrodactylus
Cestoda (tapeworms)		*Schistocephalus*
		Taenia
		Echinococcus
Digenea (internal flukes)		*Fasciola hepatica* (liverfluke)
		Phyllodistomum
		Plagiorchis
ASCHELMINTHES		
Nematoda		*Rhabditis*
(roundworms)		*Ascaris* (horse or pig roundworm)
Rotifera (rotifers)		*Asplanchna*
		Floscularia
Gastrotricha		
ECTOPROCTA (bryozoans, moss animals)		
BRACHIOPODA (lamp shells)		*Lingula*
NEMERTINA (ribbon worms)		
ECHIUROIDEA		*Bonellia*
ANNELIDA (true worms)		
Polychaeta (marine tubicolous and free-living forms)		*Diopatra*
		Endalia
		Eunice (palolo worm)
		Hydroides
		Pomatoceros
		Spirorbis

	Oligochaeta (freshwater and terrestrial worms)			*Tubifex* (sewage worm) *Lumbricus* (earth worm)
	Hirudinea (leeches)			
ARTHROPODA				
	Onychophora			*Peripatopsis*
	Trilobita			
	Crustacea			
		BRANCHIOPODA		
			ANOSTRACA	*Artemia* (brine shrimps)
			CLADOCERA	*Daphnia* (water flea)
		OSTRACODA		*Cypris* *Cypridopsis*
		COPEPODA		*Cyclops* *Calanus*
		CIRRIPEDIA		
			THORACICA (barnacles)	*Balanus and Cthalamus* (acorn barnacles) *Lepas* (goose barnacles)
			RHIZOCEPHALA	*Sacculina*
		MALACOSTRACA		
			AMPHIPODA	*Gammarus*
			STOMATOPODA (mantis shrimps)	*Squilla*
			DECAPODA	*Astacus and Procambarus* (crayfishes) *Carcinus* (shore crab) *Portunus* (swimming crab) *Homarus* (lobsters) *Palinurus* (crawfish) *Eupagurus and Clibanarius* (hermit crabs) *Cancer* (edible crab) *Uca* (fiddler crab)

Insecta		
(Paurometabola)		
	DICTYOPTERA	*Periplaneta* (American cockroach)
		Pycnoscelis (Surinam cockroach)
	MANTIDA	*Mantis* (praying mantis)
	PHASMIDA	*Dixippus* (common stick insect)
	ORTHOPTERA	*Locusta and Schistocerca* (locusts)
		Chorthippus (grasshopper)
		Gryllus (=Acheta) (cricket)
	DERMAPTERA	*Forficula* (earwig)
	ISOPTERA (termites)	*Bellicosotermes*
		Kalotermes
		Reticulotermes
	HEMIPTERA (bugs)	*Acanthochermes and Sacchiphantes and Pemphigus and Macrosiphum* (aphids or greenflies)
		Steatococcus and Pseudococcus (scale insects)
		Cimex (bed bug)
(Hemimetabola)		
	EPHEMOPTERA (may flies)	*Sialis*
	ODONATA (dragon and damsel flies)	
(Holometabola)		
	COLEOPTERA (beetles)	*Carabus* ('black', cellar and graveyard beetles)
		Brachinus (bombardier beetles)
		Tenebrio (mealworms)
		ladybirds

	LEPIDOPTERA (butterflies and moths)	*Herse* (hawk moth)
		Bombyx (silk moth)
	DIPTERA (true flies)	*Corethra and Chironomus* (midges)
		Aedes (yellow fever mosquito)
		Anopheles (malaria mosquito)
		Culex (gnats and mosquitos)
		Cecidomyids:
		Miastor
		Heteropeza
		Sciara (fungus gnat)
		Empidae
		Drosophila (fruit flies)
		Musca (house flies)
		Glossina (tsetse fly)
	SIPHONAPTERA (fleas)	
	HYMENOPTERA	alder wood wasp
		Apis (honey bee)
		Polistes
		Vespula (wasp)
		Pheidole and Formica and Anomma and Eciton (ants)
		Dahlbominus
Merostomata	XIPHOSURA	*Limulus* (king or horse-shoe crabs)
Arachnida		
	SCORPIONES (scorpions)	
	ACARI (mites and ticks)	*Argas* (cattle ticks)
		Aculus (peach mite)
	ARANEAE (spiders)	*Lycosa* (wolf spiders)
		Tegenaria (English house spider)
		Gastrotheca
Chilopoda (centipedes)		*Lithobius*

MOLLUSCA		
Gastropoda (snails)		
PROSOBRANCHIA		*Crepidula* (slipper limpet)
		Littorina (winkle)
		Paludina (=Viviparus) (river snail)
		Nassarius and Ilyanassa (dog whelks)
		Buccinum (whelk)
PULMONATA		*Lymnaea* (pond snail)
		Planorbis (ram's horn snail)
		Helix (garden snail)
		Slugs
Lamellibranchia (bivalves)		*Mytilus* (mussel)
		Crassostrea (oyster)
		Anodonta (swan mussel)
		Dreissensia
Cephalopoda		*Argonauta* (paper nautilus)
		Loligo (squid)
		Sepia (cuttlefish)
		Octopus
HEMICHORDATA		*Balanoglossus*
CHAETOGNATHA (arrow worms)		*Sagitta*
ECHINODERMATA		
Echinoidea (sea urchins)		*Lytechinus*
		Stylocidaris
		Echinus
Holothuroidea (sea cucumbers)		*Cucumaria*
Ophiuroidea (brittle stars)		
Asteroidea (starfishes)		*Asterias*
CHORDATA		
Urochordata (tunicates)		
	ASCIDIACEA (sea squirts)	*Clavelina*
		Botryllus
		Morchellium

		Styela (=Cynthia) *Ciona*
	THALIACEA (salps)	*Doliolum* *Salpa*
VERTEBRATA Agnatha		*Eptatretus* hagfish lampreys
Selachii (=Chondrichthyes) (elasmobranchs)		
	PLEUROTREMATA (sharks and dogfish)	*Scyliorhinus* (dogfish) *Mustelus* (smooth, rough etc. 'hounds') *Lamna* (porbeagle) *Squalus* = *Acanthias*
	HYPOTREMATA (skates and rays)	*Torpedo* (electric ray) *Trygon* (sting ray) *Pteroplatea*
Osteichthyes (bony fishes) *ACTINOPTERYGII* (ray-finned fishes)		
CHONDROSTEI (sturgeons)		*Acipenser* *Scaphirhynchos*
TELEOSTEI (teleosts)	ISOSPONDYLI	*Clupea* (herring) *Salmo* (trout and some salmon) *Notopterus*
	CYPRINIFORMES	*Characidae*: *Copeina* *Creagrutus* *Astyanax* *Anoptichthys* *Cyprinidae*: *Cyprinus* (carp) *Rutilus* (roach) *Alburnus* (bleak) *Leuciscus* (dace) *Gobio* (gudgeon) *Branchydanio* (zebra fish) *Misgurnus and* *Cobitis* (loaches) *Callichthyidae*: *Corydoras* (armoured cats)

APODES (eels)	*Anguilla* conger
GADIFORMES	*Gadus* (cod)
CYPRINODONTI-FORMES = MICROCYPRINI	*Cyprinodontidae*: *Epiplatys* (panchax) *Aphyosemion* *Fundulus* *Cynolebias* *Rivulus* *Poecilidae*: *Poecilia* = *Lebistes* (guppy) *Gambusia and Heterandria* (mosquito fishes) *Platypoecilus* (includes *Xiphophorus*, platies and swordtails) *Mollienisia* (mollies)
PERCOMORPHI	*Percidae:* *Perca* (perch) *Cichlidae*: *Tilapia* *Haplochromis* *Pseudotropheus* (Mbuna) *Aequidens* (*Acara*) *Anabantidae*: *Betta* (Siamese fighting fish) *Gouramis* paradise fishes
SOLENICHTHYES	*Hippocampus* (sea horse)
THORACOSTEI (sticklebacks)	*Gasterosteus* (3-spined stickleback)
PLECTOGNATHI	*Mola* (marine sunfish)
PEDICULATI (angler fish)	*Lophius* (angler fish) *Borophryne* (deep sea angler)
HETEROSOMATA (flatfishes)	*Pleuronectes* (plaice, etc.) *Xystreurys* (sole) *Scophthalmus and Pleuronichthys* (turbots)

	CROSSOPTERYGII (lobe-finned fishes)		
		DIPNOI (lungfishes)	*Lepidosiren*
	Amphibia	LABYRINTHODONTIA	*Eryops*
		GYMNOPHIONA = APODA (caecilians)	
		CAUDATA (urodeles)	*Ambystoma* (salamanders and axolotl) *Triturus* (newts) *Amphiuma* (Congo eel)
		SALIENTIA (anurans)	*Discoglossus* (painted frog) *Xenopus* (clawed toad) *Pipa* (Surinam toad) *Ascaphus* ('tailed' frog) *Sminthilus* (Cuban painted frog) *Bufo* (toads) *Rana* (some frogs and bullfrogs) *Hyla* (some tree frogs)
'Amniotes'			
	Reptilia	TESTUDINES (chelonians)	tortoises *Chelone* (marine turtle) *Chrysemys* (terrapin)
		CROCODYLIA	alligator
		SQUAMATA (lizards and snakes)	*Anolis* (American 'chameleon') *Lacerta* (green, wall, rainbow lizards) *Cnemidophorus* *Anguis* (slow worm) *Tiliqua* (blue tongue) *Eumeces* (skink) *Chalcides* *Feylinia*

			Thamnophis (garter snakes)
			black racer
			Crotalus (rattle-snake)
			Vipera (adder)
Aves (birds)			
		STRUTHIONIFORMES	*Struthio* (ostrich)
		SPHENISCIFORMES (penguins)	emperor penguin
		CHARADRIIFORMES	*Larus* (gull)
			lapwing
		COLOMBIFORMES	*Columba* (pigeon)
		CUCULIFORMES (cuckoos)	
		STRIGIFORMES (owls)	
		PSITTACIFORMES (parrots)	
		CORACIIFORMES	*Buceros* (hornbills)
		ANSERIFORMES	*Anser* (goose)
			Anas (duck)
		GALLIFORMES	*Gallus* (domestic chicken)
			Coturnix (quail)
			Perdix (partridge)
		PASSERIFORMES (includes Passeres = songbirds)	sparrow
			Sturnus (starling)
			finches
			cock-of-the-rock
			bower birds
			robins
			Dendroica (warblers)
			blackbird
			thrush
			tits
Mammalia			
	PROTOTHERIA		
		MONOTREMATA	*Echidna* (spiny ant-eater)
	METATHERIA		
		MARSUPIALIA	*Megaliea and Macropus* (kangaro
			Dasyurus
	EUTHERIA (placental mammals)		
		INSECTIVORA	*Elephantulus* (elephant shrew)

	Sorex (shrews)
CHIROPTERA (bats)	*Myotis* (brown bat)
RODENTIA (rodents)	*Sciuromorpha*:
	Sciurus (squirrel)
	Myomorpha:
	Mus (mouse)
	Rattus (rat)
	Apodemus (field mouse)
	Microtus (vole)
	Mesocricetus (golden hamster)
	Cricetulus (Chinese hamster)
	Lemmus (lemming)
	Hystricomorpha:
	Hystrix (porcupine)
	Cavia (guinea pig)
	Chinchilla
	Lagostomus (plains viscacha)
EDENTATA	*Dasypus* (armadillo)
LAGOMORPHA	*Oryctolagus* (rabbit)
	Lepus (hare)
CETACEA	*Odontoceti*: (toothed whales)
	Tursiops (dolphin)
	Physeter (sperm whale)
	Mysticeti: (whalebone whales)
	Balaenoptera (Sei whale)
	Balaenopterus (blue whale)
	humpback whale
ARTIODACTYLA	
SUIFORMES:	*Sus* (pig)
	hippopotamus
TYLOPODA:	camel
RUMINANTIA: ('ungulates')	*Bos* (cattle)
	bison
	gnu
	Rangifer (reindeer)

	antelopes
	Capra (goat)
	Ovis (sheep)
PERISSODACTYLA	*Eohippus* (ancient horselet)
	Equus (horse)
	Rhinoceros and Diceros (rhinoceroses)
	Eotitanops (early titanothere)
	Dolichorhinos and Brontotherium (later titanotheres)
PROBOSCIDEA	*Loxodonta* (African elephant)
	Elephas (Indian elephant)
CARNIVORA	*Canis* (dog)
	Felis (cat)
	Hyaena
	Panthera (lion)
	Machaerodonts:
	Smilodon (sabre tooth).
	Paradoxurus (palm civet)
	leopard
	mink, marten, sable and ferret (mustelids)
	Meles (badger)
	Lutra (otter)
	Ursus (brown bear)
	Thalarctos (polar bear)
	Vulpes (fox)
PINNIPEDIA	*Phoca* (seal)
	Pagophilius (harp seal)
	Mirounga (elephant seal)
	Odobenus (walrus)
PRIMATES	*Galago* (bush baby)
	Callithrix (marmoset)
	Saimiri (squirrel monkey)
	Macaca (rhesus monkeys, macaques)
	Papio (baboons)

Dryopithecus (ancient apės, similar to *Ramapithecus*)
Ramapithecus (ancient men, similar to *Dryopithecus*)
Pan (orang utan)
Gorilla
Homo (man and woman)

SUBJECT INDEX